Monte Carlo Method
in Statistical Physics

Monte Carlo Methods in Statistical Physics

M. E. J. NEWMAN

Santa Fe Institute

and

G. T. BARKEMA

Institute for Theoretical Physics
Utrecht University

CLARENDON PRESS · OXFORD

OXFORD
UNIVERSITY PRESS

Great Clarendon Street, Oxford OX2 6DP

Oxford University Press is a department of the University of Oxford.
It furthers the University's objective of excellence in research, scholarship,
and education by publishing worldwide in

Oxford New York

Auckland Cape Town Dar es Salaam Hong Kong Karachi
Kuala Lumpur Madrid Melbourne Mexico City Nairobi
New Delhi Shanghai Taipei Toronto

With offices in

Argentina Austria Brazil Chile Czech Republic France Greece
Guatemala Hungary Italy Japan Poland Portugal Singapore
South Korea Switzerland Thailand Turkey Ukraine Vietnam

Oxford is a registered trade mark of Oxford University Press
in the UK and in certain other countries

Published in the United States
by Oxford University Press Inc., New York

© M. E. J. Newman and G. T. Barkema, 1999

A catalogue record for this book is available from the British Library

Library of Congress Cataloging in Publication Data
Data available

ISBN 978-0-19-851797-9

Printed in Great Britain
by
the MPG Books Group, Bodmin and King's Lynn

Preface

This book is intended for those who are interested in the use of Monte Carlo simulations in classical statistical mechanics. It would be suitable for use in a course on simulation methods or on statistical physics. It would also be a good choice for those who wish to teach themselves about Monte Carlo methods, or for experienced researchers who want to learn more about some of the sophisticated new simulation techniques which have appeared in the last decade or so.

The primary goal of the book is to explain how to perform Monte Carlo simulations efficiently. For many people, Monte Carlo simulation just means applying the Metropolis algorithm to the problem in hand. Although this famous algorithm is very easy to program, it is rarely the most efficient way to perform a simulation. The Metropolis algorithm is certainly important and we do discuss it in some detail (Chapter 3 is devoted to it), but we also show that for most problems a little work with a pencil and paper can usually turn up a better algorithm, in some cases thousands or millions of times faster. In recent years there has been quite a flurry of interesting new Monte Carlo algorithms described in the literature, many of which are specifically designed to accelerate the simulation of particular classes of problems in statistical physics. Amongst others, we describe cluster algorithms, multi-grid methods, non-local algorithms for conserved-order-parameter models, entropic sampling, simulated tempering and continuous time Monte Carlo. The book is divided into parts covering equilibrium and non-equilibrium simulations, and throughout we give pointers to how the algorithms can be most efficiently implemented. At the end of the book we include a number of chapters on general implementation issues for Monte Carlo simulations. We also cover data analysis methods in some detail, including generic methods for estimating observable quantities, equilibration and correlation times, correlation functions, and standard errors, as well as a number of techniques which are specific to Monte Carlo simulation, such as the single and multiple histogram methods, finite-size scaling and the Monte Carlo renormalization group.

The *modus operandi* of this book is teaching by example. We have tried to include as many as possible of the important Monte Carlo algorithms in use today, and each one we introduce in the context of a particular model or models. For example, we illustrate the Metropolis algorithm by applying it to the simulation of the Ising model. We have not assumed however that the reader is familiar with the models studied, and give a brief outline of the physics behind each one at the start of the corresponding chapter. All we assume is that the reader has a working knowledge of basic statistical mechanics and thermodynamics at the level typical of a student beginning graduate study in physics. Reasonable fluency in a computer programming language such as FORTRAN or C will certainly be necessary if you actually want to write a Monte Carlo program yourself, but the book can be understood without it. We have avoided giving examples of actual computer code in the body of the book. There are a couple of reasons for this. Doing so would require the reader to know a particular language, or to learn that language if he or she was not already familiar with it. Furthermore, we do not believe that inclusion of the code helps much with the understanding of an algorithm. It is better to get a clear idea of the principles behind an algorithm by working through the physics and mathematics involved than to try to learn by reading someone else's program. With a clear understanding of the principles, you should be able to write your own program with little difficulty. However, inspecting other people's programs can be useful in one respect: it is a good way to learn programming tricks and techniques for writing efficient code. For this reason we have included programs for some of the more common Monte Carlo algorithms in an appendix at the end of the book. The programs are written in C, which is fast replacing FORTRAN as the most commonly used language for scientific programming.

We have also included a number of problems at the end of each chapter for the reader to work through if he or she wishes. Some of these are purely analytic and can be done on paper. Others ask the reader to write a short computer program. Answers to the analytic problems are given at the end of the book. The ones which require you to write a computer program have many equally good solutions, so we have by and large resorted to giving hints rather than answers for these problems.

There are many people and organizations who have assisted us in the writing of this book. We would like to thank our editors Sönke Adlung, Donald Degenhardt and Julia Tompson at Oxford University Press for their help and patience. We are also grateful to a number of institutions who have offered us hospitality during the four-year process of preparing the manuscript, including Cornell University, Oxford University, the Institute for Advanced Study in Princeton, Forschungszentrum Jülich, the Santa Fe Institute and Utrecht University. Finally, we would like to thank the many colleagues and friends who have offered suggestions and encouragement, including James Binney, Geoffrey Chester, Eytan Domany, Peter Grassberger, Harvey Gould, Daniel Kandel, Yongyut Laosiritaworn, Jim Louck, Jon Machta, Nick Metropolis, Cris Moore, Richard Palmer, Gunter Schütz, Jim Sethna, Kan Shen, Alan Sokal and Ben Widom. Responsibility for any mistakes which may lurk in the text rests of course with the authors. We would be very grateful to learn from our keen-eyed readers of any such problems.

June 1998

Mark Newman
Santa Fe, New Mexico, USA
Gerard Barkema
Utrecht, The Netherlands

Contents

II Out-of-equilibrium simulations

References   410

Appendices

Index   455

Part I

Equilibrium Monte Carlo simulations

1

Introduction

This book is about the use of computers to solve problems in statistical physics. In particular, it is about **Monte Carlo methods**, which form the largest and most important class of numerical methods used for solving statistical physics problems. In this opening chapter of the book we look first at what we mean by statistical physics, giving a brief overview of the discipline we call **statistical mechanics**. Whole books have been written on statistical mechanics, and our synopsis takes only a few pages, so we must necessarily deal only with the very basics of the subject. We are assuming that these basics are actually already familiar to you, but writing them down here will give us a chance to bring back to mind some of the ideas that are most relevant to the study of Monte Carlo methods. In this chapter we also look at some of the difficulties associated with solving problems in statistical physics using a computer, and outline what Monte Carlo techniques are, and why they are useful. In the last section of the chapter, purely for fun, we give a brief synopsis of the history of computational physics and Monte Carlo methods.

1.1 Statistical mechanics

Statistical mechanics is primarily concerned with the calculation of properties of condensed matter systems. The crucial difficulty associated with these systems is that they are composed of very many parts, typically atoms or molecules. These parts are usually all the same or of a small number of different types and they often obey quite simple equations of motion so that the behaviour of the entire system can be expressed mathematically in a straightforward manner. But the sheer number of equations—just the magnitude of the problem—makes it impossible to solve the mathematics exactly. A standard example is that of a volume of gas in a container. One litre of, say, oxygen at standard temperature and pressure consists of about 3×10^{22} oxygen molecules, all moving around and colliding with one another and the walls of the container. One litre of air under the same conditions contains the same number of molecules, but they are now a mixture of oxygen, nitrogen, carbon dioxide and a few other things. The atmosphere of the Earth contains 4×10^{21} litres of air, or about 1×10^{44} molecules, all moving around and colliding with each other and the ground and trees and houses and people. These are large systems. It is not feasible to solve Hamilton's equations for these systems because there are simply too many equations, and yet when we look at the macroscopic properties of the gas, they are very well-behaved and predictable. Clearly, there is something special about the behaviour of the solutions of these many equations that "averages out" to give us a predictable behaviour for the entire system. For example, the pressure and temperature of the gas obey quite simple laws although both are measures of rather gross average properties of the gas. Statistical mechanics attempts to side-step the problem of solving the equations of motion and cut straight to the business of calculating these gross properties of large systems by treating them in a probabilistic fashion. Instead of looking for exact solutions, we deal with the probabilities of the system being in one state or another, having this value of the pressure or that—hence the name *statistical* mechanics. Such probabilistic statements turn out to be extremely useful, because we usually find that for large systems the range of behaviours of the system that are anything more than phenomenally unlikely is very small; all the reasonably probable behaviours fall into a narrow range, allowing us to state with extremely high confidence that the real system will display behaviour within that range. Let us look at how statistical mechanics treats these systems and demonstrates these conclusions.

The typical paradigm for the systems we will be studying in this book is one of a system governed by a Hamiltonian function H which gives us the total energy of the system in any particular state. Most of the examples we will be looking at have discrete sets of states each with its own energy, ranging from the lowest, or ground state energy E_0 upwards, $E_1, E_2, E_3 \ldots$, possibly without limit. Statistical mechanics, and the Monte Carlo methods we will be introducing, are also applicable to systems with continuous energy spectra, and we will be giving some examples of such applications.

If our Hamiltonian system were all we had, life would be dull. Being a Hamiltonian system, energy would be conserved, which means that the system would stay in the same energy state all the time (or if there were a number of degenerate states with the same energy, maybe it would make transitions between those, but that's as far as it would get).[1] However,

---

[1] For a classical system which has a continuum of energy states there can be a continuous set of degenerate states through which the system passes, and an average over those states can sometimes give a good answer for certain properties of the system. Such sets of

there's another component to our paradigm, and that is the **thermal reservoir**. This is an external system which acts as a source and sink of heat, constantly exchanging energy with our Hamiltonian system in such a way as always to push the temperature of the system—defined as in classical thermodynamics—towards the temperature of the reservoir. In effect the reservoir is a weak perturbation on the Hamiltonian, which we ignore in our calculation of the energy levels of our system, but which pushes the system frequently from one energy level to another. We can incorporate the effects of the reservoir in our calculations by giving the system a **dynamics**, a rule whereby the system changes periodically from one state to another. The exact nature of the dynamics is dictated by the form of the perturbation that the reservoir produces in the Hamiltonian. We will discuss many different possible types of dynamics in the later chapters of this book. However, there are a number of general conclusions that we can reach without specifying the exact form of the dynamics, and we will examine these first.

Suppose our system is in a state μ. Let us define $R(\mu \to \nu)\,dt$ to be the probability that it is in state ν a time dt later. $R(\mu \to \nu)$ is the **transition rate** for the transition from μ to ν. The transition rate is normally assumed to be time-independent and we will make that assumption here. We can define a transition rate like this for every possible state ν that the system can reach. These transition rates are usually all we know about the dynamics, which means that even if we know the state μ that the system starts off in, we need only wait a short interval of time and it could be in any one of a very large number of other possible states. This is where our probabilistic treatment of the problem comes in. We define a set of weights $w_\mu(t)$ which represent the probability that the system will be in state μ at time t. Statistical mechanics deals with these weights, and they represent our entire knowledge about the state of the system. We can write a **master equation** for the evolution of $w_\mu(t)$ in terms of the rates $R(\mu \to \nu)$ thus:[2]

$$\frac{dw_\mu}{dt} = \sum_\nu \left[w_\nu(t) R(\nu \to \mu) - w_\mu(t) R(\mu \to \nu) \right]. \tag{1.1}$$

The first term on the right-hand side of this equation represents the rate at which the system is undergoing transitions into state μ; the second term is the rate at which it is undergoing transitions out of μ into other states. The probabilities $w_\mu(t)$ must also obey the sum rule

$$\sum_\mu w_\mu(t) = 1 \tag{1.2}$$

---

degenerate states are said to form a **microcanonical ensemble**. The more general case we consider here, in which there is a thermal reservoir causing the energy of the system to fluctuate, is known as a **canonical ensemble**.

[2]The master equation is really a set of equations, one for each state μ, although people always call it *the* master equation, as if there were only one equation here.

for all t, since the system must always be in *some* state. The solution of Equation (1.1), subject to the constraint (1.2), tells us how the weights w_μ vary over time.

And how are the weights w_μ related to the macroscopic properties of the system which we want to know about? Well, if we are interested in some quantity Q, which takes the value Q_μ in state μ, then we can define the **expectation** of Q at time t for our system as

$$\langle Q \rangle = \sum_\mu Q_\mu w_\mu(t). \tag{1.3}$$

Clearly this quantity contains important information about the real value of Q that we might expect to measure in an experiment. For example, if our system is definitely in one state τ then $\langle Q \rangle$ will take the corresponding value Q_τ. And if the system is equally likely to be in any of perhaps three states, and has zero probability of being in any other state, then $\langle Q \rangle$ is equal to the mean of the values of Q in those three states, and so forth. However, the precise relation of $\langle Q \rangle$ to the observed value of Q is perhaps not very clear. There are really two ways to look at it. The first, and more rigorous, is to imagine having a large number of copies of our system all interacting with their own thermal reservoirs and whizzing between one state and another all the time. $\langle Q \rangle$ is then a good estimate of the number we would get if we were to measure the instantaneous value of the quantity Q in each of these systems and then take the mean of all of them. People who worry about the conceptual foundations of statistical mechanics like to take this "many systems" approach to defining the expectation of a quantity.[3] The trouble with it however is that it's not very much like what happens in a real experiment. In a real experiment we normally only have one system and we make all our measurements of Q on that system, though we probably don't just make a single instantaneous measurement, but rather integrate our results over some period of time. There is another way of looking at the expectation value which is similar to this experimental picture, though it is less rigorous than the many systems approach. This is to envisage the expectation as a *time average* of the quantity Q. Imagine recording the value of Q every second for a thousand seconds and taking the average of those one thousand values. This will correspond roughly to the quantity calculated in Equation (1.3) as long as the system passes through a representative selection of the states in the probability distribution w_μ in those thousand seconds. And if we make ten thousand measurements of Q instead of one thousand,

---

[3]In fact the word *ensemble*, as in the "canonical ensemble" which was mentioned in a previous footnote, was originally introduced by Gibbs to describe an ensemble of *systems* like this, and not an ensemble of, say, molecules, or any other kind of ensemble. These days however, use of this word no longer implies that the writer is necessarily thinking of a many systems formulation of statistical mechanics.

or a million or more, we will get an increasingly accurate fit between our experimental average and the expectation $\langle Q \rangle$.

Why is this a less rigorous approach? The main problem is the question of what we mean by a "representative selection of the states". There is no guarantee that the system will pass through anything like a representative sample of the states of the system in our one thousand seconds. It could easily be that the system only hops from one state to another every ten thousand seconds, and so turns out to be in the same state for all of our one thousand measurements. Or maybe it changes state very rapidly, but because of the nature of the dynamics spends long periods of time in small portions of the state space. This can happen for example if the transition rates $R(\mu \rightarrow \nu)$ are only large for states of the system that differ in very small ways, so that the only way to make a large change in the state of the system is to go through very many small steps. This is a very common problem in a lot of the systems we will be looking at in this book. Another potential problem with the time average interpretation of (1.3) is that the weights $w_\mu(t)$, which are functions of time, may change considerably over the course of our measurements, making the expression invalid. This can be a genuine problem in both experiments and simulations of non-equilibrium systems, which are the topic of the second part of this book. For equilibrium systems, as discussed below, the weights are by definition not time-varying, so this problem does not arise.

Despite these problems however, this time-average interpretation of the expectation value of a quantity is the most widely used and most experimentally relevant interpretation, and it is the one that we will adopt in this book. The calculation of expectation values is one of the fundamental goals of statistical mechanics, and of Monte Carlo simulation in statistical physics, and much of our time will be concerned with it.

1.2 Equilibrium

Consider the master equation (1.1) again. If our system ever reaches a state in which the two terms on the right-hand side exactly cancel one another for all μ, then the rates of change dw_μ/dt will all vanish and the weights will all take constant values for the rest of time. This is an **equilibrium** state. Since the master equation is first order with real parameters, and since the variables w_μ are constrained to lie between zero and one (which effectively prohibits exponentially growing solutions to the equations) we can see that all systems governed by these equations must come to equilibrium in the end. A large part of this book will be concerned with Monte Carlo techniques for simulating equilibrium systems and in this section we develop some of the important statistical mechanical concepts that apply to these systems.

The transition rates $R(\mu \rightarrow \nu)$ appearing in the master equation (1.1) do not just take *any* values. They take particular values which arise out of the thermal nature of the interaction between the system and the thermal reservoir. In the later chapters of this book we will have to choose values for these rates when we simulate thermal systems in our Monte Carlo calculations, and it is crucial that we choose them so that they mimic the interactions with the thermal reservoir correctly. The important point is that we know *a priori* what the equilibrium values of the weights w_μ are for our system. We call these equilibrium values the **equilibrium occupation probabilities** and denote them by

$$p_\mu = \lim_{t \to \infty} w_\mu(t). \tag{1.4}$$

It was Gibbs (1902) who showed that for a system in thermal equilibrium with a reservoir at temperature T, the equilibrium occupation probabilities are

$$p_\mu = \frac{1}{Z} e^{-E_\mu/kT}. \tag{1.5}$$

Here E_μ is the energy of state μ and k is Boltzmann's constant, whose value is $1.38 \times 10^{-23} \, \mathrm{J K^{-1}}$. It is conventional to denote the quantity $(kT)^{-1}$ by the symbol β, and we will follow that convention in this book. Z is a normalizing constant, whose value is given by

$$Z = \sum_\mu e^{-E_\mu/kT} = \sum_\mu e^{-\beta E_\mu}. \tag{1.6}$$

Z is also known as the **partition function**, and it figures a lot more heavily in the mathematical development of statistical mechanics than a mere normalizing constant might be expected to. It turns out in fact that a knowledge of the variation of Z with temperature and any other parameters affecting the system (like the volume of the box enclosing a sample of gas, or the magnetic field applied to a magnet) can tell us virtually everything we might want to know about the macroscopic behaviour of the system. The probability distribution (1.5) is known as the **Boltzmann distribution**, after Ludwig Boltzmann, one of the pioneers of statistical mechanics. For a discussion of the origins of the Boltzmann distribution and the arguments that lead to it, the reader is referred to the exposition by Walter Grandy in his excellent book *Foundations of Statistical Mechanics* (1987). In our treatment we will take Equation (1.5) as our starting point for further developments.

From Equations (1.3), (1.4) and (1.5) the expectation of a quantity Q for a system in equilibrium is

$$\langle Q \rangle = \sum_\mu Q_\mu p_\mu = \frac{1}{Z} \sum_\mu Q_\mu e^{-\beta E_\mu}. \tag{1.7}$$

For example, the expectation value of the energy $\langle E \rangle$, which is also the quantity we know from thermodynamics as the internal energy U, is given by

$$U = \frac{1}{Z} \sum_{\mu} E_{\mu} \, e^{-\beta E_{\mu}}. \tag{1.8}$$

From Equation (1.6) we can see that this can also be written in terms of a derivative of the partition function:

$$U = -\frac{1}{Z} \frac{\partial Z}{\partial \beta} = -\frac{\partial \log Z}{\partial \beta}. \tag{1.9}$$

The specific heat is given by the derivative of the internal energy:

$$C = \frac{\partial U}{\partial T} = -k\beta^2 \frac{\partial U}{\partial \beta} = k\beta^2 \frac{\partial^2 \log Z}{\partial \beta^2}. \tag{1.10}$$

However, from thermodynamics we know that the specific heat is also related to the entropy:

$$C = T \frac{\partial S}{\partial T} = -\beta \frac{\partial S}{\partial \beta}, \tag{1.11}$$

and, equating these two expressions for C and integrating with respect to β, we find the following expression for the entropy:

$$S = -k\beta \frac{\partial \log Z}{\partial \beta} + k \log Z. \tag{1.12}$$

(There is in theory an integration constant in this equation, but it is set to zero under the convention known as the third law of thermodynamics, which fixes the arbitrary origin of entropy by saying that the entropy of a system should tend to zero as the temperature does.) We can also write an expression for the (Helmholtz) free energy F of the system, using Equations (1.9) and (1.12):

$$F = U - TS = -kT \log Z. \tag{1.13}$$

We have thus shown how U, F, C and S can all be calculated directly from the partition function Z. The last equation also tells us how we can deal with other parameters affecting the system. In classical thermodynamics, parameters and constraints and fields interacting with the system each have conjugate variables which represent the response of the system to the perturbation of the corresponding parameter. For example, the response of a gas system in a box to a change in the confining volume is a change in the pressure of the gas. The pressure p is the conjugate variable to the parameter V. Similarly, the magnetization M of a magnet changes in response to the applied magnetic field B; M and B are conjugate variables. Thermodynamics tells us that we can calculate the values of conjugate variables from derivatives of the free energy:

$$p = -\frac{\partial F}{\partial V},\tag{1.14}$$

$$M = \frac{\partial F}{\partial B}.\tag{1.15}$$

Thus, if we can calculate the free energy using Equation (1.13), then we can calculate the effects of parameter variations too.

In performing Monte Carlo calculations of the properties of equilibrium systems, it is sometimes appropriate to calculate the partition function and then evaluate other quantities from it. More often it is better to calculate the quantities of interest directly, but many times in considering the theory behind our simulations we will return to the idea of the partition function, because in principle the entire range of thermodynamic properties of a system can be deduced from this function, and any numerical method that can make a good estimate of the partition function is at heart a sound method.

1.2.1 Fluctuations, correlations and responses

Statistical mechanics can tell us about other properties of a system apart from the macroscopic ones that classical equilibrium thermodynamics deals with such as entropy and pressure. One of the most physically interesting classes of properties is **fluctuations** in observable quantities. We described in the first part of Section 1.1 how the calculation of an expectation could be regarded as a time average over many measurements of the same property of a single system. In addition to calculating the mean value of these many measurements, it is often useful also to calculate their standard deviation, which gives us a measure of the variation over time of the quantity we are looking at, and so tells us quantitatively how much of an approximation we are making by giving just the one mean value for the expectation. To take an example, let us consider the internal energy again. The mean square deviation of individual, instantaneous measurements of the energy away from the mean value $U = \langle E \rangle$ is

$$\langle (E - \langle E \rangle)^2 \rangle = \langle E^2 \rangle - \langle E \rangle^2.\tag{1.16}$$

We can calculate $\langle E^2 \rangle$ from derivatives of the partition function in a way similar to our calculation of $\langle E \rangle$:

$$\langle E^2 \rangle = \frac{1}{Z} \sum_\mu E_\mu^2 e^{-\beta E_\mu} = \frac{1}{Z} \frac{\partial^2 Z}{\partial \beta^2}.\tag{1.17}$$

So

$$\langle E^2 \rangle - \langle E \rangle^2 = \frac{1}{Z} \frac{\partial^2 Z}{\partial \beta^2} - \left[\frac{1}{Z} \frac{\partial Z}{\partial \beta} \right]^2 = \frac{\partial^2 \log Z}{\partial \beta^2}. \qquad (1.18)$$

Using Equation (1.10) to eliminate the second derivative, we can also write this as

$$\langle E^2 \rangle - \langle E \rangle^2 = \frac{C}{k\beta^2}. \qquad (1.19)$$

And the standard deviation of E, the RMS fluctuation in the internal energy, is just the square root of this expression.

This result is interesting for a number of reasons. First, it gives us the magnitude of the fluctuations in terms of the specific heat C or alternatively in terms of $\log Z = -\beta F$. In other words we can calculate the fluctuations entirely from quantities that are available within classical thermodynamics. However, this result could never have been derived within the framework of thermodynamics, since it depends on microscopic details that thermodynamics has no access to. Second, let us look at what sort of numbers we get out for the size of the energy fluctuations of a typical system. Let us go back to our litre of gas in a box. A typical specific heat for such a system is $1 \, \mathrm{J \, K^{-1}}$ at room temperature and atmospheric pressure, giving RMS energy fluctuations of about 10^{-18} J. The internal energy itself on the other hand will be around 10^2 J, so the fluctuations are only about one part in 10^{20}. This lends some credence to our earlier contention that statistical treatments can often give a very accurate estimate of the expected behaviour of a system. We see that in the case of the internal energy at least, the variation of the actual value of U around the expectation value $\langle E \rangle$ is tiny by comparison with the kind of energies we are considering for the whole system, and probably not within the resolution of our measuring equipment. So quoting the expectation value gives a very good guide to what we should expect to see in an experiment. Furthermore, note that, since the specific heat C is an extensive quantity, the RMS energy fluctuations, which are the square root of Equation (1.19), scale like $\sqrt{V}$ with the volume V of the system. The internal energy itself on the other hand scales like V, so that the relative size of the fluctuations compared to the internal energy decreases as $1/\sqrt{V}$ as the system becomes large. In the limit of a very large system, therefore, we can ignore the fluctuations altogether. For this reason, the limit of a large system is called the **thermodynamic limit**. Most of the questions we would like to answer about condensed matter systems are questions about behaviour in the thermodynamic limit. Unfortunately, in Monte Carlo simulations it is often not feasible to simulate a system large enough that its behaviour is a good approximation to a large system. Much of the effort we put into designing algorithms will be aimed at making them efficient enough that we can simulate the largest systems possible in the available computer time, in the hope of getting results which are at least a reasonable approximation to the thermodynamic limit.

What about fluctuations in other thermodynamic variables? As we discussed in Section 1.2, each parameter of the system that we fix, such as a volume or an external field, has a conjugate variable, such as a pressure or a magnetization, which is given as a derivative of the free energy by an equation such as (1.14) or (1.15). Derivatives of this general form are produced by terms in the Hamiltonian of the form $-XY$, where Y is a "field" whose value we fix, and X is the conjugate variable to which it couples. For example, the effect of a magnetic field on a magnet can be accounted for by a magnetic energy term in the Hamiltonian of the form $-MB$, where M is the magnetization of the system, and B is the applied magnetic field. We can write the expectation value of X in the form of Equations (1.14) and (1.15) thus:

$$\langle X \rangle = \frac{1}{\beta Z} \sum_\mu X_\mu e^{-\beta E_\mu} = \frac{1}{\beta Z} \frac{\partial}{\partial Y} \sum_\mu e^{-\beta E_\mu}, \qquad (1.20)$$

since E_μ now contains the term $-X_\mu Y$ which the derivative acts on. Here X_μ is the value of the quantity X in the state μ. We can then write this in terms of the free energy thus:

$$\langle X \rangle = \frac{1}{\beta} \frac{\partial \log Z}{\partial Y} = -\frac{\partial F}{\partial Y}. \qquad (1.21)$$

This is a useful technique for calculating the thermal average of a quantity, even if no appropriate field coupling to that quantity appears in the Hamiltonian. We can simply make up a fictitious field which couples to our quantity in the appropriate way—just add a term to the Hamiltonian anyway to allow us to calculate the expectation of the quantity we are interested in—and then set the field to zero after performing the derivative, making the fictitious term vanish from the Hamiltonian again. This is a very common trick in statistical mechanics.

Another derivative of $\log Z$ with respect to Y produces another factor of X_μ in the sum over states, and we find

$$-\frac{1}{\beta} \frac{\partial^2 F}{\partial Y^2} = \frac{1}{\beta} \frac{\partial \langle X \rangle}{\partial Y} = \langle X^2 \rangle - \langle X \rangle^2, \qquad (1.22)$$

which we recognize as the mean square fluctuation in the variable X. Thus we can find the fluctuations in all sorts of quantities from second derivatives of the free energy with respect to the appropriate fields, just as we can find the energy fluctuations from the second derivative with respect to β. The derivative $\partial \langle X \rangle / \partial Y$, which measures the strength of the response of X to changes in Y is called the **susceptibility** of X to Y, and is usually denoted by χ:

$$\chi \equiv \frac{\partial \langle X \rangle}{\partial Y}. \qquad (1.23)$$

Thus the fluctuations in a variable are proportional to the susceptibility of that variable to its conjugate field. This fact is known as the **linear response theorem** and it gives us a way to calculate susceptibilities within Monte Carlo calculations by measuring the size of the fluctuations of a variable.

Extending the idea of the susceptibility, and at the same time moving a step further from the realm of classical thermodynamics, we can also consider what happens when we change the value of a parameter or field at one particular position in our system and ask what effect that has on the conjugate variable at other positions. To study this question we will consider for the moment a system on a lattice. Similar developments are possible for continuous systems like gases, but most of the examples considered in this book are systems which fall on lattices, so it will be of more use to us to go through this for a lattice system here. The interested reader might like to develop the corresponding theory for a continuous system as an exercise.

Let us then suppose that we now have a field which is spatially varying and takes the value Y_i on the ith site of the lattice. The conjugate variables to this field[4] are denoted x_i, and the two are linked via a term in the Hamiltonian $-\sum_i x_i Y_i$. Clearly if we set $Y_i = Y$ and $x_i = X/N$ for all sites i, where N is the total number of sites on the lattice, then this becomes equal once more to the homogeneous situation we considered above. Now in a direct parallel with Equation (1.20) we can write the average value of x_i as

$$\langle x_i \rangle = \frac{1}{Z} \sum_\mu x_i^\mu e^{-\beta E_\mu} = \frac{1}{\beta} \frac{\partial \log Z}{\partial Y_i}, \tag{1.24}$$

where x_i^μ is the value of x_i in state μ. Then we can define a generalized susceptibility χ_{ij} which is a measure of the response of $\langle x_i \rangle$ to a variation of the field Y_j at a different lattice site:

$$\chi_{ij} = \frac{\partial \langle x_i \rangle}{\partial Y_j} = \frac{1}{\beta} \frac{\partial^2 \log Z}{\partial Y_i \partial Y_j}. \tag{1.25}$$

Again the susceptibility is a second derivative of the free energy. If we make the substitution $Z = \sum_\mu e^{-\beta E_\mu}$ again (Equation (1.6)), we see that this is also equal to

$$\chi_{ij} = \frac{\beta}{Z} \sum_\mu x_i^\mu x_j^\mu e^{-\beta E_\mu} - \beta \left[\frac{1}{Z} \sum_\mu x_i^\mu e^{-\beta E_\mu} \right] \left[\frac{1}{Z} \sum_\nu x_j^\nu e^{-\beta E_\nu} \right]$$
$$= \beta(\langle x_i x_j \rangle - \langle x_i \rangle \langle x_j \rangle) = \beta G_c^{(2)}(i, j). \tag{1.26}$$

---

[4] We use lower-case x_i to denote an intensive variable. X by contrast was extensive, i.e., its value scales with the size of the system. We will use this convention to distinguish intensive and extensive variables throughout much of this book.

The quantity $G_c^{(2)}(i,j)$ is called the **two-point connected correlation function** of x between sites i and j, or just the connected correlation, for short. The superscript (2) is to distinguish this function from higher order correlation functions, which are discussed below. As its name suggests, this function is a measure of the correlation between the values of the variable x on the two sites; it takes a positive value if the values of x on those two sites fluctuate in the same direction together, and a negative one if they fluctuate in opposite directions. If their fluctuations are completely unrelated, then its value will be zero. To see why it behaves this way consider first the simpler **disconnected correlation function** $G^{(2)}(i,j)$ which is defined to be

$$G^{(2)}(i,j) \equiv \langle x_i x_j \rangle. \tag{1.27}$$

If the variables x_i and x_j are fluctuating roughly together, around zero, both becoming positive at once and then both becoming negative, at least most of the time, then all or most of the values of the product $x_i x_j$ that we average will be positive, and this function will take a positive value. Conversely, if they fluctuate in opposite directions, then it will take a negative value. If they sometimes fluctuate in the same direction as one another and sometimes in the opposite direction, then the values of $x_i x_j$ will take a mixture of positive and negative values, and the correlation function will average out close to zero. This function therefore has pretty much the properties we desire of a correlation function, and it can tell us a lot of useful things about the behaviour of our system. However, it is not perfect, because we must also consider what happens if we apply our field Y to the system. This can have the effect that the mean value of x at a site $\langle x_i \rangle$ can be non-zero. The same thing can happen even in the absence of an external field if our system undergoes a phase transition to a **spontaneously symmetry broken state** where a variable such as x spontaneously develops a non-zero expectation value. (The Ising model of Section 1.2.2, for instance, does this.) In cases like these, the disconnected correlation function above can have a large positive value simply because the values of the variables x_i and x_j are always either both positive or both negative, even though this has nothing to do with them being correlated to one another. The *fluctuations* of x_i and x_j can be completely unrelated and still the disconnected correlation function takes a non-zero value. To obviate this problem we define the connected correlation function as above:

$$\begin{aligned} G_c^{(2)}(i,j) &\equiv \langle x_i x_j \rangle - \langle x_i \rangle \langle x_j \rangle \\ &= \langle (x_i - \langle x_i \rangle) \times (x_j - \langle x_j \rangle) \rangle. \end{aligned} \tag{1.28}$$

When the expectations $\langle x_i \rangle$ and $\langle x_j \rangle$ are zero and x_i and x_j are just fluctuating around zero, this function is exactly equal to the disconnected correlation function. But when the expectations are non-zero, the connected correlation function correctly averages only the fluctuations about those expectations—the term we subtract exactly takes care of any trivial contribution arising because of external fields or spontaneous symmetry breaking. If such trivial contributions are the only reason why $G^{(2)}$ is non-zero then $G_c^{(2)}$ will be zero, which is what we would like. If it is not zero, then we have a genuine correlation between the fluctuations of x_i and x_j.

Although they are not often used in the sorts of systems we will be studying in this book and we will not have call to calculate their values in any of the calculations we will describe here, it is worth mentioning, in case you ever need to use them, that there are also higher-order connected correlation functions, defined by generalizing Equation (1.25) like this:

$$
\begin{aligned}
G_c^{(3)}(i,j,k) &= \frac{1}{\beta^3} \frac{\partial^3 \log Z}{\partial Y_i \partial Y_j \partial Y_k}, \\
G_c^{(4)}(i,j,k,l) &= \frac{1}{\beta^4} \frac{\partial^4 \log Z}{\partial Y_i \partial Y_j \partial Y_k \partial Y_l},
\end{aligned}
\tag{1.29}
$$

and so on. These are measures of the correlation between simultaneous fluctuations on three and four sites respectively. For a more detailed discussion of these correlation functions and other related ones, see for example Binney *et al.* (1992).

1.2.2   An example: the Ising model

To try to make all of this a bit more concrete, we now introduce a particular model which we can try these concepts out on. That model is the Ising model, which is certainly the most thoroughly researched model in the whole of statistical physics. Without doubt more person-hours have been spent investigating the properties of this model than any other, and although an exact solution of its properties in three dimensions still eludes us, despite many valiant and increasingly sophisticated attempts, a great deal about it is known from computer simulations, and also from approximate methods such as series expansions and ϵ-expansions. We will spend three whole chapters of this book (Chapters 3, 4 and 10) discussing Monte Carlo techniques for studying the model's equilibrium and non-equilibrium properties. Here we will just introduce it briefly and avoid getting too deeply into the discussion of its properties.

The Ising model is a model of a magnet. The essential premise behind it, and behind many magnetic models, is that the magnetism of a bulk material is made up of the combined magnetic dipole moments of many atomic spins within the material. The model postulates a lattice (which can be of any geometry we choose—the simple cubic lattice in three dimensions is a common choice) with a magnetic dipole or spin on each site. In the Ising model these spins assume the simplest form possible, which is not particularly realistic, of scalar variables s_i which can take only two values ± 1, representing up-pointing or down-pointing dipoles of unit magnitude. In a real magnetic material the spins interact, for example through exchange interactions or RKKY interactions (see, for instance, Ashcroft and Mermin 1976), and the Ising model mimics this by including terms in the Hamiltonian proportional to products $s_i s_j$ of the spins. In the simplest case, the interactions are all of the same strength, denoted by J which has the dimensions of an energy, and are only between spins on sites which are nearest neighbours on the lattice. We can also introduce an external magnetic field B coupling to the spins. The Hamiltonian then takes the form

$$H = -J \sum_{\langle ij \rangle} s_i s_j - B \sum_i s_i, \qquad (1.30)$$

where the notation $\langle ij \rangle$ indicates that the sites i and j appearing in the sum are nearest neighbours.[5] The minus signs here are conventional. They merely dictate the choice of sign for the interaction parameter J and the external field B. With the signs as they are here, a positive value of J makes the spins want to line up with one another—a ferromagnetic model as opposed to an anti-ferromagnetic one which is what we get if J is negative—and the spins also want to line up in the same direction as the external field—they want to be positive if $B > 0$ and negative if $B < 0$.

The states of the Ising system are the different sets of values that the spins can take. Since each spin can take two values, there are a total of 2^N states for a lattice with N spins on it. The partition function of the model is the sum

$$Z = \sum_{s_1 = \pm 1} \sum_{s_2 = \pm 1} \cdots \sum_{s_N = \pm 1} \exp\left[\beta J \sum_{\langle ij \rangle} s_i s_j + \beta B \sum_i s_i\right]. \qquad (1.31)$$

To save the eyes, we'll write this in the shorter notation

$$Z = \sum_{\{s_i\}} \mathrm{e}^{-\beta H}. \qquad (1.32)$$

If we can perform this sum, either analytically or using a computer, then we can apply all the results of the previous sections to find the internal energy, the entropy, the free energy, the specific heat, and so forth. We can also calculate the mean magnetization $\langle M \rangle$ of the model from the partition

---

[5]This notation is confusingly similar to the notation for a thermal average, but unfortunately both are sufficiently standard that we feel compelled to use them here. In context it is almost always possible to tell them apart because one involves site labels and the other involves physical variables appearing in the model.

function using Equation (1.15), although as we will see it is usually simpler to evaluate $\langle M \rangle$ directly from an average over states:

$$\langle M \rangle = \left\langle \sum_i s_i \right\rangle. \tag{1.33}$$

Often, in fact, we are more interested in the mean magnetization per spin $\langle m \rangle$, which is just

$$\langle m \rangle = \frac{1}{N} \left\langle \sum_i s_i \right\rangle. \tag{1.34}$$

(In the later chapters of this book, we frequently use the letter m alone to denote the average magnetization per spin, and omit the brackets $\langle \ldots \rangle$ around it indicating the average. This is also the common practice of many other authors. In almost all cases it is clear from the context when an average over states is to be understood.)

We can calculate fluctuations in the magnetization or the internal energy by calculating derivatives of the partition function. Or, as we mentioned in Section 1.2.1, if we have some way of calculating the size of the fluctuations in the magnetization, we can use those to evaluate the **magnetic suscep-tibility**

$$\frac{\partial \langle M \rangle}{\partial B} = \beta(\langle M^2 \rangle - \langle M \rangle^2). \tag{1.35}$$

(See Equation (1.22).) Again, it is actually more common to calculate the magnetic susceptibility per spin:

$$\chi = \frac{\beta}{N}(\langle M^2 \rangle - \langle M \rangle^2) = \beta N(\langle m^2 \rangle - \langle m \rangle^2). \tag{1.36}$$

(Note the leading factor of N here, which is easily overlooked when calculating χ from Monte Carlo data.) Similarly we can calculate the specific heat per spin c from the energy fluctuations thus:

$$c = \frac{k\beta^2}{N}(\langle E^2 \rangle - \langle E \rangle^2). \tag{1.37}$$

(See Equation (1.19).)

We can also introduce a spatially varying magnetic field into the Hamiltonian thus:

$$H = -J \sum_{\langle ij \rangle} s_i s_j - \sum_i B_i s_i. \tag{1.38}$$

This gives us a different mean magnetization on each site:

$$\langle m_i \rangle = \langle s_i \rangle = \frac{1}{\beta} \frac{\partial \log Z}{\partial B_i}, \tag{1.39}$$

and allows us to calculate the connected correlation function

$$G_c^{(2)}(i,j) = \frac{1}{\beta^2} \frac{\partial^2 \log Z}{\partial B_i \partial B_j}. \tag{1.40}$$

When we look at the equilibrium simulation of the Ising model in Chapters 3 and 4, all of these will be quantities of interest, and relations like these between them give us useful ways of extracting good results from our numerical data.

1.3   Numerical methods

While the formal developments of statistical mechanics are in many ways very elegant, the actual process of calculating the properties of a particular model is almost always messy and taxing. If we consider calculating the partition function Z, from which, as we have shown, a large number of interesting properties of a system can be deduced, we see that we are going to have to perform a sum over a potentially very large number of states. Indeed, if we are interested in the thermodynamic limit, the sum is over an infinite number of states, and performing such sums is a notoriously difficult exercise. It has been accomplished exactly for a number of simple models with discrete energy states, most famously the Ising model in two dimensions (Onsager 1944). This and other exact solutions are discussed at some length by Baxter (1982). However, for the majority of models of interest today, it has not yet proved possible to find an exact analytic expression for the partition function, or for any other equivalent thermodynamic quantity. In the absence of such exact solutions a number of approximate techniques have been developed including series expansions, field theoretical methods and computational methods. The focus of this book is on the last of these, the computational methods.

The most straightforward computational method for solving problems in statistical physics is to take the model we are interested in and put it on a lattice of finite size, so that the partition function becomes a sum with a finite number of terms. (Or in the case of a model with a continuous energy spectrum it becomes an integral of finite dimension.) Then we can employ our computer to evaluate that sum (or integral) numerically, by simply evaluating each term in turn and adding them up. Let's see what happens when we apply this technique to the Ising model of Section 1.2.2.

If we were really interested in tackling an unsolved problem, we might look at the Ising model in three dimensions, whose exact properties have not yet been found by any method. However, rather than jump in at the deep end, let's first look at the two-dimensional case. For a system of a given linear dimension, this model will have fewer energy states than the three-dimensional one, making the sum over states simpler and quicker to perform, and the model has the added pedagogical advantage that its behaviour has been solved exactly, so we can compare our numerical calculations with the exact solution. Let's take a smallish system to start with, of 25 spins on a square lattice in a 5 × 5 arrangement. By convention we apply periodic boundary conditions, so that there are interactions between spins on the border of the array and the opposing spins on the other side. We will also set the external magnetic field B to zero, to make things simpler still.

With each spin taking two possible states, represented by ±1, our 25 spin system has a total of $2^{25} = 33\,554\,432$ possible states. However, we can save ourselves from summing over half of these, because the system has up/down symmetry, which means that for every state there is another one in which every spin is simply flipped upside down, which has exactly the same energy in zero magnetic field. So we can simplify the calculation of the partition function by just taking one out of every pair of such states, for a total of $16\,777\,216$ states, and summing up the corresponding terms in the partition function, Equation (1.6), and then doubling the sum.[6]

In Figure 1.1 we show the mean magnetization per spin and the specific heat per spin for this 5 × 5 system, calculated from Equations (1.10) and (1.34). On the same axes we show the exact solutions for these quantities on an infinite lattice, as calculated by Onsager. The differences between the two are clear, and this is precisely the difference between our small finite-sized system and the infinite thermodynamic-limit system which we discussed in Section 1.2.1. Notice in particular that the exact solution has a non-analytic point at about $kT = 2.3J$ which is not reproduced even moderately accurately by our small numerical calculation. This point is the so-called "critical temperature" at which the length-scale ξ of the fluctuations in the magnetization, also called the "correlation length", diverges. (This point is discussed in more detail in Section 3.7.1.) Because of this divergence of the length-scale, it is never possible to get good results for the behaviour of the system at the critical temperature out of any calculation performed on a finite lattice—the lattice is never large enough to include all of the important physics of the critical point. Does this mean that calculations on finite lattices are useless? No, it certainly does not. To start with, at temperatures well away from the critical point the problems are much less severe, and the numerical calculation and the exact solution agree better,

---

[6] If we were really serious about this, we could save ourselves further time by making use of other symmetries too. For example the square system we are investigating here also has a reflection symmetry and a four-fold rotational symmetry (the symmetry group is C_4), meaning that the states actually group into sets of 16 states (including the up–down symmetry pairs), all of which have the same energy. This would reduce the number of terms we have to evaluate to $2\,105\,872$. (The reader may like to ponder why this number is not exactly $2^{25}/16$, as one might expect.) However, such efforts are not really worthwhile, since, as we will see very shortly, this direct evaluation of the partition function is not a promising method for solving models.

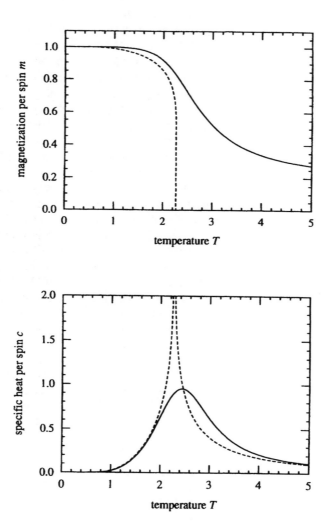

FIGURE 1.1   Top: the mean magnetization per spin m of a 5×5 Ising model on a square lattice in two dimensions (solid line) and the same quantity on an infinitely big square lattice (dashed line). Bottom: the specific heat per spin c for the same two cases.

as we can see in the figure. If we are interested in physics in this regime, then a calculation on a small lattice may well suffice. Second, the technique of "finite size scaling", which is discussed in Section 8.3, allows us to extrapolate results for finite lattices to the limit of infinite system size, and extract good results for the behaviour in the thermodynamic limit. Another technique, that of "Monte Carlo renormalization", discussed in Section 8.4, provides us with a cunning indirect way of calculating some of the features of the critical regime from just the short length-scale phenomena that we get out of a calculation on a small lattice, even though the direct cause of the features that we are interested in is the large length-scale fluctuations that we mentioned.

However, although these techniques can give answers for the critical properties of the system, the accuracy of the answers they give still depends on the size of the system we perform the calculation on, with the answers improving steadily as the system size grows. Therefore it is in our interest to study the largest system we can. However, the calculation which appears as the solid lines in Figure 1.1 took eight hours on a moderately powerful computer. The bulk of this time is spent running through the terms in the sum (1.6). For a system of N spins there are 2^N terms, of which, as we mentioned, we only need actually calculate a half, or 2^{N-1}. This number increases exponentially with the size of the lattice, so we can expect the time taken by the program to increase very rapidly with lattice size. The next size of square lattice up from the present one would be 6×6 or $N = 36$, which should take about $2^{36-1}/2^{25-1} = 2048$ times as long as the previous calculation, or about two years. Clearly this is an unacceptably long time to wait for the answer to this problem. If we are interested in results for any system larger than 5×5, we are going to have to find other ways of getting them.

1.3.1 Monte Carlo simulation

There is essentially only one known numerical method for calculating the partition function of a model such as the Ising model on a large lattice, and that method is Monte Carlo simulation, which is the subject of this book. The basic idea behind Monte Carlo simulation is to simulate the random thermal fluctuation of the system from state to state over the course of an experiment. In Section 1.1 we pointed out that for our purposes it is most convenient to regard the calculation of an expectation value as a time average over the states that a system passes through. In a Monte Carlo calculation we directly simulate this process, creating a model system on our computer and making it pass through a variety of states in such a way that the probability of it being in any particular state μ at a given time t is equal to the weight $w_\mu(t)$ which that state would have in a real system.

In order to achieve this we have to choose a dynamics for our simulation—a
rule for changing from one state to another during the simulation—which
results in each state appearing with exactly the probability appropriate to
it. In the next chapter we will discuss at length a number of strategies for
doing this, but the essential idea is that we try to simulate the physical
processes that give rise to the master equation, Equation (1.1). We choose
a set of rates $R(\mu \rightarrow \nu)$ for transitions from one state to another, and we
choose them in such a way that the equilibrium solution to the corresponding
master equation is precisely the Boltzmann distribution (1.5). Then we use
these rates to choose the states which our simulated system passes through
during the course of a simulation, and from these states we make estimates
of whatever observable quantities we are interested in.

The advantage of this technique is that we need only sample quite a
small fraction of the states of the system in order to get accurate estimates
of physical quantities. For example, we do not need to include every state
of the system in order to get a decent value for the partition function, as
we would if we were to evaluate it directly from Equation (1.6). The principal disadvantage of the technique is that there are statistical errors in the
calculation due to this same fact that we don't include every state in our
calculation, but only some small fraction of the states. In particular this
means that there will be statistical noise in the partition function. Taking
the derivative of a noisy function is always problematic, so that calculating
expectation values from derivatives of the partition function as discussed in
Section 1.2 is usually not a good way to proceed. Instead it is normally better in Monte Carlo simulations to calculate as many expectations as we can
directly, using equations such as (1.34). We can also make use of relations
such as (1.36) to calculate quantities like susceptibilities without having to
evaluate a derivative.

In the next chapter we will consider the theory of Monte Carlo simulation
in equilibrium thermal systems, and the rest of the first part of the book
will deal with the design of algorithms to investigate these systems. In the
second part of the book we look at algorithms for non-equilibrium systems.

1.4   A brief history of the Monte Carlo method

In this section we outline the important historical developments in the evolution of the Monte Carlo method. This section is just for fun; feel free to
skip over it to the next chapter if you're not interested.

The idea of Monte Carlo calculation is a lot older than the computer. The
name "Monte Carlo" is relatively recent—it was coined by Nicolas Metropolis
in 1949—but under the older name of "statistical sampling" the method
has a history stretching back well into the last century, when numerical
calculations were performed by hand using pencil and paper and perhaps

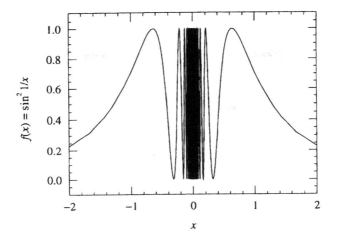

FIGURE 1.2   The pathological function $f(x) \equiv \sin^2 \frac{1}{x}$, whose integral with respect to x, though hard to evaluate analytically, can be evaluated in a straightforward manner using the Monte Carlo integration technique described in the text.

a slide-rule. As first envisaged, Monte Carlo was not a method for solving problems in physics, but a method for estimating integrals which could not be performed by other means. Integrals over poorly-behaved functions and integrals in high-dimensional spaces are two areas in which the method has traditionally proved profitable, and indeed it is still an important technique for problems of these types. To give an example, consider the function

$$f(x) \equiv \sin^2 \frac{1}{x} \qquad (1.41)$$

which is pictured in Figure 1.2. The values of this function lie entirely between zero and one, but it is increasingly rapidly varying in the neighbourhood of $x = 0$. Clearly the integral

$$I(x) \equiv \int_0^x f(x') \, \mathrm{d}x' \qquad (1.42)$$

which is the area under this curve between 0 and x, takes a finite value somewhere in the range $0 < I(x) < x$, but it is not simple to calculate this value exactly because of the pathologies of the function near the origin. However, we can make an estimate of it by the following method. If we choose a random real number h, uniformly distributed between zero and x, and another v between zero and one and plot on Figure 1.2 the point for which these are the horizontal and vertical coordinates, the probability that

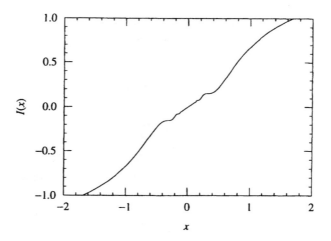

FIGURE 1.3   The function $I(x)$, calculated by Monte Carlo integration as described in the text.

this point will be below the line of $f(x)$ is just $I(x)/x$. It is easy to determine whether the point is in fact below the line: it is below it if $h < f(v)$. Thus if we simply pick a large number N of these random points and count up the number M which fall below the line, we can estimate $I(x)$ from

$$I(x) = \lim_{N \to \infty} \frac{Mx}{N}. \tag{1.43}$$

You can get an answer accurate to one figure by taking a thousand points, which would be about the limit of what one could have reasonably done in the days before computers. Nowadays, even a cheap desktop computer can comfortably run through a million points in a few seconds, giving an answer accurate to about three figures. In Figure 1.3 we have plotted the results of such a calculation for a range of values of x. The errors in this calculation are smaller than the width of the line in the figure.[7]

A famous early example of this type of calculation is the experiment known as "Buffon's needle" (Dörrie 1965), in which the mathematical constant π is determined by repeatedly dropping a needle onto a sheet of paper ruled with evenly spaced lines. The experiment is named after Georges-Louis Leclerc, Comte de Buffon who in 1777 was the first to show that if we throw a needle of length l completely at random onto a sheet of paper ruled with lines a distance d apart, then the chances that the needle will fall so as to

---

[7]In fact there exist a number of more sophisticated Monte Carlo integration techniques which give more accurate answers than the simple "hit or miss" method we have described here. A discussion can be found in the book by Kalos and Whitlock (1986).

intersect one of the lines is $2l/\pi d$, provided that $d \geq l$. It was Laplace in 1820 who then pointed out that if the needle is thrown down N times and is observed to land on a line M of those times, we can make an estimate of π from

$$\pi = \lim_{N \to \infty} \frac{2Nl}{Md}. \qquad (1.44)$$

(Perhaps the connection between this and the Monte Carlo evaluation of integrals is not immediately apparent, but it will certainly become clear if you try to derive Equation (1.44) for yourself, or if you follow Dörrie's derivation.) A number of investigators made use of this method over the years to calculate approximate values for π. The most famous of these is Mario Lazzarini, who in 1901 announced that he had calculated a value of 3.1415929 for π from an experiment in which a $2\frac{1}{2}$ cm needle was dropped 3408 times onto a sheet of paper ruled with lines 3 cm apart. This value, accurate to better than three parts in ten million, would be an impressive example of the power of the statistical sampling method were it not for the fact that it is almost certainly faked. Badger (1994) has demonstrated extremely convincingly that, even supposing Lazzarini had the technology at his disposal to measure the length of his needle and the spaces between his lines to a few parts in 10^7 (a step necessary to ensure the accuracy of Equation (1.44)), still the chances of his finding the results he did were poorer than three in a million; Lazzarini was imprudent enough to publish details of the progress of the experiment through the 3408 castings of the needle, and it turns out that the statistical "fluctuations" in the numbers of intersections of the needle with the ruled lines are much smaller than one would expect in a real experiment. All indications are that Lazzarini forged his results. However, other, less well known attempts at the experiment were certainly genuine, and yielded reasonable figures for π: 3.1596 (Wolf 1850), 3.1553 (Smith 1855). Apparently, performing the Buffon's needle experiment was for a while quite a sophisticated pastime amongst Europe's intellectual gentry.

With the advent of mechanical calculating machines at the end of the nineteenth century, numerical methods took a large step forward. These machines increased enormously the number and reliability of the arithmetic operations that could be performed in a numerical "experiment", and made the application of statistical sampling techniques to research problems in physics a realistic possibility for the first time. An early example of what was effectively a Monte Carlo calculation of the motion and collision of the molecules in a gas was described by William Thomson (later Lord Kelvin) in 1901. Thomson's calculations were aimed at demonstrating the truth of the equipartition theorem for the internal energy of a classical system. However, after the fashion of the time, he did not perform the laborious analysis himself, and a lot of the credit for the results must go to Thomson's secretary, William Anderson, who apparently solved the kinetic equations for more than five thousand molecular collisions using nothing more than a pencil and a mechanical adding machine.

Aided by mechanical calculators, numerical methods, particularly the method of finite differences, became an important tool during the First World War. The authors recently heard the intriguing story of the Herculean efforts of French mathematician Henri Soudée, who in 1916 calculated firing tables for the new 400 mm cannons being set up at Verdun, directly from his knowledge of the hydrodynamic properties of gases. The tables were used when the cannons were brought to bear on the German-occupied Fort de Douaumont, and as a result the fort was taken by the allies. Soudée was later honoured by the French. By the time of the Second World War the mechanical calculation of firing angles for large guns was an important element of military technology. The physicist Richard Feynman tells the story of his employment in Philadelphia during the summer of 1940 working for the army on a mechanical device for predicting the trajectories of planes as they flew past (Feynman 1985). The device was to be used to guide anti-aircraft guns in attacking the planes. Despite some success with the machine, Feynman left the army's employ after only a few months, joking that the subject of mechanical computation was too difficult for him. He was shrewd enough to realize he was working on a dinosaur, and that the revolution of electronic computing was just around the corner. It was some years however before that particular dream would become reality, and before it did Feynman had plenty more chance to spar with the mechanical calculators. As a group leader during the Manhattan Project at Los Alamos he created what was effectively a highly pipelined human CPU, by employing a large number of people armed with Marchant mechanical adding machines in an arithmetic assembly line in which little cards with numbers on were passed from one worker to the next for processing on the machines. A number of numerical calculations crucial to the design of the atomic bomb were performed in this way.

The first real applications of the statistical sampling method to research problems in physics seem to have been those of Enrico Fermi, who was working on neutron diffusion in Rome in the early 1930s. Fermi never published his numerical methods—apparently he considered only the results to be of interest, not the methods used to obtain them—but according to his influential student and collaborator Emilio Segrè those methods were, in everything but name, precisely the Monte Carlo methods later employed by Ulam and Metropolis and their collaborators in the construction of the hydrogen bomb (Segrè 1980).

So it was that when the Monte Carlo method finally caught the attention of the physics community, it was again as the result of armed conflict. The important developments took place at the Los Alamos National Laboratory in New Mexico, where Nick Metropolis, Stanislaw Ulam and John von Neumann gathered in the last months of the Second World War shortly after the epochal bomb test at Alamagordo, to collaborate on numerical calculations to be performed on the new ENIAC electronic computer, a mammoth, room-filing machine containing some 18 000 triode valves, whose construction was nearing completion at the University of Pennsylvania. Metropolis (1980) has remarked that the technology that went into the ENIAC existed well before 1941, but that it took the pressure of America's entry into the war to spur the construction of the machine.

It seems to have been Stan Ulam who was responsible for reinventing Fermi's statistical sampling methods. He tells of how the idea of calculating the average effect of a frequently repeated physical process by simply simulating the process over and over again on a digital computer came to him whilst huddled over a pack of cards, playing patience[8] one day. The game he was playing was "Canfield" patience, which is one of those forms of patience where the goal is simply to turn up every card in the pack, and he wondered how often on average one could actually expect to win the game. After abandoning the hopelessly complex combinatorics involved in answering this question analytically, it occurred to him that you could get an approximate answer simply by playing a very large number of games and seeing how often you win. With his mind never far from the exciting new prospect of the ENIAC computer, the thought immediately crossed his mind that he might be able to get the machine to play these games for him far faster than he ever could himself, and it was only a short conceptual leap to applying the same idea to some of the problems of the physics of the hydrogen bomb that were filling his work hours at Los Alamos. He later described his idea to John von Neumann who was very enthusiastic about it, and the two of them began making plans to perform actual calculations. Though Ulam's idea may appear simple and obvious to us today, there are actually many subtle questions involved in this idea that a physical problem with an exact answer can be approximately solved by studying a suitably chosen random process. It is a tribute to the ingenuity of the early Los Alamos workers that, rather than plunging headlong into the computer calculations, they considered most of these subtleties right from the start.

The war ended before the first Monte Carlo calculations were performed on the ENIAC. There was some uncertainty about whether the Los Alamos laboratory would continue to exist in peacetime, and Edward Teller, who was leading the project to develop the hydrogen bomb, was keen to apply the power of the computer to the problems of building the new bomb, in order to show that significant work was still going on at Los Alamos. Von Neumann developed a detailed plan of how the Monte Carlo method could be

---

[8]Also called "solitaire" in the USA.

implemented on the ENIAC to solve a number of problems concerned with neutron transport in the bomb, and throughout 1947 worked with Metropolis on preparations for the calculations. They had to wait to try their ideas out however, because the ENIAC was to be moved from Philadelphia where it was built to the army's Ballistics Research Laboratory in Maryland. For a modern computer this would not be a problem, but for the gigantic ENIAC, with its thousands of fragile components, it was a difficult task, and there were many who did not believe the computer would survive the journey. It did, however, and by the end of the year it was working once again in its new home. Before von Neumann and the others put it to work on the calculations for the hydrogen bomb, Richard Clippinger of the Ballistics Lab suggested a modification to the machine which allowed it to store programs in its electronic memory. Previously a program had to be set up by plugging and unplugging cables at the front of the machine, an arduous task which made the machine inflexible and inconvenient to use. Von Neumann was in favour of changing to the new "stored program" model, and Nick Metropolis and von Neumann's wife, Klari, made the necessary modifications to the computer themselves. It was the end of 1947 before the machine was at last ready, and Metropolis and von Neumann set to work on the planned Monte Carlo calculations.

The early neutron diffusion calculations were an impressive success, but Metropolis and von Neumann were not able to publish their results, because they were classified as secret. Over the following two years however, they and others, including Stan Ulam and Stanley Frankel, applied the new statistical sampling method to a variety of more mundane problems in physics, such as the calculation of the properties of hard-sphere gases in two and three dimensions, and published a number of papers which drew the world's attention to this emerging technique. The 1949 paper by Metropolis and Ulam on statistical techniques for studying integro-differential equations is of interest because it contained in its title the first use of the term "Monte Carlo" to describe this type of calculation. Also in 1949 the first conference on Monte Carlo methods was held in Los Alamos, attracting more than a hundred participants. It was quickly followed by another similar meeting in Gainesville, Florida.

The calculations received a further boost in 1948 with the arrival at Los Alamos of a new computer, humorously called the MANIAC. (Apparently the name was suggested by Enrico Fermi, who was tiring of computers with contrived acronyms for names—he claimed that it stood for "Metropolis and Neumann Invent Awful Contraption". Nowadays, with all our computers called things like XFK-23/z we would no doubt appreciate a few pronounceable names.) Apart from the advantage of being in New Mexico rather than Maryland, the MANIAC was a significant technical improvement over the ENIAC which Presper Eckert (1980), its principal architect, refers to as a "hastily built first try". It was faster and contained a larger memory (40 kilobits, or 5 kilobytes in modern terms). It was built under the direction of Metropolis, who had been lured back to Los Alamos after a brief stint on the faculty at Chicago by the prospect of the new machine. The design was based on ideas put forward by John von Neumann and incorporated a number of technical refinements proposed by Jim Richardson, an engineer working on the project. A still more sophisticated computer, the MANIAC 2, was built at Los Alamos two years later, and both machines remained in service until the late fifties, producing a stream of results, many of which have proved to be seminal contributions to the field of Monte Carlo simulation. Of particular note to us is the publication in 1953 of the paper by Nick Metropolis, Marshall and Arianna Rosenbluth, and Edward and Mici Teller, in which they describe for the first time the Monte Carlo technique that has come to be known as the Metropolis algorithm. This algorithm was the first example of a thermal "importance sampling" method, and it is to this day easily the most widely used such method. We will be discussing it in some detail in Chapter 3. Also of interest are the Monte Carlo studies of nuclear cascades performed by Antony Turkevich and Nick Metropolis, and Edward Teller's work on phase changes in interacting hard-sphere gases using the Metropolis algorithm.

The exponential growth in computer power since those early days is by now a familiar story to us all, and with this increase in computational resources Monte Carlo techniques have looked deeper and deeper into the subject of statistical physics. Monte Carlo simulations have also become more accurate as a result of the invention of new algorithms. Particularly in the last twenty years, many new ideas have been put forward, of which we describe a good number in the rest of this book.

Problems

1.1 "If a system is in equilibrium with a thermal reservoir at temperature T, the probability of its having a total energy E varies with E in proportion to $e^{-\beta E}$." True or false?

1.2 A certain simple system has only two energy states, with energies E_0 and E_1, and transitions between the two states take place at rates $R(0 \rightarrow 1) = R_0 \exp[-\beta(E_1 - E_0)]$ and $R(1 \rightarrow 0) = R_0$. Solve the master equation (1.1) for the probabilities w_0 and w_1 of occupation of the two states as a function of time with the initial conditions $w_0 = 0$, $w_1 = 1$. Show that as $t \rightarrow \infty$ these solutions tend to the Boltzmann probabilities, Equation (1.5).

1.3 A slightly more complex system contains N distinguishable particles, each of which can be in one of two boxes. The particles in the first box have energy $E_0 = 0$ and the particles in the second have energy E_1, and particles are allowed to move back and forward between the boxes under the influence of thermal excitations from a reservoir at temperature T. Find the partition function for this system and then use this result to calculate the internal energy.

1.4 Solve the Ising model, whose Hamiltonian is given in Equation (1.30), in one dimension for the case where $B = 0$ as follows. Define a new set of variables σ_i which take values 0 and 1 according to $\sigma_i = \frac{1}{2}(1 - s_i s_{i+1})$ and rewrite the Hamiltonian in terms of these variables for a system of N spins with periodic boundary conditions. Show that the resulting system is equivalent to the one studied in Problem 1.3 in the limit of large N and hence calculate the internal energy as a function of temperature.

2

The principles of equilibrium thermal Monte Carlo simulation

In Section 1.3.1 we looked briefly at the general ideas behind equilibrium thermal Monte Carlo simulations. In this chapter we discuss these ideas in more detail in preparation for the discussion in the following chapters of a variety of specific algorithms for use with specific problems. The three crucial ideas that we introduce in this chapter are "importance sampling", "detailed balance" and "acceptance ratios". If you know what these phrases mean, you can understand most of the thermal Monte Carlo simulations that have been performed in the last thirty years.

2.1 The estimator

The usual goal in the Monte Carlo simulation of a thermal system is the calculation of the expectation value $\langle Q \rangle$ of some observable quantity Q, such as the internal energy in a model of a gas, or the magnetization in a magnetic model. As we showed in Section 1.3, the ideal route to calculating such an expectation, that of averaging the quantity of interest over all states μ of the system, weighting each with its own Boltzmann probability

$$\langle Q \rangle = \frac{\sum_\mu Q_\mu \, \mathrm{e}^{-\beta E_\mu}}{\sum_\mu \mathrm{e}^{-\beta E_\mu}} \tag{2.1}$$

is only tractable in the very smallest of systems. In larger systems, the best we can do is average over some subset of the states, though this necessarily introduces some inaccuracy into the calculation. Monte Carlo techniques work by choosing a subset of states at random from some probability distribution p_μ which we specify. Suppose we choose M such states $\{\mu_1 \ldots \mu_M\}$.

Our best estimate of the quantity Q will then be given by

$$Q_M = \frac{\sum_{i=1}^{M} Q_{\mu_i} p_{\mu_i}^{-1} \mathrm{e}^{-\beta E_{\mu_i}}}{\sum_{j=1}^{M} p_{\mu_j}^{-1} \mathrm{e}^{-\beta E_{\mu_j}}}. \tag{2.2}$$

Q_M is called the **estimator** of Q. It has the property that, as the number M of states sampled increases, it becomes a more and more accurate estimate of $\langle Q \rangle$, and when $M \to \infty$ we have $Q_M = \langle Q \rangle$.

The question we would like to answer now is how should we choose our M states in order that Q_M be an accurate estimate of $\langle Q \rangle$? In other words, how should we choose the probability distribution p_μ? The simplest choice is to pick all states with equal probability; in other words make all p_μ equal. Substituting this choice into Equation (2.2), we get

$$Q_M = \frac{\sum_{i=1}^{M} Q_{\mu_i} \mathrm{e}^{-\beta E_{\mu_i}}}{\sum_{j=1}^{M} \mathrm{e}^{-\beta E_{\mu_j}}}. \tag{2.3}$$

It turns out however, that this is usually a rather poor choice to make. In most numerical calculations it is only possible to sample a very small fraction of the total number of states. Consider, for example, the Ising model of Section 1.2.2 again. A small three-dimensional cubic system of $10 \times 10 \times 10$ Ising spins would have $2^{1000} \simeq 10^{300}$ states, and a typical numerical calculation could only hope to sample up to about 10^8 of those in a few hours on a good computer, which would mean we were only sampling one in every 10^{292} states of the system, a very small fraction indeed. The estimator given above is normally a poor guide to the value of $\langle Q \rangle$ under these circumstances. The reason is that one or both of the sums appearing in Equation (2.1) may be dominated by a small number of states, with all the other states, the vast majority, contributing a negligible amount even when we add them all together. This effect is often especially obvious at low temperatures, where these sums may be dominated by a hundred states, or ten states, or even one state, because at low temperatures there is not enough thermal energy to lift the system into the higher excited states, and so it spends almost all of its time sitting in the ground state, or one of the lowest of the excited states. In the example described above, the chances of one of the 10^8 random states we sample in our simulation being the ground state are one in 10^{292}, which means there is essentially no chance of our picking it, which makes Q_M a very inaccurate estimate of $\langle Q \rangle$ if the sums are dominated by the contribution from this state.

On the other hand, if we had some way of knowing which states made the important contributions to the sums in Equation (2.1) and if we could pick our sample of M states from just those states and ignore all the others, we could get a very good estimate of $\langle Q \rangle$ with only a small number of terms. This is the essence of the idea behind thermal Monte Carlo methods. The technique for picking out the important states from amongst the very large number of possibilities is called **importance sampling**.

2.2   Importance sampling

As discussed in Section 1.1, we can regard an expectation value as a time average over the states that a system passes through during the course of a measurement. We do not assume that the system passes through every state during the measurement, even though every state appears in the sums of Equation (2.1). When you count how many states a typical system has you realize that this would never be possible. For instance, consider again the example we took in the last chapter of a litre container of gas at room temperature and atmospheric pressure. Such a system contains on the order of 10^{22} molecules. Typical speeds for these molecules are in the region of 100 m s^{-1}, giving them a de Broglie wavelength of around 10^{-10} m. Each molecule will then have about 10^{27} different quantum states within the one litre box, and the complete gas will have around $(10^{27})^{10^{22}}$ states, which is a spectacularly large number.[1] The molecules will change from one state to another when they undergo collisions with one another or with the walls of the container, which they do at a rate of about 10^9 collisions per second, or 10^{31} changes of state per second for the whole gas. At this rate, it will take about $10^{10^{23}}$ times the lifetime of the universe for our litre of gas to move through every possible state. Clearly then, our laboratory systems are only sampling the tiniest portion of their state spaces during the time that we conduct our experiments on them. In effect, real systems are carrying out a sort of Monte Carlo calculation of their own properties; they are "analogue computers" which evaluate expectations by taking a small but representative sample of their own states and averaging over that sample.[2] So it should not come as a great surprise to learn that we can also perform a reasonable calculation of the properties of a system using a simulation which only samples a small fraction of its states.

In fact, our calculations are often significantly better than this simple argument suggests. In Section 1.2.1 we showed that the range of energies of the states sampled by a typical system is very small compared with the total

---

[1] Actually, this is probably an overestimate, since it counts states which are classically distinguishable but quantum mechanically identical. For the purpose of the present rough estimation however, it will do fine.

[2] There are some systems which, because they have certain conservation laws, will not in fact sample their state spaces representatively, and this can lead to discrepancies between theory and experiment. Special Monte Carlo techniques exist for simulating these "conservative" systems, and we will touch on one or two of them in the coming chapters. For the moment, however, we will make the assumption that our system takes a representative sample of its own states.

energy of the system—the ratio was about 10^{-20} in the case of our litre of gas, for instance. Similar arguments can be used to show that systems sample very narrow ranges of other quantities as well. The reason for this, as we saw, is that the system is not sampling all states with equal probability, but instead sampling them according to the Boltzmann probability distribution, Equation (1.5). If we can mimic this effect in our simulations, we can exploit these narrow ranges of energy and other quantities to make our estimates of such quantities very accurate. For this reason, we normally try to take a sample of the states of the system in which the likelihood of any particular one appearing is proportional to its Boltzmann weight. This is the most common form of importance sampling, and most of the algorithms in this book make use of this idea in one form or another.

Our strategy then is this: instead of picking our M states in such a way that every state of the system is as likely to get chosen as every other, we pick them so that the probability that a particular state μ gets chosen is $p_\mu = Z^{-1}e^{-\beta E_\mu}$. Then our estimator for $\langle Q \rangle$, Equation (2.2), becomes just

$$Q_M = \frac{1}{M}\sum_{i=1}^{M} Q_{\mu_i}. \tag{2.4}$$

Notice that the Boltzmann factors have now cancelled out of the estimator, top and bottom, leaving a particularly simple expression. This definition of Q_M works much better than (2.3), especially when the system is spending the majority of its time in a small number of states (such as, for example, the lowest-lying ones when we are at low temperatures), since these will be precisely the states that we pick most often, and the relative frequency with which we pick them will exactly correspond to the amount of time the real system would spend in those states.

The only remaining question is *how* exactly we pick our states so that each one appears with its correct Boltzmann probability. This is by no means a simple task. In the remainder of this chapter we describe the standard solution to the problem, which makes use of a "Markov process".

2.2.1 Markov processes

The tricky part of performing a Monte Carlo simulation is the generation of an appropriate random set of states according to the Boltzmann probability distribution. For a start, one cannot simply choose states at random and accept or reject them with a probability proportional to $e^{-\beta E_\mu}$. That would be no better than our original scheme of sampling states at random; we would end up rejecting virtually all states, since the probabilities for their acceptance would be exponentially small. Instead, almost all Monte Carlo schemes rely on **Markov processes** as the generating engine for the set of states used.

For our purposes, a Markov process is a mechanism which, given a system in one state μ, generates a new state of that system ν. It does so in a random fashion; it will not generate the same new state every time it is given the initial state μ. The probability of generating the state ν given μ is called the **transition probability** $P(\mu \rightarrow \nu)$ for the transition from μ to ν, and for a true Markov process all the transition probabilities should satisfy two conditions: (1) they should not vary over time, and (2) they should depend only on the properties of the current states μ and ν, and not on any other states the system has passed through. These conditions mean that the probability of the Markov process generating the state ν on being fed the state μ is the same every time it is fed the state μ, irrespective of anything else that has happened. The transition probabilities $P(\mu \rightarrow \nu)$ must also satisfy the constraint

$$\sum_{\nu} P(\mu \rightarrow \nu) = 1, \tag{2.5}$$

since the Markov process must generate *some* state ν when handed a system in the state μ. Note however, that the transition probability $P(\mu \rightarrow \mu)$, which is the probability that the new state generated will be the same as the old one, need not be zero. This amounts to saying there may be a finite probability that the Markov process will just stay in state μ.

In a Monte Carlo simulation we use a Markov process repeatedly to generate a **Markov chain** of states. Starting with a state μ, we use the process to generate a new one ν, and then we feed *that* state into the process to generate another λ, and so on. The Markov process is chosen specially so that when it is run for long enough starting from any state of the system it will eventually produce a succession of states which appear with probabilities given by the Boltzmann distribution. (We call the process of reaching the Boltzmann distribution "coming to equilibrium", since it is exactly the process that a real system goes through with its "analogue computer" as it reaches equilibrium at the ambient temperature.) In order to achieve this, we place two further conditions on our Markov process, in addition to the ones specified above, the conditions of "ergodicity" and "detailed balance"

2.2.2 Ergodicity

The **condition of ergodicity** is the requirement that it should be possible for our Markov process to reach any state of the system from any other state, if we run it for long enough. This is necessary to achieve our stated goal of generating states with their correct Boltzmann probabilities. Every state ν appears with some non-zero probability p_ν in the Boltzmann distribution, and if that state were inaccessible from another state μ no matter how long we continue our process for, then our goal is thwarted if we start in state μ: the probability of finding ν in our Markov chain of states will be zero, and not p_ν as we require it to be.

The condition of ergodicity tells us that we are allowed to make some of the transition probabilities of our Markov process zero, but that there must be at least one path of non-zero transition probabilities between any two states that we pick. In practice, most Monte Carlo algorithms set almost all of the transition probabilities to zero, and we must be careful that in so doing we do not create an algorithm which violates ergodicity. For most of the algorithms we describe in this book we will explicitly prove that ergodicity is satisfied before making use of the algorithm.

2.2.3   Detailed balance

The other condition we place on our Markov process is the **condition of detailed balance**. This condition is the one which ensures that it is the Boltzmann probability distribution which we generate after our system has come to equilibrium, rather than any other distribution. Its derivation is quite subtle. Consider first what it means to say that the system is in equilibrium. The crucial defining condition is that the rate at which the system makes transitions into and out of any state μ must be equal. Mathematically we can express this as[3]

$$\sum_\nu p_\mu P(\mu \to \nu) = \sum_\nu p_\nu P(\nu \to \mu). \tag{2.6}$$

Making use of the sum rule, Equation (2.5), we can simplify this to

$$p_\mu = \sum_\nu p_\nu P(\nu \to \mu). \tag{2.7}$$

For any set of transition probabilities satisfying this equation, the probability distribution p_μ will be an equilibrium of the dynamics of the Markov process. Unfortunately, however, simply satisfying this equation is not sufficient to guarantee that the probability distribution will tend to p_μ from any state of the system if we run the process for long enough. We can demonstrate this as follows.

The transition probabilities $P(\mu \to \nu)$ can be thought of as the elements of a matrix $\mathbf{P}$. This matrix is called the **Markov matrix** or the **stochastic matrix** for the Markov process. Let us return to the notation of Section 1.1, in which we denoted by $w_\mu(t)$, the probability that our system is in a state μ at time t. If we measure time in steps along our Markov chain, then the

---

[3]This equation is essentially just a discrete-time version of the one we would get if we were to set the derivative in the master equation, Equation (1.1), to zero.

probability $w_\nu(t+1)$ of being in state ν at time $t+1$ is given by[4]

$$w_\nu(t+1) = \sum_\mu P(\mu \to \nu)\, w_\mu(t). \qquad (2.8)$$

In matrix notation, this becomes

$$\mathbf{w}(t+1) = \mathbf{P} \cdot \mathbf{w}(t), \qquad (2.9)$$

where $\mathbf{w}(t)$ is the vector whose elements are the weights $w_\mu(t)$. If the Markov process reaches a simple equilibrium state $\mathbf{w}(\infty)$ as $t \to \infty$, then that state satisfies

$$\mathbf{w}(\infty) = \mathbf{P} \cdot \mathbf{w}(\infty). \qquad (2.10)$$

However, it is also possible for the process to reach a **dynamic equilibrium** in which the probability distribution $\mathbf{w}$ rotates around a number of different values. Such a rotation is called a **limit cycle**. In this case $\mathbf{w}(\infty)$ would satisfy

$$\mathbf{w}(\infty) = \mathbf{P}^n \cdot \mathbf{w}(\infty), \qquad (2.11)$$

where n is the length of the limit cycle. If we choose our transition probabilities (or equivalently our Markov matrix) to satisfy Equation (2.7) we guarantee that the Markov chain will have a simple equilibrium probability distribution p_μ, but it may also have any number of limit cycles of the form (2.11). This means that there is no guarantee that the actual states generated will have anything like the desired probability distribution.

We get around this problem by applying an additional condition to our transition probabilities thus:

$$p_\mu P(\mu \to \nu) = p_\nu P(\nu \to \mu). \qquad (2.12)$$

This is the condition of detailed balance. It is clear that any set of transition probabilities which satisfy this condition also satisfy Equation (2.6). (To prove it, simply sum both sides of Equation (2.12) over ν.) We can also show that this condition eliminates limit cycles. To see this, look first at the left-hand side of the equation, which is the probability of being in a state μ multiplied by the probability of making a transition from that state to another state ν. In other words, it is the overall rate at which transitions from μ to ν happen in our system. The right-hand side is the overall rate for the reverse transition. The condition of detailed balance tells us that on average the system should go from μ to ν just as often as it goes from ν to μ. In a limit cycle, in which the probability of occupation of some or all of the states changes in a cyclic fashion, there must be states for which this

---

[4]This equation is also closely related to Equation (1.1). The reader may like to work out how the one can be transformed into the other.

condition is violated on any particular step of the Markov chain; in order for the probability of occupation of a particular state to increase, for instance, there must be more transitions into that state than out of it, on average. The condition of detailed balance forbids dynamics of this kind and hence forbids limit cycles.

Once we remove the limit cycles in this way, it is straightforward to show that the system will always tend to the probability distribution p_μ as $t \to \infty$. As $t \to \infty$, $\mathbf{w}(t)$ will tend exponentially towards the eigenvector corresponding to the largest eigenvalue of $\mathbf{P}$. This may be obvious to you if you are familiar with stochastic matrices. If not, we prove it in Section 3.3.2. For the moment, let us take it as given. Looking at Equation (2.10) we see that the largest eigenvalue of the Markov matrix must in fact be one.[5] If limit cycles of the form (2.11) were present, then we could also have eigenvalues which are complex roots of one, but the condition of detailed balance prevents this from happening. Now look back at Equation (2.7) again. We can express this equation in matrix notation as

$$\mathbf{p} = \mathbf{P} \cdot \mathbf{p}. \tag{2.13}$$

In other words, if Equation (2.7) (or equivalently the condition of detailed balance) holds for our Markov process, then the vector $\mathbf{p}$ whose elements are the probabilities p_μ is precisely the one correctly normalized eigenvector of the Markov matrix which has eigenvalue one. Putting this together with Equation (2.10) we see that the equilibrium probability distribution over states $\mathbf{w}(\infty)$ is none other than $\mathbf{p}$, and hence $\mathbf{w}(t)$ must tend exponentially to $\mathbf{p}$ as $t \to \infty$.

There is another reason why detailed balance makes sense for Monte Carlo simulations: the "analogue computers" which constitute the real physical systems we are trying to mimic almost always obey the condition of detailed balance. The reason is that they are based on standard quantum or classical mechanics, which is time-reversal symmetric. If they did not obey detailed balance, then in equilibrium they could have one or more limit cycles around which the system passes in one particular direction. If we take such a system and reverse it in time, the motion around this cycle is also reversed, and it becomes clear that the dynamics of the system in equilibrium is not the same forward as it is in reverse. Such a violation of time-reversal symmetry is forbidden for most systems, implying that they must satisfy detailed balance. Although this does not mean that we are necessarily obliged

---

[5] All Markov matrices have at least one eigenvector with corresponding eigenvalue one, a fact which is easily proven since Equation (2.5) implies that the vector $(1, 1, 1, \ldots)$ is a left eigenvector of $\mathbf{P}$ with eigenvalue one. It is possible to have more than one eigenvector with eigenvalue one if the states of the system divide into two or more mutually inaccessible subsets. However, if the condition of ergodicity is satisfied then such subsets are forbidden and hence there is only one such eigenvector.

to enforce detailed balance in our simulations as well, it is helpful if we do, because it makes the behaviour of our model system more similar to that of the real one we are trying to understand.

So, we have shown that we can arrange for the probability distribution of states generated by our Markov process to tend to any distribution p_μ we please by choosing a set of transition probabilities which satisfy Equation (2.12). Given that we wish the equilibrium distribution to be the Boltzmann distribution,[6] clearly we want to choose the values of p_μ to be the Boltzmann probabilities, Equation (1.5). The detailed balance equation then tells us that the transition probabilities should satisfy

$$\frac{P(\mu \to \nu)}{P(\nu \to \mu)} = \frac{p_\nu}{p_\mu} = e^{-\beta(E_\nu - E_\mu)}. \qquad (2.14)$$

This equation and Equation (2.5) are the constraints on our choice of transition probabilities $P(\mu \to \nu)$. If we satisfy these, as well as the condition of ergodicity, then the equilibrium distribution of states in our Markov process will be the Boltzmann distribution. Given a suitable set of transition probabilities, our plan is then to write a computer program which implements the Markov process corresponding to these transition probabilities so as to generate a chain of states. After waiting a suitable length of time[7] to allow the probability distribution of states $w_\mu(t)$ to get sufficiently close to the Boltzmann distribution, we average the observable Q that we are interested in over M states and we have calculated the estimator Q_M defined in Equation (2.4). A number of refinements on this outline are possible and we will discuss some of those in the remainder of this chapter and in later chapters of the book, but this is the basic principle on which virtually all modern equilibrium Monte Carlo calculations are based.

Our constraints still leave us a good deal of freedom over how we choose the transition probabilities. There are many ways in which to satisfy them. One simple choice for example is

$$P(\mu \to \nu) \propto e^{-\frac{1}{2}\beta(E_\mu - E_\nu)}, \qquad (2.15)$$

although as we will show in Section 3.1 this choice is not a very good one. There are some other choices which are known to work well in many cases, such as the "Metropolis algorithm" proposed by Metropolis and co-workers in 1953, and we will discuss the most important of these in the coming chapters. However, it must be stressed—and this is one of the most important

---

[6]Occasionally, in fact, we want to generate equilibrium distributions other than the Boltzmann distribution. An example is the entropic sampling algorithm of Section 6.3. In this case the arguments here still apply. We simply feed our required distribution into the condition of detailed balance.

[7]Exactly how long we have to wait can be a difficult thing to decide. A number of possible criteria are discussed in Section 3.2.

things this book has to say—that the standard algorithms are *very rarely* the best ones for solving new problems with. In most cases they will work, and in some cases they will even give quite good answers, but you can almost always do a better job by giving a little extra thought to choosing the best set of transition probabilities to construct an algorithm that will answer the particular questions that you are interested in. A purpose-built algorithm can often give a much faster simulation than an equivalent standard algorithm, and the improvement in efficiency can easily make the difference between finding an answer to a problem and not finding one.

2.3 Acceptance ratios

Our little summary above makes rather light work of the problems of constructing a Monte Carlo algorithm. Given a desired set of transition probabilities $P(\mu \to \nu)$ satisfying the conditions (2.5) and (2.14), we say, we simply concoct some Markov process that generates states with exactly those transition probabilities, and *presto!* we produce a string of states of our system with exactly their correct Boltzmann probabilities. However, it is often very far from obvious what the appropriate Markov process is that has the required transition probabilities, and finding one can be a haphazard, trial-and-error process. For some problems we can use known algorithms such as the Metropolis method (see Section 3.1), but for many problems the standard methods are far from ideal, and we will do much better if we can tailor a new algorithm to our specific needs. But though we may be able to suggest many candidate Markov processes—different ways of creating a new state ν from an old one μ—still we may not find one which gives exactly the right set of transition probabilities. The good news however is that we don't have to. In fact it turns out that we can choose any algorithm we like for generating the new states, and still have our desired set of transition probabilities, by introducing something called an **acceptance ratio**. The idea behind the trick is this.

We mentioned in Section 2.2.1 that we are allowed to make the "stay-at-home" transition probability $P(\mu \to \mu)$ non-zero if we want. If we set $\nu = \mu$ in Equation (2.14), we get the simple tautology $1 = 1$, which means that the condition of detailed balance is always satisfied for $P(\mu \to \mu)$, no matter what value we choose for it. This gives us some flexibility about how we choose the other transition probabilities with $\mu \neq \nu$. For a start, it means that we can adjust the value of any $P(\mu \to \nu)$ and keep the sum rule (2.5) satisfied, by simply compensating for that adjustment with an equal but opposite adjustment of $P(\mu \to \mu)$. The only thing we need to watch is that $P(\mu \to \mu)$ never passes out of its allowed range between zero and one. If we make an adjustment like this in $P(\mu \to \nu)$, we can also arrange for Equation (2.14) to remain satisfied, by simultaneously making a change in

$P(\nu \to \mu)$, so that the ratio of the two is preserved.

It turns out that these considerations actually give us enough freedom that we can make the transition probabilities take any set of values we like by tweaking the values of the probabilities $P(\mu \to \mu)$. To see this, we break the transition probability down into two parts:

$$P(\mu \to \nu) = g(\mu \to \nu)\, A(\mu \to \nu). \tag{2.16}$$

The quantity $g(\mu \to \nu)$ is the **selection probability**, which is the probability, given an initial state μ, that our algorithm will generate a new target state ν, and $A(\mu \to \nu)$ is the acceptance ratio (sometimes also called the "acceptance probability"). The acceptance ratio says that if we start off in a state μ and our algorithm generates a new state ν from it, we should accept that state and change our system to the new state ν a fraction of the time $A(\mu \to \nu)$. The rest of the time we should just stay in the state μ. We are free to choose the acceptance ratio to be any number we like between zero and one; choosing it to be zero for all transitions is equivalent to choosing $P(\mu \to \mu) = 1$, which is the largest value it can take, and means that we will never leave the state μ. (Not a very desirable situation. We would never choose an acceptance ratio of zero for an actual calculation.)

This gives us complete freedom about how we choose the selection probabilities $g(\mu \to \nu)$, since the constraint (2.14) only fixes the ratio

$$\frac{P(\mu \to \nu)}{P(\nu \to \mu)} = \frac{g(\mu \to \nu)A(\mu \to \nu)}{g(\nu \to \mu)A(\nu \to \mu)}. \tag{2.17}$$

The ratio $A(\mu \to \nu)/A(\nu \to \mu)$ can take any value we choose between zero and infinity, which means that both $g(\mu \to \nu)$ and $g(\nu \to \mu)$ can take any values we like.

Our other constraint, the sum rule of Equation (2.5), is still satisfied, since the system must end up in *some* state after each step in the Markov chain, even if that state is just the state we started in.

So, in order to create our Monte Carlo algorithm what we actually do is think up an algorithm which generates random new states ν given old ones μ, with some set of probabilities $g(\mu \to \nu)$, and then we accept or reject those states with acceptance ratios $A(\mu \to \nu)$ which we choose to satisfy Equation (2.17). This will then satisfy all the requirements for the transition probabilities, and so produce a string of states which, when the algorithm reaches equilibrium, will each appear with their correct Boltzmann probability.

This all seems delightful, but there is a catch which we must always bear in mind, and which is one of the most important considerations in the design of Monte Carlo algorithms. If the acceptance ratios for our moves are low, then the algorithm will on most time steps simply stay in the state that it is in, and not go anywhere. The step on which it actually accepts a change to a new state will be rare, and this is wasteful of time. We want an algorithm that moves nimbly about state space and samples a wide selection of different states. We don't want to take a million time steps and find that our algorithm has only sampled a dozen states. The solution to this problem is to make the acceptance ratio as close to unity as possible. One way to do this is to note that Equation (2.17) fixes only the ratio $A(\mu \rightarrow \nu)/A(\nu \rightarrow \mu)$ of the acceptance ratios for the transitions in either direction between any two states. Thus we are free to multiply both $A(\mu \rightarrow \nu)$ and $A(\nu \rightarrow \mu)$ by the same factor, and the equation will still be obeyed. The only constraint is that both acceptance ratios should remain between zero and one. In practice then, what we do is to set the larger of the two acceptance ratios to one, and have the other one take whatever value is necessary for the ratio of the two to satisfy (2.17). This ensures that the acceptance ratios will be as large as they can be while still satisfying the relevant conditions, and indeed that the ratio in one direction will be unity, which means that in that direction at least, moves will always be accepted.

However, the best thing we can do to keep the acceptance ratios large is to try to embody in the selection probabilities $g(\mu \rightarrow \nu)$ as much as we can of the dependence of $P(\mu \rightarrow \nu)$ on the characteristics of the states μ and ν, and put as little as we can in the acceptance ratio. The ideal algorithm is one in which the new states are selected with exactly the correct transition probabilities all the time, and the acceptance ratio is always one. A good algorithm is one in which the acceptance probability is usually close to one. Much of the effort invested in the algorithms described in this book is directed at making the acceptance ratios large.

2.4   Continuous time Monte Carlo

There is a another twist we can add to our Markov process to allow ourselves further freedom about the way in which we choose states, without letting the acceptance ratios get too low. It is called **continuous time Monte Carlo**, or sometimes the **BKL algorithm**, after Bortz, Kalos and Lebowitz (1975), who invented it. Continuous time Monte Carlo is not nearly as widely used as it ought to be; it is an important and powerful technique and many calculations can be helped enormously by making use of it.

Consider a system at low temperature. Such systems are always a problem where Monte Carlo methods are concerned—cool systems move from state to state very slowly in real life and the problem is no less apparent in simulations. A low-temperature system is a good example of the sort of problem system that was described in the last section. Once it reaches equilibrium at its low temperature it will spend a lot of its time in the ground state. Maybe it will spend a hundred consecutive time-steps of the simulation in the ground state, then move up to the first excited state for one time-step and then relax back to the ground state again. Such behaviour is not unreasonable for a cold system but we waste a lot of computer time simulating it. Time-step after time-step our algorithm selects a possible move to some excited state, but the acceptance ratio is very low and virtually all of these possible moves are rejected, and the system just ends up spending most of its time in the ground state.

Well, what if we were to accept that this is the case, and take a look at the acceptance ratio for a move from the ground state to the first excited state, and say to ourselves, "Judging by this acceptance ratio, this system is going to spend a hundred time-steps in the ground state before it accepts a move to the first excited state". Then we could jump the gun by *assuming* that the system will do this, miss out the calculations involved in the intervening useless one hundred time-steps, and progress straight to the one time-step in which something interesting happens. This is the essence of the idea behind the continuous time method. In this technique, we have a time-step which corresponds to a varying length of time, depending on how long we expect the system to remain in its present state before moving to a new one. Then when we come to take the average of our observable Q over many states, we weight the states in which the system spends longest the most heavily—the calculation of the estimator of Q is no more than a time average, so each value Q_μ for Q in state μ should be weighted by how long the system spends in that state.

How can we adapt our previous ideas concerning the transition probabilities for our Markov process to take this new idea into account? Well, assuming that the system is in some state μ, we can calculate how long a time Δt (measured in steps of the simulation) it will stay there for before a move to another state is accepted by considering the "stay-at-home" probability $P(\mu \to \mu)$. The probability that it is still in this same state μ after t time-steps is just

$$[P(\mu \to \mu)]^t = e^{t \log P(\mu \to \mu)}, \tag{2.18}$$

and so the time-scale Δt is

$$\Delta t = -\frac{1}{\log P(\mu \to \mu)} = -\frac{1}{\log[1 - \sum_{\nu \neq \mu} P(\mu \to \nu)]}$$
$$\simeq \frac{1}{\sum_{\nu \neq \mu} P(\mu \to \nu)}. \tag{2.19}$$

So, if we can calculate this quantity Δt, then rather than wait this many time-steps for a Monte Carlo move to get accepted, we can simply pretend that we have done the waiting and go right ahead and change the state of the system to a new state $\nu \neq \mu$. Which state should we choose for ν? We should choose one at random, but in proportion to $P(\mu \to \nu)$. Thus our continuous time Monte Carlo algorithm consists of the following steps:

1. We calculate the probabilities $P(\mu \to \nu)$ for transitions to all states which can be reached in one Monte Carlo step from the current state μ. We choose a new state ν with probability proportional to $P(\mu \to \nu)$ and change the state of the system to ν.

2. Using our values for the $P(\mu \to \nu)$ we also calculate the time interval Δt. Notice that we have to recalculate Δt at each step, since in general it will change from one step to the next.

3. We increment the time t by Δt, to mimic the effect of waiting Δt Monte Carlo steps. The variable t keeps a record of how long the simulation has gone on for in "equivalent Monte Carlo steps".

While this technique is in many respects a very elegant solution to the problem of simulating a system at low temperatures (or any other system which has a low acceptance ratio), it does suffer from one obvious drawback, which is that step (1) above involves calculating $P(\mu \to \nu)$ for every possible state ν which is accessible from μ. There may be very many such states (for some systems the number goes up exponentially with the size of the system), and so this step may take a very long time. However, in some cases, it turns out that the set of transition probabilities is very similar from one step of the algorithm to the next, only a few of them changing at each step, and hence it is possible to keep a table of probabilities and update only a few entries at each step to keep the table current. In cases such as these the continuous time method becomes very efficient and can save us a great deal of CPU time, despite being more complex than the accept/reject method discussed in the previous section. One example of a continuous time Monte Carlo algorithm is presented in Section 5.2.1 for the conserved-order-parameter Ising model.

In the next few chapters, we will examine a number of common models used for calculating the equilibrium properties of condensed matter systems, and show how the general ideas presented in this chapter can be used to find efficient numerical solutions to these physical problems.

Problems

2.1 Derive Equation (2.8) from Equation (1.1).

2.2 Consider a system which has just three energy states, with energies $E_0 < E_1 < E_2$. Suppose that the only allowed transitions are ones of the form $\mu \to \nu$, where $\nu = (\mu + 1) \bmod 3$. Such a system cannot satisfy detailed balance. Show nonetheless that it is possible to choose the transition probabilities $P(\mu \to \nu)$ so that the Boltzmann distribution is an equilibrium of the dynamics.

3

The Ising model and the Metropolis algorithm

In Section 1.2.2 we introduced the Ising model, which is one of the simplest and best studied of statistical mechanical models. In this chapter and the next we look in detail at the Monte Carlo methods that have been used to investigate the properties of this model. As well as demonstrating the application of the basic principles described in the last chapter, the study of the Ising model provides an excellent introduction to the most important Monte Carlo algorithms in use today. Along the way we will also look at some of the tricks used for implementing Monte Carlo algorithms in computer programs and at some of the standard techniques used to analyse the data those programs generate.

To recap briefly, the Ising model is a simple model of a magnet, in which dipoles or "spins" s_i are placed on the sites i of a lattice. Each spin can take either of two values: $+1$ and -1. If there are N sites on the lattice, then the system can be in 2^N states, and the energy of any particular state is given by the Ising Hamiltonian:

$$H = -J \sum_{\langle ij \rangle} s_i s_j - B \sum_i s_i, \qquad (3.1)$$

where J is an interaction energy between nearest-neighbour spins $\langle ij \rangle$, and B is an external magnetic field. We are interested in simulating an Ising system of finite size using Monte Carlo methods, so that we can estimate the values of quantities such as the magnetization m (Equation (1.34)) or the specific heat c (Equation (1.37)) at any given temperature. Most of the interesting questions concerning the Ising model can be answered by performing simulations in zero magnetic field $B = 0$, so for the moment at least we will concentrate on this case.

3.1   The Metropolis algorithm

The very first Monte Carlo algorithm we introduce in this book is the most famous and widely used algorithm of them all, the **Metropolis algorithm**, which was introduced by Nicolas Metropolis and his co-workers in a 1953 paper on simulations of hard-sphere gases (Metropolis *et al.* 1953). We will use this algorithm to illustrate many of the general concepts involved in a real Monte Carlo calculation, including equilibration, measurement of expectation values, and the calculation of errors. First however, let us see how the algorithm is arrived at, and how one might go about implementing it on a computer.

The derivation of the Metropolis algorithm follows exactly the plan we outlined in Section 2.3. We choose a set of selection probabilities $g(\mu \to \nu)$, one for each possible transition from one state to another, $\mu \to \nu$, and then we choose a set of acceptance probabilities $A(\mu \to \nu)$ such that Equation (2.17) satisfies the condition of detailed balance, Equation (2.14). The algorithm works by repeatedly choosing a new state ν, and then accepting or rejecting it at random with our chosen acceptance probability. If the state is accepted, the computer changes the system to the new state ν. If not, it just leaves it as it is. And then the process is repeated again and again.

The selection probabilities $g(\mu \to \nu)$ should be chosen so that the condition of ergodicity—the requirement that every state be accessible from every other in a finite number of steps—is fulfilled (see Section 2.2.2). This still leaves us a good deal of latitude about how they are chosen; given an initial state μ we can generate any number of candidate states ν simply by flipping different subsets of the spins on the lattice. However, as we demonstrated in Section 1.2.1, the energies of systems in thermal equilibrium stay within a very narrow range—the energy fluctuations are small by comparison with the energy of the entire system. In other words, the real system spends most of its time in a subset of states with a narrow range of energies and rarely makes transitions that change the energy of the system dramatically. This tells us that we probably don't want to spend much time in our simulation considering transitions to states whose energy is very different from the energy of the present state. The simplest way of achieving this in the Ising model is to consider only those states which differ from the present one by the flip of a single spin. An algorithm which does this is said to have **single-spin-flip dynamics**. The algorithm we describe in this chapter has single-spin-flip dynamics, although this is not what makes it the Metropolis algorithm. (As discussed below, it is the particular choice of acceptance ratio that characterizes the Metropolis algorithm. Our algorithm would still be a Metropolis algorithm even if it flipped many spins at once.)

Using single-spin-flip dynamics guarantees that the new state ν will have an energy E_ν differing from the current energy E_μ by at most $2J$ for each bond between the spin we flip and its neighbours. For example, on a square lattice in two dimensions each spin has four neighbours, so the maximum difference in energy would be $8J$. The general expression is $2zJ$, where z is the **lattice coordination number**, i.e., the number of neighbours that each site on the lattice has.[1] Using single-spin-flip dynamics also ensures that our algorithm obeys ergodicity, since it is clear that we can get from any state to any other on a finite lattice by flipping one by one each of the spins by which the two states differ.

In the Metropolis algorithm the selection probabilities $g(\mu \to \nu)$ for each of the possible states ν are all chosen to be equal. The selection probabilities of all other states are set to zero. Suppose there are N spins in the system we are simulating. With single-spin-flip dynamics there are then N different spins that we could flip, and hence N possible states ν which we can reach from a given state μ. Thus there are N selection probabilities $g(\mu \to \nu)$ which are non-zero, and each of them takes the value

$$g(\mu \to \nu) = \frac{1}{N}. \tag{3.2}$$

With these selection probabilities, the condition of detailed balance, Equation (2.14), takes the form

$$\frac{P(\mu \to \nu)}{P(\nu \to \mu)} = \frac{g(\mu \to \nu)A(\mu \to \nu)}{g(\nu \to \mu)A(\nu \to \mu)} = \frac{A(\mu \to \nu)}{A(\nu \to \mu)} = e^{-\beta(E_\nu - E_\mu)}. \tag{3.3}$$

Now we have to choose the acceptance ratios $A(\mu \to \nu)$ to satisfy this equation. As we pointed out in Section 2.2.3, one possibility is to choose

$$A(\mu \to \nu) = A_0 e^{-\frac{1}{2}\beta(E_\nu - E_\mu)}. \tag{3.4}$$

The constant of proportionality A_0 cancels out in Equation (3.3), so we can choose any value for it that we like, except that $A(\mu \to \nu)$, being a probability, should never be allowed to become greater than one. As we mentioned above, the largest difference in energy $E_\nu - E_\mu$ that we can have between our two states is $2zJ$, where z is the lattice coordination number. That means that the largest value of $e^{-\frac{1}{2}\beta(E_\nu - E_\mu)}$ is $e^{\beta zJ}$. Thus, in order to make sure $A(\mu \to \nu) \leq 1$ we want to choose

$$A_0 \leq e^{-\beta zJ}. \tag{3.5}$$

To make the algorithm as efficient as possible, we want the acceptance probabilities to be as large as possible, so we make A_0 as large as it is allowed to

---

[1] This is not the same thing as the "spin coordination number" which we introduce in Chapter 5. The spin coordination number is the number of spins j neighbouring i which have the same value as spin i: $s_j = s_i$.

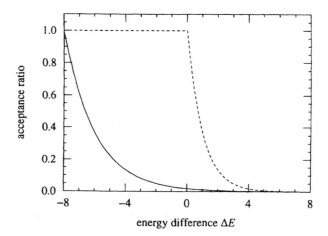

FIGURE 3.1   Plot of the acceptance ratio given in Equation (3.6) (solid line). This acceptance ratio gives rise to an algorithm which samples the Boltzmann distribution correctly, but is very inefficient, since it rejects the vast majority of the moves it selects for consideration. The Metropolis acceptance ratio (dashed line) is much more efficient.

be, which gives us

$$A(\mu \rightarrow \nu) = e^{-\frac{1}{2}\beta(E_\nu - E_\mu + 2zJ)}. \tag{3.6}$$

This is not the Metropolis algorithm (we are coming to that), but using this acceptance probability we can perform a Monte Carlo simulation of the Ising model, and it will correctly sample the Boltzmann distribution. However, the simulation will be very inefficient, because the acceptance ratio, Equation (3.6), is very small for almost all moves. Figure 3.1 shows the acceptance ratio (solid line) as a function of the energy difference $\Delta E = E_\nu - E_\mu$ over the allowed range of values for a simulation with $\beta = J = 1$ and a lattice coordination number $z = 4$, as on a square lattice for example. As we can see, although $A(\mu \rightarrow \nu)$ starts off at 1 for $\Delta E = -8$, it quickly falls to only about 0.13 at $\Delta E = -4$, and to only 0.02 when $\Delta E = 0$. The chances of making any move for which $\Delta E > 0$ are pitifully small, and in practice this means that an algorithm making use of this acceptance ratio would be tremendously slow, spending most of its time rejecting moves and not flipping any spins at all. The solution to this problem is as follows.

In Equation (3.4) we have assumed a particular functional form for the acceptance ratio, but the condition of detailed balance, Equation (3.3), doesn't actually require that it take this form. Equation (3.3) only specifies the ratio of pairs of acceptance probabilities, which still leaves us quite a lot of room to manœuvre. In fact, as we pointed out in Section 2.3, when given a constraint like (3.3) the way to maximize the acceptance ratios (and therefore produce the most efficient algorithm) is always to give the larger of the two ratios the largest value possible—namely 1—and then adjust the other to satisfy the constraint. To see how that works out in this case, suppose that of the two states μ and ν we are considering here, μ has the lower energy and ν the higher: $E_\mu < E_\nu$. Then the larger of the two acceptance ratios is $A(\nu \rightarrow \mu)$, so we set that equal to one. In order to satisfy Equation (3.3), $A(\mu \rightarrow \nu)$ must then take the value $e^{-\beta(E_\nu - E_\mu)}$. Thus the optimal algorithm is one in which

$$A(\mu \rightarrow \nu) = \begin{cases} e^{-\beta(E_\nu - E_\mu)} & \text{if } E_\nu - E_\mu > 0 \\ 1 & \text{otherwise.} \end{cases} \tag{3.7}$$

In other words, if we select a new state which has an energy lower than or equal to the present one, we should always accept the transition to that state. If it has a higher energy then we maybe accept it, with the probability given above. This is the Metropolis algorithm for the Ising model with single-spin-flip dynamics. It is Equation (3.7) which makes it the Metropolis algorithm. This is the part that was pioneered by Metropolis and co-workers in their paper on hard-sphere gases, and any algorithm, applied to any model, which chooses selection probabilities according to a rule like (3.7) can be said to be a Metropolis algorithm. At first, this rule may seem a little strange, especially the part about how we *always* accept a move that will lower the energy of the system. The first algorithm we suggested, Equation (3.6), seems much more natural in this respect, since it sometimes rejects moves to lower energy. However, as we have shown, the Metropolis algorithm satisfies detailed balance, and is by far the more efficient algorithm, so, natural or not, it has become the algorithm of choice in the overwhelming majority of Monte Carlo studies of simple statistical mechanical models in the last forty years. We have also plotted Equation (3.7) in Figure 3.1 (dashed line) for comparison between the two algorithms.

3.1.1 Implementing the Metropolis algorithm

Let us look now at how we would actually go about writing a computer program to perform a simulation of the Ising model using the Metropolis algorithm. For simplicity we will continue to focus on the case of zero magnetic field $B = 0$, although the generalization to the case $B \neq 0$ is not hard (see Problem 3.1). In fact almost all the past studies of the Ising model, including Onsager's exact solution in two dimensions, have looked only at the zero-field case.

First, we need an actual lattice of spins to work with, so we would define a set of N variables—an array—which can take the values ± 1. Probably we would use integer variables, so it would be an integer array. Normally, we apply **periodic boundary conditions** to the array. That is, we specify that the spins on one edge of the lattice are neighbours of the corresponding spins on the other edge. This ensures that all spins have the same number of neighbours and local geometry, and that there are no special edge spins which have different properties from the others; all the spins are equivalent and the system is completely translationally invariant. In practice this considerably improves the quality of the results from our simulation.

A variation on the idea of periodic boundary conditions is to use "helical boundary conditions" which are only very slightly different from periodic ones and possess all the same benefits but are usually considerably simpler to implement and can make our simulation significantly faster. The various types of boundary conditions and their implementation are described in detail in Section 13.1, along with methods for representing most common lattice geometries using arrays.

Next we need to decide at what temperature, or alternatively at what value of β we want to perform our simulation, and we need to choose some starting value for each of the spins—the initial state of the system. In a lot of cases, the initial state we choose is not particularly important, though sometimes a judicious choice can reduce the time taken to come to equilibrium (see Section 3.2). The two most commonly used initial states are the zero-temperature state and the infinite temperature state. At $T = 0$ the Ising model will be in its ground state. When the interaction energy J is greater than zero and the external field B is zero (as is the case in the simulations we will present in this chapter) there are actually two ground states. These are the states in which the spins are all up or all down. It is easy to see that these must be ground states, since in these states each pair of spins in the first term of Equation (3.1) contributes the lowest possible energy $-J$ to the Hamiltonian. In any other state there will be pairs of spins which contribute $+J$ to the Hamiltonian, so that its overall value will be higher. (If $B \neq 0$ then there will only be one ground state—the field ensures that one of the two is favoured over the other.) The other commonly used initial state is the $T = \infty$ state. When $T = \infty$ the thermal energy kT available to flip the spins is infinitely larger than the energy due to the spin–spin interaction J, so the spins are just oriented randomly up or down in an uncorrelated fashion.

These two choices of initial state are popular because they each correspond to a known, well defined temperature, and they are easy to generate. There is, however, one other initial state which can sometimes be very useful, which we should mention. Often we don't just perform one simulation at a single temperature, but rather a set of simulations one after another at a range of different values of T, to probe the behaviour of the model with varying temperature. In this case it is often advantageous to us to choose as the initial state of our system the *final* state of the system for a simulation at a nearby temperature. For example, suppose we are interested in probing a range of temperatures between $T = 1.0$ and $T = 2.0$ in steps of 0.1. (Here and throughout much of the rest of this book, we measure temperature in energy units, so that $k = 1$. Thus when we say $T = 2.0$ we mean that $\beta^{-1} = 2.0$.) Then we might start off by performing a simulation at $T = 1.0$ using the zero-temperature state with all spins aligned as our initial state. At the end of the simulation, the system will be in equilibrium at $T = 1.0$, and we can use the final state of that simulation as the initial state for the simulation at $T = 1.1$, and so on. The justification for doing this is clear: we hope that the equilibrium state at $T = 1.0$ will be more similar to that at $T = 1.1$ than will the zero-temperature state. In most cases this is a correct assumption and our system will come to equilibrium quicker with this initial state than with either a $T = 0$ or a $T = \infty$ one.

Now we start our simulation. The first step is to generate a new state—the one we called ν in the discussion above. The new state should differ from the present one by the flip of just one spin, and every such state should be exactly as likely as every other to be generated. This is an easy task to perform. We just pick a single spin k at random from the lattice to be flipped. Next, we need to calculate the difference in energy $E_\nu - E_\mu$ between the new state and the old one, in order to apply Equation (3.7). The most straightforward way to do this would be to calculate E_μ directly by substituting the values s_i^μ of the spins in state μ into the Hamiltonian (3.1), then flip spin k and calculate E_ν, and take the difference. This, however, is not a very efficient way to do it. Even in zero magnetic field $B = 0$ we still have to perform the sum in the first term of (3.1), which has as many terms as there are bonds on the lattice, which is $\frac{1}{2}Nz$. But most of these terms don't change when we flip our single spin. The only ones that change are those that involve the flipped spin. The others stay the same and so cancel out when we take the difference $E_\nu - E_\mu$. The *change* in energy between the two states is thus

$$E_\nu - E_\mu = -J \sum_{\langle ij \rangle} s_i^\nu s_j^\nu + J \sum_{\langle ij \rangle} s_i^\mu s_j^\mu$$

$$= -J \sum_{i \text{ n.n. to } k} s_i^\mu (s_k^\nu - s_k^\mu). \tag{3.8}$$

In the second line the sum is over only those spins i which are nearest neighbours of the flipped spin k and we have made use of the fact that all of these spins do not themselves flip, so that $s_i^\nu = s_i^\mu$. Now if $s_k^\mu = +1$, then after spin k has been flipped we must have $s_k^\nu = -1$, so that $s_k^\nu - s_k^\mu = -2$. On the other hand, if $s_k^\mu = -1$ then $s_k^\nu - s_k^\mu = +2$. Thus we can write

$$s_k^\nu - s_k^\mu = -2s_k^\mu, \tag{3.9}$$

and so

$$E_\nu - E_\mu = 2J \sum_{i \text{ n.n. to } k} s_i^\mu s_k^\mu$$

$$= 2J s_k^\mu \sum_{i \text{ n.n. to } k} s_i^\mu. \qquad (3.10)$$

This expression only involves summing over z terms, rather than $\frac{1}{2}Nz$, and it doesn't require us to perform any multiplications for the terms in the sum, so it is much more efficient than evaluating the change in energy directly. What's more, it involves only the values of the spins in state μ, so we can evaluate it *before* we actually flip the spin k.

The algorithm thus involves calculating $E_\nu - E_\mu$ from Equation (3.10) and then following the rule given in Equation (3.7): if $E_\nu - E_\mu \leq 0$ we definitely accept the move and flip the spin $s_k \rightarrow -s_k$. If $E_\nu - E_\mu > 0$ we still may want to flip the spin. The Metropolis algorithm tells us to flip it with probability $A(\mu \rightarrow \nu) = e^{-\beta(E_\nu - E_\mu)}$. We can do this as follows. We evaluate the acceptance ratio $A(\mu \rightarrow \nu)$ using our value of $E_\nu - E_\mu$ from Equation (3.10), and then we choose a random number r between zero and one. (Strictly the number can be equal to zero, but it must be less than one: $0 \leq r < 1$.) If that number is *less* than our acceptance ratio, $r < A(\mu \rightarrow \nu)$, then we flip the spin. If it isn't, we leave the spin alone.

And that is our complete algorithm. Now we just keep on repeating the same calculations over and over again, choosing a spin, calculating the energy change we would get if we flipped it, and then deciding whether to flip it according to Equation (3.7). Actually, there is one other trick that we can pull that makes our algorithm a bit faster still. (In fact, on most computers it will make it a lot faster.) One of the slowest parts of the algorithm as we have described it is the calculation of the exponential, which we have to perform if the energy of the new state we choose is greater than that of the current state. Calculating exponentials on a computer is usually done using a polynomial approximation which involves performing a number of floating-point multiplications and additions, and can take a considerable amount of time. We can save ourselves this effort, and thereby speed up our simulation, if we notice that the quantity, Equation (3.10), which we are calculating the exponential of, can only take a rather small number of values. Each of the terms in the sum can only take the values $+1$ and -1. So the entire sum, which has z terms, can only take the values $-z, -z+2, -z+4 \ldots$ and so on up to $+z$—a total of $z+1$ possible values. And we only actually need to calculate the exponential when the sum is negative (see Equation (3.7) again), so in fact there are only $\frac{1}{2}z$ values of $E_\mu - E_\nu$ for which we ever need to calculate exponentials. Thus, it makes good sense to calculate the values of these $\frac{1}{2}z$ exponentials before we start the calculation proper, and store them in the computer's memory (usually in an array), where we can simply look them up when we need them during the simulation. We pay the one-time cost of evaluating them at the beginning, and save a great deal more by never having to evaluate any exponentials again during the rest of the simulation. Not only does this save us the effort of evaluating all those exponentials, it also means that we hardly have to perform any floating-point arithmetic during the simulation. The only floating-point calculations will be in the generation of the random number r. (We discuss techniques for doing this in Chapter 16.) All the other calculations involve only integers, which on most computers are much quicker to deal with than real numbers.

3.2 Equilibration

So what do we do with our Monte Carlo program for the Ising model, once we have written it? Well, we probably want to know the answer to some questions like "What is the magnetization at such-and-such a temperature?", or "How does the internal energy behave with temperature over such-and-such a range?" To answer these questions we have to do two things. First we have to run our simulation for a suitably long period of time until it has come to equilibrium at the temperature we are interested in—this period is called the **equilibration time** τ_{eq}—and then we have to measure the quantity we are interested in over another suitably long period of time and average it, to evaluate the estimator of that quantity (see Equation (2.4)). This leads us to several other questions. What exactly do we mean by "allowing the system to come to equilibrium"? And how long is a "suitably long" time for it to happen? How do we go about measuring our quantity of interest, and how long do we have to average over to get a result of a desired degree of accuracy? These are very general questions which we need to consider every time we do a Monte Carlo calculation. Although we will be discussing them here using our Ising model simulation as an example, the conclusions we will draw in this and the following sections are applicable to all equilibrium Monte Carlo calculations. These sections are some of the most important in this book.

As we discussed in Section 1.2, "equilibrium" means that the average probability of finding our system in any particular state μ is proportional to the Boltzmann weight $e^{-\beta E_\mu}$ of that state. If we start our system off in a state such as the $T = 0$ or $T = \infty$ states described in the last section and we want to perform a simulation at some finite non-zero temperature, it will take a little time before we reach equilibrium. To see this, recall that, as we demonstrated in Section 1.2.1, a system at equilibrium spends the overwhelming majority of its time in a small subset of states in which its internal energy and other properties take a narrow range of values. In order to get a good estimate of the equilibrium value of any property of the system therefore, we need to wait until it has found its way to one of the states that

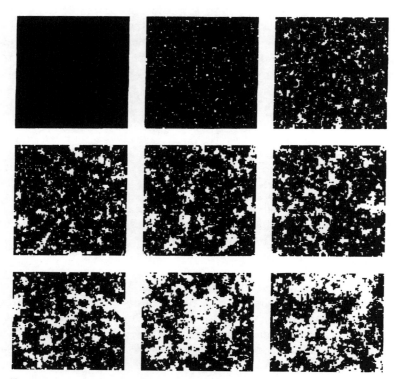

FIGURE 3.2 Nine snapshots of a 100×100 Ising model on a square lattice with $J = 1$ coming to equilibrium at a temperature $T = 2.4$ using the Metropolis algorithm. In these pictures the up-spins ($s_i = +1$) are represented by black squares and the down-spins ($s_i = -1$) by white ones. The starting configuration is one in which all the spins are pointing up. The progression of the figures is horizontally across the top row, then the middle row, then the bottom one. They show the lattice after 0, 1, 2, 4, 6, 10, 20, 40 and 100 times 100 000 steps of the simulation. In the last frame the system has reached equilibrium according to the criteria given in this section.

fall in this narrow range. Then, we assume, the Monte Carlo algorithm we have designed will ensure that it stays roughly within that range for the rest of the simulation—it should do since we designed the algorithm specifically to simulate the behaviour of the system at equilibrium. But it may take some time to find a state that lies within the correct range. In the version of the Metropolis algorithm which we have described here, we can only flip one spin at a time, and since we are choosing the spins we flip at random, it could take quite a while before we hit on the correct sequence of spins to

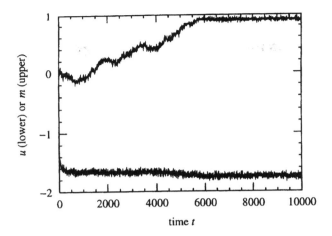

FIGURE 3.3   The magnetization (upper curve) and internal energy
(lower curve) per site of a two-dimensional Ising model on a square
lattice of 100×100 sites with $J = 1$ simulated using the Metropolis algorithm of Section 3.1. The simulation was started at $T = \infty$
(i.e., the initial states of the spins were chosen completely at random)
and "cooled" to equilibrium at $T = 2.0$. Time is measured in Monte
Carlo steps per lattice site, and equilibrium is reached after about 6000
steps per site (in other words, 6×10^7 steps altogether).

flip in order to get us to one of the states we want to be in. At the very
least we can expect it to take about N Monte Carlo steps to reach a state
in the appropriate energy range, where N is the number of spins on the
lattice, since we need to allow every spin the chance to flip at least once. In
Figure 3.2 we show a succession of states of a two-dimension Ising model on
a square lattice of 100×100 spins with $J = 1$, as it is "warmed" up to a
temperature $T = 2.4$ from an initial $T = 0$ state in which all the spins are
aligned. In these pictures the $+1$ and -1 spins are depicted as black and
white squares. By the time we reach the last frame out of nine, the system
has equilibrated. The whole process takes on the order of 10^7 steps in this
case.

However looking at pictures of the lattice is not a reliable way of gauging
when the system has come to equilibrium. A better way, which takes very
little extra effort, is to plot a graph of some quantity of interest, like the
magnetization per spin m of the system or the energy of the system E, as
a function of time from the start of the simulation. We have done this in
Figure 3.3. (We will discuss the best ways of measuring these quantities
in the next section, but for the moment let's just assume that we calculate
them directly. For example, the energy of a given state can be calculated by

feeding all the values of the spins s_i into the Hamiltonian, Equation (3.1).)
It is not hard to guess simply by looking at this graph that the system came
to equilibrium at around time $t = 6000$. Up until this point the energy
and the magnetization are changing, but after this point they just fluctuate
around a steady average value.

The horizontal axis in Figure 3.3 measures time in Monte Carlo steps *per
lattice site*, which is the normal practice for simulations of this kind. The
reason is that if time is measured in this way, then the average frequency
with which any particular spin is selected for flipping is independent of the
total number of spins N on the lattice. This average frequency is called
the "attempt frequency" for the spin. In the simulation we are considering
here the attempt frequency has the value 1. It is natural that we should
arrange for the attempt frequency to be independent of the lattice size; in
an experimental system, the rate at which spins or atoms or molecules change
from one state to another does not depend on how many there are in the
whole system. An atom in a tiny sample will change state as often as one in
a sample the size of a house. Attempt frequencies are discussed further in
Section 11.1.1.

When we perform N Monte Carlo steps—one for each spin in the system,
on average—we say we have completed one **sweep** of the lattice. We could
therefore also say that the time axis of Figure 3.3 was calibrated in sweeps.

Judging the equilibration of a system by eye from a plot such as Figure 3.3
is a reasonable method, provided we know that the system will come to
equilibrium in a smooth and predictable fashion as it does in this case.
The trouble is that we usually know no such thing. In many cases it is
possible for the system to get stuck in some metastable region of its state
space for a while, giving roughly constant values for all the quantities we
are observing and so appearing to have reached equilibrium. In statistical
mechanical terms, there can be a **local energy minimum** in which the
system can remain temporarily, and we may mistake this for the **global
energy minimum**, which is the region of state space that the equilibrium
system is most likely to inhabit. (These ideas are discussed in more detail
in the first few sections of Chapter 6.) To avoid this potential pitfall, we
commonly adopt a different strategy for determining the equilibration time,
in which we perform two different simulations of the same system, starting
them in different initial states. In the case of the Ising model we might, for
example, start one in the $T = 0$ state with all spins aligned, and one in the
$T = \infty$ state with random spins. Or we could choose two different $T = \infty$
random-spin states. We should also run the two simulations with different
"seeds" for the random number generator (see Section 16.1.2), to ensure
that they take different paths to equilibrium. Then we watch the value of
the magnetization or energy or other quantity in the two systems and when
we see them reach the same approximately constant value, we deduce that

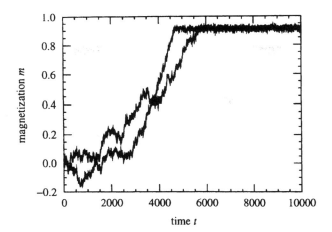

FIGURE 3.4   The magnetization of our 100×100 Ising model as a function of time (measured in Monte Carlo steps per lattice site) for two different simulations using the Metropolis algorithm.   The two simulations were started off in two different $T = \infty$ (random-spin) states.   By about time $t = 6000$ the two simulations have converged to the same value of the mean magnetization, within the statistical errors due to fluctuations, and so we conclude that both have equilibrated.

both systems have reached equilibrium. We have done this for two 100×100 Ising systems in Figure 3.4. Again, we clearly see that it takes about 6000 Monte Carlo steps for the two systems to reach a consensus about the value of the magnetization. This technique avoids the problem mentioned above, since if one of the systems finds itself in some metastable region, and the other reaches equilibrium or gets stuck in another metastable region, this will be apparent from the graph, because the magnetization (or other quantity) will take different values for the two systems. Only in the unlikely event that the two systems coincidentally become trapped in the same metastable region (for example, if we choose two initial states that are too similar to one another) will we be misled into thinking they have reached equilibrium when they haven't. If we are worried about this possibility, we can run *three* different simulations from different starting points, or four, or five. Usually, however, two is sufficient.

3.3   Measurement

Once we are sure the system has reached equilibrium, we need to measure whatever quantity it is that we are interested in. The most likely candidates for the Ising model are the energy and the magnetization of the system. As we pointed out above, the energy E_μ of the current state μ of the system can be evaluated directly from the Hamiltonian by substituting in the values of the spins s_i from our array of integers. However, this is not an especially efficient way of doing it, and there is a much better way. As part of our implementation of the Metropolis algorithm, you will recall we calculated the energy difference $\Delta E = E_\nu - E_\mu$ in going from state μ to state ν (see Equation (3.10)). So, if we know the energy of the current state μ, we can calculate the new energy when we flip a spin, using only a single addition:

$$E_\nu = E_\mu + \Delta E. \tag{3.11}$$

So the clever thing to do is to calculate the energy of the system from the Hamiltonian at the very start of the simulation, and then every time we flip a spin calculate the new energy from Equation (3.11) using the value of ΔE, which we have to calculate anyway.

Calculating the magnetization is even easier. The total magnetization M_μ of the whole system in state μ (as opposed to the magnetization *per spin*—we'll calculate that in a moment), is given by the sum

$$M_\mu = \sum_i s_i^\mu. \tag{3.12}$$

As with the energy, it is not a shrewd idea to evaluate the magnetization directly from this sum every time we want to know it. It is much better to notice that only one spin k flips at a time in the Metropolis algorithm, so the change of magnetization from state μ to state ν is

$$\Delta M = M_\nu - M_\mu = \sum_i s_i^\nu - \sum_i s_i^\mu = s_k^\nu - s_k^\mu = 2s_k^\nu, \tag{3.13}$$

where the last equality follows from Equation (3.9). Thus, the clever way to evaluate the magnetization is to calculate its value at the beginning of the simulation, and then make use of the formula

$$M_\nu = M_\mu + \Delta M = M_\mu + 2s_k^\nu \tag{3.14}$$

every time we flip a spin.[2]

---

[2] However, to be absolutely fair, we should point out that doing this involves performing at least one addition operation every time we flip a spin, or one addition every $\bar{A}^{-1}$ steps, where $\bar{A}$ is the mean acceptance ratio. Direct evaluation of Equation (3.12) on the other hand involves N additions every time we want to know the magnetization. Thus, if we want to make measurements less often than once every $N/\bar{A}$ steps, it may pay to use the direct method rather than employing Equation (3.14). Similar considerations apply to the measurement of the energy also.

Given the energy and the magnetization of our Ising system at a selection of times during the simulation, we can average them to find the estimators of the internal energy and average magnetization. Then dividing these figures by the number of sites N gives us the internal energy and average magnetization per site.

We can also average the squares of the energy and magnetization to find quantities like the specific heat and the magnetic susceptibility:

$$c = \frac{\beta^2}{N}(\langle E^2 \rangle - \langle E \rangle^2), \tag{3.15}$$

$$\chi = \beta N(\langle m^2 \rangle - \langle m \rangle^2). \tag{3.16}$$

(See Equations (1.36) and (1.37). Note that we have set $k = 1$ again.)

In order to average quantities like E and M, we need to know how long a run we have to average them over to get a good estimate of their expectation values. One simple solution would again be just to look at a graph like Figure 3.3 and guess how long we need to wait. However (as you might imagine) this is not a very satisfactory solution. What we really need is a measure of the **correlation time** τ of the simulation. The correlation time is a measure of how long it takes the system to get from one state to another one which is significantly different from the first, i.e., a state in which the number of spins which are the same as in the initial state is no more than what you would expect to find just by chance. (We will give a more rigorous definition in a moment.) There are a number of ways to estimate the correlation time. One that is sometimes used is just to assume that it is equal to the equilibration time. This is usually a fairly safe assumption:[3] usually the equilibration time is considerably longer than the correlation time, $\tau_{eq} > \tau$, because two states close to equilibrium are qualitatively more similar than a state far from equilibrium (like the $T = 0$ or $T = \infty$ states we suggested for starting this simulation with) and one close to equilibrium. However, this is again a rather unrigorous supposition, and there are more watertight ways to estimate τ. The most direct of these is to calculate the "time-displaced autocorrelation function" of some property of the model.

3.3.1 Autocorrelation functions

Let us take the example of the magnetization m of our Ising model. The **time-displaced autocorrelation** $\chi(t)$ of the magnetization is given by

$$\chi(t) = \int dt' \, [m(t') - \langle m \rangle][m(t' + t) - \langle m \rangle]$$

$$= \int dt' \, [m(t')m(t' + t) - \langle m \rangle^2]. \tag{3.17}$$

---

[3]In particular, it works fine for the Metropolis simulation of the Ising model which we are considering here.

where $m(t)$ is the instantaneous value of the magnetization at time t and $\langle m \rangle$ is the average value. This is rather similar to the connected correlation function which we defined in Equation (1.26). That measured the correlation between the values of a quantity (such as the magnetization) on two different sites, i and j, on the lattice. The autocorrelation gives us a similar measure of the correlation at two different times, one an interval t later than the other. If we measure the difference between the magnetization $m(t')$ at time t' and its mean value, and then we do the same thing at time $t' + t$, and we multiply them together, we will get a positive value if they were fluctuating in the same direction at those two times, and a negative one if they were fluctuating in opposite directions. If we then integrate over time as in Equation (3.17), then $\chi(t)$ will take a non-zero value if on average the fluctuations are correlated, or it will be zero if they are not. For our Metropolis simulation of the Ising model it is clear that if we measure the magnetization at two times just a single Monte Carlo step apart, the values we get will be very similar, so we will have a large positive autocorrelation. On the other hand, for two times a long way apart the magnetizations will probably be totally unrelated, and their autocorrelation will be close to zero. Ideally, we should calculate $\chi(t)$ by integrating over an infinite time, but this is obviously impractical in a simulation, so we do the best we can and just sum over all the measurements of m that we have, from beginning to end of our run. Figure 3.5 shows the magnetization autocorrelation of our 100×100 Ising model at temperature $T = 2.4$ and interaction energy $J = 1$, calculated in exactly this manner using results from our Metropolis Monte Carlo simulation. As we can see, the autocorrelation does indeed drop from a significant non-zero value at short times t towards zero at very long times. In this case, we have divided $\chi(t)$ by its value $\chi(0)$ at $t = 0$, so that its maximum value is one. The typical time-scale (if there is one) on which it falls off is a measure of the correlation time τ of the simulation. In fact, this is the *definition* of the correlation time. It is the typical time-scale on which the autocorrelation drops off; the autocorrelation is expected to fall off exponentially at long times thus:

$$\chi(t) \sim e^{-t/\tau}. \tag{3.18}$$

(In the next section we show why it should take this form.) With this definition, we see that in fact there is still a significant correlation between two samples taken a correlation time apart: at time $t = \tau$ the autocorrelation function, which is a measure of the similarity of the two states, is only a factor of $1/e$ down from its maximum value at $t = 0$. If we want truly independent samples then, we may want to draw them at intervals of greater than one correlation time. In fact, the most natural definition of statistical independence turns out to be samples drawn at intervals of 2τ. We discuss this point further in Section 3.4.1.

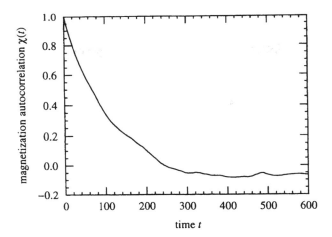

FIGURE 3.5 The magnetization autocorrelation function $\chi(t)$ for a two-dimensional Ising model at temperature $T = 2.4$ on a square lattice of 100×100 sites with $J = 1$ simulated using the Metropolis algorithm of Section 3.1. Time is measured in Monte Carlo steps per site.

We can make a reasonable estimate of τ by eye from Figure 3.5. At a guess we'd probably say τ was about 100 in this case. This is an accurate enough figure for estimating how long a Monte Carlo run we need to do in order to get decent statistics. It tells us that we expect to get a new independent spin configuration about once every $2\tau = 200$ sweeps of the lattice in our simulation. So if we want, say, 10 independent measurements of the magnetization, we need to run our Metropolis algorithm for about 2000 sweeps after equilibration, or 2×10^7 Monte Carlo steps. If we want 100 measurements we need to do 2×10^8 steps. In general, if a run lasts a time $t_{\max}$, then the number of independent measurements we can get out of the run, after waiting a time τ_{eq} for equilibration, is on the order of

$$n = \frac{t_{\max}}{2\tau}. \tag{3.19}$$

It is normal practice in a Monte Carlo simulation to make measurements at intervals of less than the correlation time. For example, in the case of the Metropolis algorithm we might make one measurement every sweep of the lattice. Thus the total number of measurements we make of magnetization or energy (or whatever) during the run is usually greater than the number of independent measurements. There are a number of reasons why we do it this way. First of all, we usually don't know what the correlation time is until after the simulation has finished, or at least until after it has run

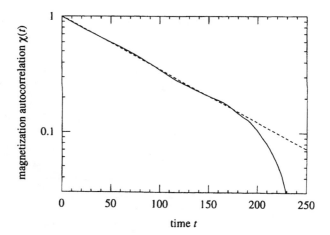

FIGURE 3.6   The autocorrelation function of Figure 3.5 replotted on semi-logarithmic axes. The dashed line is a straight line fit which yields a figure of $\tau = 95 \pm 5$ in this case. Note that the horizontal scale is not the same as in Figure 3.5.

for a certain amount of time, and we want to be sure of having at least one measurement every two correlation times. Another reason is that we want to be able to calculate the autocorrelation function for times less than a correlation time, so that we can use it to make an accurate estimate of the correlation time. If we only had one measurement every 2τ, we wouldn't be able to calculate τ with any accuracy at all.

If we want a more reliable figure for τ, we can replot our autocorrelation function on semi-logarithmic axes as we have done in Figure 3.6, so that the slope of the line gives us the correlation time. Then we can estimate τ by fitting the straight-line portion of the plot using a least-squares method. The dotted line in the figure is just such a fit and its slope gives us a figure of $\tau = 95 \pm 5$ for the correlation time in this case.

An alternative is to calculate the **integrated correlation time**.[4] If we assume that Equation (3.18) is accurate for all times t then

$$\int_0^\infty \frac{\chi(t)}{\chi(0)} \, dt = \int_0^\infty e^{-t/\tau} \, dt = \tau. \tag{3.20}$$

This form has a number of advantages. First, it is often easier to apply Equation (3.20) than it is to perform the exponential fit to the autocorrelation

---

[4]This is a rather poor name for this quantity, since it is not the correlation time that is integrated but the autocorrelation function. However, it is the name in common use so we use it here too.

function. Second, the method for estimating τ illustrated in Figure 3.6 is rather sensitive to the range over which we perform the fit. In particular, the very long time behaviour of the autocorrelation function is often noisy, and it is important to exclude this portion from the fitted data. However, the exact point at which we truncate the fit can have quite a large effect on the resulting value for τ. The integrated correlation time is much less sensitive to the way in which the data are truncated, although it is by no means perfect either, since, as we will demonstrate in Section 3.3.2, Equation (3.18) is only strictly correct for long times t, and we introduce an uncontrolled error by assuming it to be true for all times. On the other hand, the direct fitting method also suffers from this problem, unless we only perform our fit over the exact range of times for which true exponential behaviour is present. Normally, we don't know what this range is, so the fitting method is no more accurate than calculating the integrated correlation time. Usually then, Equation (3.20) is the method of choice for calculating τ. Applying it to the data from Figure 3.6 gives a figure of $\tau = 86 \pm 5$, which is in moderately good agreement with our previous figure.

The autocorrelation function in Figure 3.5 was calculated directly from a discrete form of Equation (3.17). If we have a set of samples of the magnetization $m(t)$ measured at evenly-spaced times up to some maximum time t_{max}, then the correct formula for the autocorrelation function is[5]

$$\chi(t) = \frac{1}{t_{\mathrm{max}} - t} \sum_{t'=0}^{t_{\mathrm{max}}-t} m(t')\, m(t'+t)$$

$$-\frac{1}{t_{\mathrm{max}} - t} \sum_{t'=0}^{t_{\mathrm{max}}-t} m(t') \times \frac{1}{t_{\mathrm{max}} - t} \sum_{t'=0}^{t_{\mathrm{max}}-t} m(t'+t). \quad (3.21)$$

Notice how we have evaluated the mean magnetization m in the second term using the same subsets of the data that we used in the first term. This is not strictly speaking necessary, but it makes $\chi(t)$ a little better behaved. In Figure 3.5 we have also normalized $\chi(t)$ by dividing throughout by $\chi(0)$, but this is optional. We've just done it for neatness.

Note that one should be careful about using Equation (3.21) to evaluate $\chi(t)$ at long times. When t gets close to t_{max}, the upper limit of the sums becomes small and we end up integrating over a rather small time interval to get our answer. This means that the statistical errors in $\chi(t)$ due to the random nature of the fluctuations in $m(t)$ may become large. A really satisfactory simulation would always run for many correlation times, in which case we will probably not be interested in the very tails of $\chi(t)$, since the correlations will have died away by then, by definition. However, it is not

---

[5]In fact, this formula differs from (3.17) by a multiplicative constant, but this makes no difference as far as the calculation of the correlation time is concerned.

always possible, because we only have limited computer resources, to perform simulations as long as we would like, and one should always be aware that errors of this type can crop up with shorter data sets.

Calculating the autocorrelation function from Equation (3.21) takes time of order n^2, where n is the number of samples. For most applications, this is not a problem. Even for simulations where several thousand samples are taken, the time needed to evaluate the autocorrelation is only a few seconds on a modern computer, and the simplicity of the formulae makes their programming very straightforward, which is a big advantage—people-time is usually more expensive than computer-time. However, sometimes it is desirable to calculate the autocorrelation function more quickly. This is particularly the case when one needs to do it very often during the course of a calculation, for some reason. If you need to calculate an autocorrelation a thousand times, and each time takes a few seconds on the computer, then the seconds start to add up. In this case, at the expense of rather greater programming effort, we can often speed up the process by the following trick. Instead of calculating $\chi(t)$ directly, we calculate the Fourier transform $\tilde{\chi}(\omega)$ of the autocorrelation and then we invert the Fourier transform to get $\chi(t)$. The Fourier transform is related to the magnetization as follows:

$$
\begin{aligned}
\tilde{\chi}(\omega) &= \int dt \, e^{i\omega t} \int dt' \, [m(t') - \langle m \rangle][m(t' + t) - \langle m \rangle] \\
&= \int dt \int dt' \, e^{-i\omega t'} [m(t') - \langle m \rangle] \, e^{i\omega(t'+t)} [m(t' + t) - \langle m \rangle] \\
&= \tilde{m}'(\omega) \, \tilde{m}'(-\omega) = |\tilde{m}'(\omega)|^2,
\end{aligned}
\tag{3.22}
$$

where $\tilde{m}'(\omega)$ is the Fourier transform of $m'(t) \equiv m(t) - \langle m \rangle$.[6]

So all we need to do is calculate the Fourier transform of $m'(t)$ and feed it into this formula to get $\tilde{\chi}(\omega)$. The advantage in doing this is that the Fourier transform can be evaluated using the so-called **fast Fourier transform** or FFT algorithm, which was given by Cooley and Tukey in 1965. This is a standard algorithm which can evaluate the Fourier transform in a time which goes like $n \log n$, where n is the number of measurements of the magnetization.[7] Furthermore, there exist a large number of ready-made computer software packages that perform FFTs. These packages have

---

[6]Since $m(t)$ and $m'(t)$ differ only by a constant, $\tilde{m}(\omega)$ and $\tilde{m}'(\omega)$ differ only in their $\omega = 0$ component (which is zero in the latter case, but may be non-zero in the former). For this reason, it is often simplest to calculate $\tilde{m}'(\omega)$ by first calculating $\tilde{m}(\omega)$ and then just setting the $\omega = 0$ component to zero.

[7]On a technical note, if we simply apply the FFT algorithm directly to our magnetization data, the result produced is the Fourier transform of an infinite periodic repetition of the data set, which is not quite what we want in Equation (3.22). A simple way of getting around this problem is to add n zeros to the end of the data set before we perform the transform. It can be shown that this then gives a good estimate of the autocorrelation function (Futrelle and McGinty 1971).

been written very carefully by people who understand the exact workings of the computer and are designed to be as fast as possible at performing this particular calculation, so it usually saves us much time and effort to make use of the programs in these packages. Having calculated $\widetilde{\chi}(\omega)$, we can then invert the Fourier transform, again using a streamlined inverse FFT routine which also runs in time proportional to $n \log n$, and so recover the function $\chi(t)$.

We might imagine that, if we wanted to calculate the integrated correlation time, Equation (3.20), we could avoid inverting the Fourier transform, since

$$\widetilde{\chi}(0) = \int_0^\infty \chi(t) \, dt, \tag{3.23}$$

and thus

$$\tau = \frac{\widetilde{\chi}(0)}{\chi(0)}, \tag{3.24}$$

where $\chi(0)$ is simply the magnetization fluctuation

$$\chi(0) = \langle m^2 \rangle - \langle m \rangle^2. \tag{3.25}$$

However, you should avoid calculating τ this way because, as footnote 6 on the previous page makes clear, $\widetilde{\chi}(0)$ is zero when calculated directly for finite datasets. Equation (3.24) is only applicable if $\langle m \rangle$ is calculated over a much longer run than $\widetilde{\chi}(\omega)$.

3.3.2 Correlation times and Markov matrices

The techniques outlined in the previous section are in most cases quite sufficient for estimating correlation times in Monte Carlo simulations. However, we have simplified the discussion somewhat by supposing there to be only one correlation time in the system. In real life there are as many correlation times as there are states of the system, and the interplay between these different times can sometimes cause the methods of the last section to give inaccurate results. In this section we look at these issues in more detail. The reader who is interested only in how to calculate a rough measure of τ could reasonably skip this section.

In Section 2.2.3 we showed that the probabilities $w_\mu(t)$ and $w_\mu(t+1)$ of being in a particular state μ at consecutive Monte Carlo steps are related by the Markov matrix $\mathbf{P}$ for the algorithm. In matrix notation we wrote

$$\mathbf{w}(t+1) = \mathbf{P} \cdot \mathbf{w}(t), \tag{3.26}$$

where $\mathbf{w}$ is the vector whose elements are the probabilities w_μ (see Equation (2.9)). By iterating this equation from time $t = 0$ we can then show that

$$\mathbf{w}(t) = \mathbf{P}^t \cdot \mathbf{w}(0). \tag{3.27}$$

Now $\mathbf{w}(0)$ can be expressed as a linear combination of the right eigenvectors $\mathbf{v}_i$ of $\mathbf{P}$ thus:[8]

$$\mathbf{w}(0) = \sum_i a_i \mathbf{v}_i, \tag{3.28}$$

where the quantities a_i are coefficients whose values depend on the configuration of the system at $t = 0$. Then

$$\mathbf{w}(t) = \mathbf{P}^t \cdot \sum_i a_i \mathbf{v}_i = \sum_i a_i \lambda_i^t \mathbf{v}_i, \tag{3.29}$$

where λ_i is the eigenvalue of $\mathbf{P}$ corresponding to the eigenvector $\mathbf{v}_i$. As $t \to \infty$, the right-hand side of this equation will be dominated by the term involving the largest eigenvalue λ_0 of the Markov matrix. This means that in the limit of long times, the probability distribution $\mathbf{w}(t)$ becomes proportional to $\mathbf{v}_0$, the eigenvector corresponding to the largest eigenvalue. We made use of this result in Section 2.2.3 to demonstrate that $\mathbf{w}(t)$ tends to the Boltzmann distribution at long times.

Now suppose that we are interested in knowing the value of some observable quantity Q, such as the magnetization. The expectation value of this quantity at time t can be calculated from the formula

$$Q(t) = \sum_\mu w_\mu(t) \, Q_\mu \tag{3.30}$$

or

$$Q(t) = \mathbf{q} \cdot \mathbf{w}(t), \tag{3.31}$$

where $\mathbf{q}$ is the vector whose elements are the values Q_μ of the quantity in the various states of the system. Substituting Equation (3.29) into (3.31) we then get

$$Q(t) = \sum_i a_i \lambda_i^t \, \mathbf{q} \cdot \mathbf{v}_i = \sum_i a_i \lambda_i^t \, q_i. \tag{3.32}$$

Here $q_i \equiv \mathbf{q} \cdot \mathbf{v}_i$ is the expectation value of Q in the ith eigenstate. The long time limit $Q(\infty)$ of this expression is also dominated by the largest eigenvalue λ_0 and is proportional to q_0. If we now define a set of quantities τ_i thus:

$$\tau_i = -\frac{1}{\log \lambda_i} \tag{3.33}$$

for all $i \neq 0$, then Equation (3.32) can be written

$$Q(t) = Q(\infty) + \sum_{i \neq 0} a_i q_i e^{-t/\tau_i}. \tag{3.34}$$

---

[8]$\mathbf{P}$ is in general not symmetric, so its right and left eigenvectors are not the same.

The quantities τ_i are the correlation times for the system, as we can demonstrate by showing that they also govern the decay of the autocorrelation function. Noting that the long-time limit $Q(\infty)$ of the expectation of Q is none other than the equilibrium expectation $\langle Q \rangle$, we can write the autocorrelation of Q as the correlation between the expectations at zero time and some later time t thus:

$$\chi(t) = [Q(0) - Q(\infty)][Q(t) - Q(\infty)] = \sum_{i \neq 0} b_i e^{-t/\tau_i}, \qquad (3.35)$$

with

$$b_i = \sum_{j \neq 0} a_i a_j q_i q_j. \qquad (3.36)$$

Equation (3.35) is the appropriate generalization of Equation (3.18) to all times t (not just long ones).

As we said, there are as many correlation times as there are states of the system since that is the rank of the matrix **P**. (Well, strictly there are as many of them as the rank of the matrix *less one*, since there is no τ_0 corresponding to the highest eigenvalue. There are $2^N - 1$ correlation times in the case of the Ising model, for example. However, the rank of the matrix is usually very large, so let's not quibble over one correlation time.) The longest of these correlation times is τ_1, the one which corresponds to the second largest eigenvalue of the matrix. This is the correlation time we called τ in the last section. Clearly, for large enough times t, this will be the only correlation time we need to worry about, since all the other terms in Equation (3.35) will have decayed away to insignificance. (This is how Equation (3.18) is derived.) However, depending on how close together the higher eigenvalues of the Markov matrix are, and how long our simulation runs for, we may or may not be able to extract reliable results for τ_1 by simply ignoring all the other terms. In general the most accurate results are obtained by fitting our autocorrelation function to a sum of a small number of decaying exponentials, of the form of Equation (3.35), choosing values for the quantities b_i by a least-squares or similar method. In work on the three-dimensional Ising model, for example, Wansleben and Landau (1991) showed that including three terms was sufficient to get a good fit to the magnetization autocorrelation function, and thus get an accurate measure of the longest correlation time $\tau \equiv \tau_1$. In studies of the dynamical properties of statistical mechanical models this is the most correct way to measure the correlation time. Strictly speaking it gives only a lower bound on τ since it is always possible that correlations exist beyond the longest times that one can measure. However, in practice it usually gives good results.

3.4 Calculation of errors

Normally, as well as measuring expectation values, we also want to calculate the errors on those values, so that we have an idea of how accurate they are. As with experiments, the errors on Monte Carlo results divide into two classes: **statistical errors** and **systematic errors**.[9] Statistical errors are errors which arise as a result of random changes in the simulated system from measurement to measurement—thermal fluctuations, for example—and they can be estimated simply by taking many measurements of the quantity we are interested in and calculating the spread of the values. Systematic errors on the other hand, are errors due to the procedure we have used to make the measurements, and they affect the whole simulation. An example is the error introduced by waiting only a finite amount of time for our system to equilibrate. (Ideally, we should allow an infinite amount of time for this, in order to be sure the system has completely equilibrated. However, this, of course, is not practical.)

3.4.1 Estimation of statistical errors

In a Monte Carlo calculation the principal source of statistical error in the measured value of a quantity is usually the fluctuation of that quantity from one time step to the next. This error is inherent in the Monte Carlo method. As the name "Monte Carlo" itself makes clear, there is an innate randomness and statistical nature to Monte Carlo calculations. (In Chapter 6 on glassy spin models, we will see another source of statistical error: "sample-to-sample" fluctuations in the actual system being simulated. However, for the simple Ising model we have been considering, thermal fluctuations are the only source of statistical error. All other errors fall into the category of systematic errors.) It is often straightforward to estimate the statistical error in a measured quantity, since the assumption that the error is statistical— i.e., that it arises through random deviations in the measured value of the quantity—implies that we can estimate the true value by taking the mean of several different measurements, and that the error on that estimate is simply the error on the mean. Thus, if we are performing the Ising model simulation described in the last section, and we make n measurements m_i of the magnetization of the system during a particular run, then our best estimate of the true thermal average of the magnetization is the mean $\overline{m}$ of those n measurements (which is just the estimator of m, as defined in Section 2.1),

---

[9]Monte Carlo simulations are in many ways rather similar to experiments. It often helps to regard them as "computer experiments", and analyse the results in the same way as we would analyse the results of a laboratory experiment.

and our best estimate of the standard deviation on the mean is given by[10]

$$\sigma = \sqrt{\frac{\frac{1}{n}\sum_{i=0}^{n}(m_i - \overline{m})^2}{n-1}} = \sqrt{\frac{1}{n-1}(\overline{m^2} - \overline{m}^2)}. \tag{3.37}$$

This expression assumes that our samples m_i are statistically independent, which in general they won't be. As we pointed out in Section 3.3.1, it is normal to sample at intervals of less than a correlation time, which means that successive samples will in general be correlated. A simple and usually perfectly adequate solution to this problem is to use the value of n given by Equation (3.19), rather than the actual number of samples taken. In fact, it can be shown (Müller-Krumbhaar and Binder 1973) that the correct expression for σ in terms of the actual number of samples is

$$\sigma = \sqrt{\frac{1 + 2\tau/\Delta t}{n-1}(\overline{m^2} - \overline{m}^2)}, \tag{3.38}$$

where τ is the correlation time and Δt is the time interval at which the samples were taken. Clearly this becomes equal to Equation (3.37) when $\Delta t \gg \tau$, but more often we have $\Delta t \ll \tau$. In this case, we can ignore the 1 in the numerator of Equation (3.38). Noting that for a run of length $t_{\max}$ (after equilibration) the interval Δt is related to the total number of samples by

$$n = \frac{t_{\max}}{\Delta t}, \tag{3.39}$$

we then find that for large n

$$\sigma = \sqrt{\frac{2\tau}{t_{\max}}(\overline{m^2} - \overline{m}^2)}, \tag{3.40}$$

which is the same result as we would get by simply using Equation (3.19) for n in Equation (3.37). This in fact was the basis for our assertion in Section 3.3.1 that the appropriate sampling interval for getting independent samples was twice the correlation time. Note that the value of σ in Equation (3.40) is independent of the value of Δt, which means we are free to choose Δt in whatever way is most convenient.

3.4.2 The blocking method

There are some cases where it is either not possible or not straightforward to estimate the error in a quantity using the direct method described in the last

---

[10]The origin of the $n - 1$ in this and following expressions for error estimates is a little obscure. The curious reader is referred to any good book on data analysis for an explanation, such as Bevington and Robinson (1992).

section. This happens when the result we want is not merely the average of some measurement repeated many times over the course of the simulation, as the magnetization is, but is instead derived in some more complex way from measurements we make during the run. An example is the specific heat c, Equation (3.15), which is inherently an average macroscopic quantity. Unlike the magnetization, the specific heat is not defined at a single time step in the simulation. It is only defined in terms of averages of many measurements of E and E^2 over a longer period of time. We might imagine that we could calculate the error on $\langle E \rangle$ and the error on $\langle E^2 \rangle$ using the techniques we employed for the magnetization, and then combine them in some fashion to give an estimate of the error in c. But this is not as straightforward as it seems at first, since the errors in these two quantities are correlated—when $\langle E \rangle$ goes up, so does $\langle E^2 \rangle$. It is possible to do the analysis necessary to calculate the error on c in this fashion. However, it is not particularly simple, and there are other more general methods of error estimation which lend themselves to this problem. As the quantities we want to measure become more complex, these methods—"blocking", the "bootstrap" and the "jackknife"—will save us a great deal of effort in estimating errors. We illustrate these methods here for the case of the specific heat, though it should be clear that they are applicable to almost any quantity that can be measured in a Monte Carlo simulation.

The simplest of our general-purpose error estimation methods is the **blocking method**. Applied to the specific heat, the idea is that we take the measurements of E that we made during the simulation and divide them into several groups, or **blocks**. We then calculate c separately for each block, and the spread of values from one block to another gives us an estimate of the error. To see how this works, suppose we make 200 measurements of the energy during our Ising model simulation, and then split those into 10 groups of 20 measurements. We can evaluate the specific heat from Equation (3.15) for each group and then find the mean of those 10 results exactly as we did for the magnetization above. The error on the mean is given again by Equation (3.37), except that n is now replaced by the number n_b of blocks, which would be 10 in our example. This method is intuitive, and will give a reasonable estimate of the order of magnitude of the error in a quantity such as c. However, the estimates it gives vary depending on the number of different blocks you divide your data up into, with the smallest being associated with large numbers of blocks, and the largest with small numbers of blocks, so it is clearly not a very rigorous method. A related but more reliable method, which can be used for error estimation in a wide variety of different circumstances, is the **bootstrap method**, which we now describe.

3.4.3   The bootstrap method

The bootstrap method is a **resampling method**. Applied to our problem
of calculating the specific heat for the Ising model, it would work like this.
We take our list of measurements of the energy of the model and from the
n numbers in this list we pick out n at random. (Strictly, n should be the
number of *independent* measurements. In practice the measurements made
are usually not all independent, but luckily it transpires that the bootstrap
method is not much affected by this difference, a point which is discussed
further below.) We specifically allow ourselves to pick the same number
twice from the list, so that we end up with n numbers each of which appears
on the original list, and some of which may be duplicates of one another. (In
fact, if you do your sums, you can show that about a fraction $1 - 1/e \simeq 63\%$
of the numbers will be duplicates.) We calculate the specific heat from
these n numbers just as we would normally, and then we repeat the process,
picking (or **resampling**) another n numbers at random from the original
measurements. It can be shown (Efron 1979) that after we have repeated
this calculation several times, the standard deviation of the distribution in
the results for c is a measure of the error in the value of c. In other words
if we make several of these "bootstrap" calculations of the specific heat, our
estimate of the error σ is given by

$$\sigma = \sqrt{\overline{c^2} - \bar{c}^2}. \tag{3.41}$$

Notice that there is no extra factor of $1/(n-1)$ here as there was in Equation (3.37). (It is clear that the latter would not give a correct result, since
it would imply that our estimate of the error could be reduced by simply
resampling our data more times.)

As we mentioned, it is not necessary for the working of the bootstrap
method that all the measurements made be independent in the sense of Section 3.3.1 (i.e., one every two correlation times or more). As we pointed
out earlier, it is more common in a Monte Carlo simulation to make measurements at comfortably short intervals throughout the simulation so as to
be sure of making at least one every correlation time or so and then calculate the number of independent measurements made using Equation (3.19).
Thus the number of samples taken usually exceeds the number which actually constitute independent measurements. One of the nice things about
the bootstrap method is that it is not necessary to compensate for this difference in applying the method. You get fundamentally the same estimate
of the error if you simply resample your n measurements from the entire
set of measurements that were made. In this case, still about 63% of the
samples will be duplicates of one another, but many others will effectively
be duplicates as well because they will be measurements taken at times less
than a correlation time apart. Nonetheless, the resulting estimate of σ is the

same.

The bootstrap method is a good general method for estimating errors in quantities measured by Monte Carlo simulation. Although the method initially met with some opposition from mathematicians who were not convinced of its statistical validity, it is now widely accepted as giving good estimates of errors (Efron 1979).

3.4.4   The jackknife method

A slightly different but related method of error estimation is the **jackknife**. For this method, unlike the bootstrap method, we really do need to choose n independent samples out of those that were made during the run, taking one approximately every two correlation times or more. Applying the jackknife method to the case of the specific heat, we would first use these samples to calculate a value c for the specific heat. Now however, we also calculate n other estimates c_i as follows. We take our set of n measurements, and we remove the first one, leaving $n - 1$, and we calculate the specific heat c_1 from that subset. Then we put the first one back, but remove the second and calculate c_2 from that subset, and so forth. Each c_i is the specific heat calculated with the ith measurement of the energy removed from the set, leaving $n - 1$ measurements. It can then be shown that an estimate of the error in our value of c is

$$\sigma = \sqrt{\sum_{i=1}^{n}(c_i - c)^2}, \qquad (3.42)$$

where c is our estimate of the specific heat using all the data.[11]

Both the jackknife and the bootstrap give good estimates of errors for large data sets, and as the size of the data set becomes infinite they give exact estimates. Which one we choose in a particular case usually depends on how much work is involved applying them. In order to get a decent error estimate from the bootstrap method we usually need to take at least 100 resampled sets of data, and 1000 would not be excessive. (100 would give the error to a bit better than 10% accuracy.) With the jackknife we have to recalculate the quantity we are interested in exactly n times to get the error estimate. So, if n is much larger than 100 or so, the bootstrap is probably the more efficient. Otherwise, we should use the jackknife.[12]

---

[11] In fact, it *is* possible to use the jackknife method with samples taken at intervals Δt less than 2τ. In this case we just reduce the sum inside the square root by a factor of $\Delta t/2\tau$ to get an estimate of σ which is independent of the sampling interval.

[12] For a more detailed discussion of these two methods, we refer the interested reader to the review article by Efron (1979).

3.4.5 Systematic errors

Just as in experiments, systematic errors are much harder to gauge than statistical ones; they don't show up in the fluctuations of the individual measurements of a quantity. (That's why they are systematic errors.) The main source of systematic error in the Ising model simulation described in this chapter is the fact that we wait only a finite amount of time for the system to equilibrate. There is no good general method for estimating systematic errors; each source of error has to be considered separately and a strategy for estimating it evolved. This is essentially what we were doing when we discussed ways of estimating the equilibration time for the Metropolis algorithm. Another possible source of systematic error would be not running the simulation for a long enough time after equilibration to make good independent measurements of the quantities of interest. When we discussed methods for estimating the correlation time τ, we were dealing with this problem. In the later sections of this chapter, and indeed throughout this book, we will discuss methods for estimating and controlling systematic errors as they crop up in various situations.

3.5 Measuring the entropy

We have described how we go about measuring the internal energy, specific heat, magnetization and magnetic susceptibility from our Metropolis Monte Carlo simulation of the Ising model. However, there are three quantities of interest which were mentioned in Chapter 1 which we have not yet discussed: the free energy F, the entropy S and the correlation function $G_c^{(2)}(i,j)$. The correlation function we will consider in the next section. The other two we consider now.

The free energy and the entropy are related by

$$F = U - TS \tag{3.43}$$

so that if we can calculate one, we can easily find the other using the known value of the total internal energy U. Normally, we calculate the entropy, which we do by integrating the specific heat over temperature as follows.

We can calculate the specific heat of our system from the fluctuations in the internal energy as described in Section 3.3. Moreover, we know that the specific heat C is equal to

$$C = T\frac{dS}{dT}. \tag{3.44}$$

Thus the entropy $S(T)$ at temperature T is

$$S(T) = S(T_0) + \int_{T_0}^{T} \frac{C}{T} \, dT. \tag{3.45}$$

If we are only interested in how S varies with T, then it is not necessary to know the value of the integration constant $S(T_0)$ and we can give it any value we like. If we want to know the absolute value of S, then we have to fix $S(T_0)$ by choosing T_0 to be some temperature at which we know the value of the entropy. The conventional choice, known as the third law of thermodynamics, is to make the entropy zero[13] when $T_0 = 0$. In other words

$$S(T) = \int_0^T \frac{C}{T} \, dT. \qquad (3.46)$$

As with the other quantities we have discussed, we often prefer to calculate the entropy per spin $s(T)$ of our system, which is given in terms of the specific heat per spin c by

$$s(T) = \int_0^T \frac{c}{T} \, dT. \qquad (3.47)$$

Of course, evaluating either of these expressions involves calculating the specific heat over a range of temperatures up to the temperature we are interested in, at sufficiently small intervals that its variation with T is well approximated. Then we have to perform a numerical integral using, for example, the trapezium rule or any of a variety of more sophisticated integration techniques (see, for instance, Press *et al.* 1988). Calculating the entropy (or equivalently the free energy) of a system is therefore a more complex task than calculating the internal energy, and may use up considerably more computer time. On the other hand, if we are interested in probing the behaviour of our system over a range of temperatures anyway (as we often are), we may as well make use of the data to calculate the entropy; the integration is a small extra computational effort to make by comparison with the simulation itself.

The integration involved in the entropy calculation can give problems. In particular, if there is any very sharp behaviour in the curve of specific heat as a function of temperature, we may miss it in our simulation, which would give the integral the wrong value. This is a particular problem near "phase transitions", where the specific heat often diverges (see Section 3.7.1).

3.6   Measuring correlation functions

One other quantity which we frequently want to measure is the two-point connected correlation function $G_c^{(2)}(i,j)$. Let us see how we would go about

---

[13]Systems which have more than one ground state may violate the third law. This point is discussed in more detail in Section 7.1.2.

calculating this correlation function for the Ising model. The most straight-forward way is to evaluate it directly from the definition, Equation (1.26), using the values of the spins s_i from our simulation:

$$G_c^{(2)}(i,j) = \langle s_i s_j \rangle - \langle s_i \rangle \langle s_j \rangle = \langle s_i s_j \rangle - m^2. \tag{3.48}$$

(Here m denotes the expectation value of the magnetization.) In fact, since our Ising system is translationally invariant, $G_c^{(2)}(i,j)$ is dependent only on the displacement $\mathbf{r}$ between the sites i and j, and not on exactly where they are. In other words, if $\mathbf{r}_i$ is the position vector of the ith spin, then we should have

$$G_c^{(2)}(\mathbf{r}_i, \mathbf{r}_i + \mathbf{r}) = G_c^{(2)}(\mathbf{r}) \tag{3.49}$$

independent of the value of $\mathbf{r}_i$. This means that we can improve our estimate of the correlation function $G_c^{(2)}(\mathbf{r})$ by averaging its value over the whole lattice for all pairs of spins separated by a displacement $\mathbf{r}$:

$$G_c^{(2)}(\mathbf{r}) = \frac{1}{N} \sum_{\substack{i,j \text{ with} \\ \mathbf{r}_j - \mathbf{r}_i = \mathbf{r}}} \left[\langle s_i s_j \rangle - m^2 \right]. \tag{3.50}$$

If, as with our Ising model simulation, the system we are simulating has periodic boundary conditions (see Section 3.1), then $G_c^{(2)}(\mathbf{r})$ will not die away for very large values of $\mathbf{r}$. Instead, it will be periodic, dying away for values of $\mathbf{r}$ up to half the width of the lattice, and then building up again to another maximum when we have gone all the way across the lattice and got back to the spin we started with.

In order to evaluate $G_c^{(2)}(\mathbf{r})$ using Equation (3.50) we have to record the value of every single spin on the lattice at intervals during the simulation. This is not usually a big problem given the generous amounts of storage space provided by modern computers. However, if we want to calculate $G_c^{(2)}(\mathbf{r})$ for every value of $\mathbf{r}$ on the lattice, this kind of direct calculation does take an amount of time which scales with the number N of spins on the lattice as N^2. As with the calculation of the autocorrelation function in Section 3.3.1, it actually turns out to be quicker to calculate the Fourier transform of the correlation function instead.

The spatial Fourier transform $\widetilde{G}_c^{(2)}(\mathbf{k})$ is defined by

$$\begin{aligned}
\widetilde{G}_c^{(2)}(\mathbf{k}) &= \sum_{\mathbf{r}} e^{i\mathbf{k}\cdot\mathbf{r}} G_c^{(2)}(\mathbf{r}) \\
&= \frac{1}{N} \sum_{\mathbf{r}} \sum_{\substack{i,j \text{ with} \\ \mathbf{r}_j - \mathbf{r}_i = \mathbf{r}}} e^{i\mathbf{k}\cdot(\mathbf{r}_j - \mathbf{r}_i)} \left[\langle s_i s_j \rangle - m^2 \right] \\
&= \frac{1}{N} \left\langle \sum_{\mathbf{r}_i} e^{-i\mathbf{k}\cdot\mathbf{r}_i} (s_i - m) \sum_{\mathbf{r}_j} e^{i\mathbf{k}\cdot\mathbf{r}_j} (s_j - m) \right\rangle
\end{aligned}$$

$$= \frac{1}{N} \langle |\tilde{s}'(\mathbf{k})|^2 \rangle, \tag{3.51}$$

where $\tilde{s}'(\mathbf{k})$ is the Fourier transform of $s_i' \equiv s_i - m$. In other words, in order to calculate $\tilde{G}_c^{(2)}(\mathbf{k})$, we just need to perform a Fourier transform of the spins at a succession of different times throughout the simulation and then feed the results into Equation (3.51). As in the case of the autocorrelation function of Section 3.2, this can be done using a standard FFT algorithm. To get the correlation function in real space we then have to use the algorithm again to Fourier transform back, but the whole process still only takes a time which scales as $N \log N$, and so for the large lattices of today's Monte Carlo studies it is usually faster than direct calculation from Equation (3.50).[14]

Occasionally we also need to calculate the disconnected correlation function defined in Section 1.2.1. The equivalent of (3.51) in this case is simply

$$\tilde{G}^{(2)}(\mathbf{k}) = \frac{1}{N} \langle |\tilde{s}(\mathbf{k})|^2 \rangle. \tag{3.52}$$

Note that s_i and s_i' differ only by the average magnetization m, which is a constant. As a result, $\tilde{G}^{(2)}(\mathbf{k})$ and $\tilde{G}_c^{(2)}(\mathbf{k})$ are in fact identical, except at $\mathbf{k} = 0$. For this reason, it is often simpler to calculate $\tilde{G}_c^{(2)}(\mathbf{k})$ by first calculating $\tilde{G}^{(2)}(\mathbf{k})$ and then just setting the $\mathbf{k} = 0$ component to zero.

3.7   An actual calculation

In this section we go through the details of an actual Monte Carlo simulation and demonstrate how the calculation proceeds. The example that we take is that of the simulation of the two-dimensional Ising model on a square lattice using the Metropolis algorithm. This system has the advantage that its properties in the thermodynamic limit are known exactly, following the analytic solution given by Onsager (1944). Comparing the results from our simulation with the exact solution will give us a feel for the sort of accuracy one can expect to achieve using the Monte Carlo method. Some of the results have already been presented (see Figures 3.3 and 3.5 for example). Here we

---

[14] Again we should point out that this does not necessarily mean that one should always calculate the correlation function this way. As with the calculation of the autocorrelation function, using the Fourier transform is a more complicated method than direct calculation of the correlation function, and if your goal is to get an estimate quickly, and your lattice is not very large, you may be better advised to go the direct route. However, the Fourier transform method is more often of use in the present case of the two-point correlation function, since in order to perform the thermal average appearing in Equations (3.50) and (3.51) we need to repeat the calculation about once every two correlation times throughout the entire simulation, which might mean doing it a hundred or a thousand times in one run. Under these circumstances the FFT method may well be advantageous.

describe in detail how these and other results are arrived at, and discuss what conclusions we can draw from them.

The first step in performing any Monte Carlo calculation, once we have decided on the algorithm we are going to use, is to write the computer program to implement that algorithm. The code used for our Metropolis simulation of the Ising model is given in Appendix B. It is written in the computer language C.

As a test of whether the program is correct, we have first used it to simulate a small 5×5 Ising model for a variety of temperatures between $T = 0$ and $T = 5.0$ with J set equal to 1. For such a small system our program runs very fast, and the entire simulation only took about a second at each temperature. In Section 1.3 we performed an exact calculation of the magnetization and specific heat for the 5×5 Ising system by directly evaluating the partition function from a sum over all the states of the system. This gives us something to compare our Monte Carlo results with, so that we can tell if our program is doing the right thing. At this stage, we are not interested in doing a very rigorous calculation, only in performing a quick check of the program, so we have not made much effort to ensure the equilibration of the system or to measure the correlation time. Instead, we simply ran our program for 20 000 Monte Carlo steps per site (i.e., 20 000 × 25 = 500 000 steps in all), and averaged over the last 18 000 of these to measure the magnetization and the energy. Then we calculated m from Equation (3.12) and c from Equation (3.15). If the results do not agree with our exact calculation then it could mean either that there is a problem with the program, or that we have not waited long enough in either the equilibration or the measurement sections of the simulation. However, as shown in Figure 3.7, the numerical results agree rather well with the exact ones. Even though we have not calculated the statistical errors on our data in order to determine the degree of agreement, these results still give us enough confidence in our program to proceed with a more thorough calculation on a larger system.

For our large-scale simulation, we have chosen to examine a system of 100×100 spins on a square lattice. We started the program with randomly chosen values of all the spins—the $T = \infty$ state of Section 3.1.1—and ran the simulations at a variety of temperatures from $T = 0.2$ to $T = 5.0$ in steps of 0.2, for a total of 25 simulations in all.[15] Again we ran our simulations

---

[15] Note that it is not possible to perform a simulation at $T = 0$ because the acceptance ratio, Equation (3.7), for spin flips which increase the energy of the system becomes zero in this limit. This means that it is not possible to guarantee that the system will come to equilibrium, because the requirement of ergodicity is violated; there are some states which it is not possible to get to in a finite number of moves. It is true in general of thermal Monte Carlo methods that they break down at $T = 0$, and often they become very slow close to $T = 0$. The continuous time Monte Carlo method of Section 2.4 can sometimes by used to overcome this problem in cases where we are particularly interested

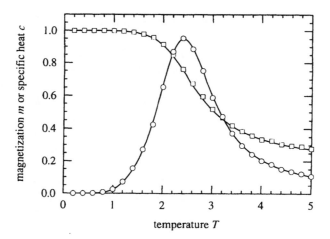

FIGURE 3.7   The magnetization (squares) and specific heat (circles) per spin of an Ising model in two dimensions on a 5 × 5 square lattice. The points are the results of the Metropolis Monte Carlo calculation described in the text. The lines are the exact calculations performed in Section 1.3, in which we evaluated the partition function by summing over all the states of the system.

for 20 000 Monte Carlo steps per lattice site. This is a fairly generous first run, and is only possible because we are looking at quite a small system still. In the case of larger or more complex models, one might well first perform a shorter run to get a rough measure of the equilibration and correlation times for the system, before deciding how long a simulation to perform. A still more sophisticated approach is to perform a short run and then store the configuration of the spins on the lattice before the program ends. Then, after deciding how long the entire calculation should last on the basis of the measurements during that short run, we can pick up exactly where we left off using the stored configuration, thus saving ourselves the effort of equilibrating the system twice.

Taking the data from our 25 simulations at different temperatures, we first estimate the equilibration times τ_{eq} at each temperature using the methods described in Section 3.2. In this case we found that all the equilibration times were less than about 1000 Monte Carlo steps per site, except for the simulations performed at $T = 2.0$ and $T = 2.2$, which both had equilibration times on the order of 6000 steps per site. (The reason for this anomaly is explained in the next section.) Allowing ourselves a margin for error in these estimates, we therefore took the data from time 2000 onwards as our equi-

in the behaviour of a model close to $T = 0$.

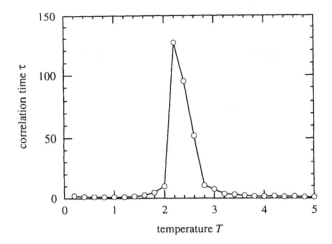

FIGURE 3.8 The correlation time for the 100×100 Ising model simulated using the Metropolis algorithm. The correlation time is measured in Monte Carlo steps per lattice site (i.e., in multiples of 10 000 Monte Carlo steps in this case). The straight lines joining the points are just to guide the eye.

librium measurements for all the temperatures except the two slower ones, for which we took the data for times 10 000 onwards.

Next we need to estimate how many independent measurements these data constitute, which means estimating the correlation time. To do this, we calculate the magnetization autocorrelation function at each temperature from Equation (3.21), for times t up to 1000. (We must be careful only to use our equilibrium data for this calculation since the autocorrelation function is an equilibrium quantity. That is, we should not use the data from the early part of the simulation during which the system was coming to equilibrium.) Performing a fit to these functions as in Figure 3.6, we make an estimate of the correlation time τ at each temperature. The results are shown in Figure 3.8. Note the peak in the correlation time around $T = 2.2$. This effect is called "critical slowing down", and we will discuss it in more detail in Section 3.7.2. Given the length of the simulation $t_{\max}$ and our estimates of τ_{eq} and τ for each temperature, we can calculate the number n of independent measurements to which our simulations correspond using Equation (3.19).

Using these figures we can now calculate the equilibrium properties of the 100×100 Ising model in two dimensions. As an example, we have calculated the magnetization and the specific heat again. Our estimate of the magnetization is calculated by averaging over the magnetization measurements from the simulation, again excluding the data from the early portion

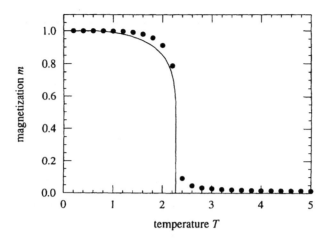

FIGURE 3.9   The magnetization per spin of the two-dimensional Ising model. The points are the results from our Monte Carlo simulation using the Metropolis algorithm. The errors are actually smaller than the points in this figure because the calculation is so accurate. The solid line is the known exact solution for the Ising model on an infinite two-dimensional square lattice.

of the run where the system was not equilibrated. The results are shown in Figure 3.9, along with the known exact solution for the infinite system. Calculating the errors on the magnetization from Equation (3.37), we find that the errors are so small that the error bars would be completely covered by the points themselves, so we have not bothered to put them in the figure. The agreement between the numerical calculation and the exact one for the infinite lattice is much better than it was for the smaller system in Figure 1.1, although there is still some discrepancy between the two. This discrepancy arises because the quantity plotted in the figure is in fact the average $\langle |m| \rangle$ of the *magnitude* of the magnetization, and not the average magnetization itself; we discuss our reasons for doing this in Section 3.7.1 when we examine the spontaneous magnetization of the Ising model.

The figure clearly shows the benefits of the Monte Carlo method. The calculation on the 5×5 system which we performed in Section 1.3 was exact, whereas the Monte Carlo calculation on the 100×100 system is not. However, the Monte Carlo calculation still gives a better estimate of the magnetization of the infinite system. The errors due to statistical fluctuations in our measurements of the magnetization are much smaller than the inaccuracy of working with a tiny 5×5 system.

Using the energy measurements from our simulation, we have also calculated the specific heat for our Ising model from Equation (3.15). To calculate

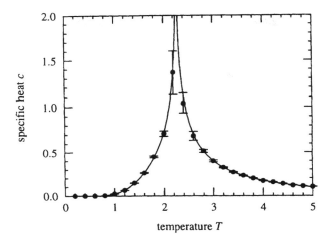

FIGURE 3.10 The specific heat per spin of the two-dimensional Ising model calculated by Monte Carlo simulation (points with error bars) and the exact solution for the same quantity (solid line). Note how the error bars get bigger close to the peak in the specific heat. This phenomenon is discussed in detail in the next section.

the errors on the resulting numbers we could use the blocking method of Section 3.4.2 to get a rough estimate. Here we are interested in doing a more accurate calculation, and for that we should use either the bootstrap or the jackknife method (see Sections 3.4.3 and 3.4.4). The number of independent samples n is for most temperatures considerably greater than 100, so by the criterion given in Section 3.4.4, the bootstrap method is the more efficient one to use. In Figure 3.10 we show our results for the specific heat with error bars calculated from 200 bootstrap resamplings of the data (giving errors accurate to about 5%). The agreement here with the known exact result for the specific heat is excellent—better than for the magnetization—though the errors are larger, especially in the region close to the peak. If we were particularly interested to know the value of c in this region it would make sense for us to go back and do a longer run of our program in this region to get more accurate data. For example, if we wanted to calculate the entropy difference from one side of the peak to the other using Equation (3.45), then large error bars in this region would make the integral inaccurate and we might well benefit from expending a little more effort in this region to get more accurate results.

3.7.1   The phase transition

It is not really the intention of this book to discuss the properties of the Ising model in detail, or the properties of any other model. However, there are a few things about the Ising model which we need to look into in a little more detail in order to understand the strengths and weaknesses of the Metropolis algorithm, and to explain why other algorithms may be better for certain calculations.

If we look at Figures 3.9 and 3.10, the most obvious feature that strikes us is the behaviour of the model around temperature $T = 2.2$ or so. The results shown in Figure 3.9 indicate that above this temperature the mean magnetization per site m is quite small, whereas below it the magnetization is definitely non-zero and for the most part quite close to its maximum possible value of 1. This seems like a sensible way for the model to behave, since we know (see Section 3.1.1) that the $T = \infty$ state is one in which all the spins are randomly oriented up or down so that the net magnetization will be zero on average, and we know that the $T = 0$ state is one in which all the spins line up with one another, either all up or all down, so that the magnetization per site is either $+1$ or -1. If we only had results for a small Ising system to go by, like the ones depicted in Figure 3.7, we might imagine that the true behaviour of the system was simply that the magnetization rose smoothly from zero in the $T \to \infty$ limit to 1 as the temperature tends to zero. However, our results for the larger 100×100 system indicate that the transition from small m to large m becomes sharper as we go to larger systems, and in fact we know in the case of this particular model, because we have an exact solution, that the change from one regime to the other is actually infinitely sharp in the thermodynamic limit. This kind of change is called a **phase transition**. The Ising model is known to have a phase transition in two or more dimensions. The two regimes we observe are called the **phases** of the model. The phase transition between them takes place at a temperature T_c, which we call the **critical temperature**, whose value in the particular case of the two-dimensional Ising model is known to be

$$T_c = \frac{2J}{\log(1 + \sqrt{2})} \simeq 2.269J. \tag{3.53}$$

Above this temperature the system is in the **paramagnetic phase**, in which the average magnetization is zero. Below it, the system is in the **ferromagnetic phase** and develops a **spontaneous magnetization** (i.e., most of the spins align one way or the other and the magnetization becomes non-zero all of its own accord without the application of a magnetic field to the model). This spontaneous magnetization rises from zero at the phase transition to unity at absolute zero. The magnetization is referred to as the **order parameter** of the Ising model because of this behaviour. In general, an order parameter is any quantity which is zero on one side of a phase transition and non-zero on the other. A phase transition in which the order parameter is continuous at T_c, as it is here, is called a **continuous phase transition**.

In fact, to be strictly correct, the *mean* magnetization of the Ising model below the critical temperature is still zero, since the system is equally happy to have most of its spins pointing either down or up. Thus if we average over a long period of time we will find that the magnetization is close to +1 half the time and close to −1 for the other half, with occasional transitions between the two, so that the average is still close to zero. However, the average of the magnitude $|m|$ of the magnetization will be close to +1, whereas it will be close to zero above the phase transition. In Figure 3.9 we therefore actually plotted the average of $|m|$, and not m. This explains why the magnetization above the transition temperature is still slightly greater than zero. The average magnetization in this phase is definitely zero (give or take the statistical error) but the average of the *magnitude* of m is always greater than zero, since we are taking the average of a number which is never negative. Still, as we go to the thermodynamic limit we expect this quantity to tend to zero, so that the numerical result and the exact solution should agree.[16]

We can look in detail at what happens to the spins in our Ising system as we pass through the phase transition from high to low temperatures by examining pictures such as those in Figure 3.2. At high temperatures the spins are random and uncorrelated, but as the temperature is lowered the interactions between them encourage nearby spins to point in the same direction, giving rise to correlations in the system. Groups of adjacent spins which are correlated in this fashion and tend to point in the same direction are called **clusters**.[17] As we approach T_c, the typical size ξ of these clusters—also called the **correlation length**—diverges, so that when we are precisely at the transition, we may encounter arbitrarily large areas in which the spins are pointing mostly up or mostly down. Then, as we pass below the transition temperature, the system spontaneously chooses to have the majority of its spins in either the up or the down direction, and develops a non-zero magnetization in that direction. Which direction it chooses depends solely

---

[16]This also provides an explanation of why the agreement between the analytic solution and the Monte Carlo calculation was better for the specific heat, Figure 3.10, than it was for the magnetization. The process of taking the mean of the magnitude $|m|$ of the magnetization means that we consistently overestimate the magnetization above the critical temperature, and in fact this problem extends to temperatures a little below T_c as well (see Figure 3.9). No such adjustments are necessary when calculating the specific heat, and as a result our simulation agrees much better with the known values for c, even though the error bars are larger in this case.

[17]A number of different mathematical definitions of a cluster are possible. Some of them require that all spins in the cluster point in the same direction whilst others are less strict. We discuss these definitions in detail in the next chapter, particularly in Sections 4.2 and 4.4.2.

on the random details of the thermal fluctuations it was going through as we passed the critical temperature, and so is itself completely random. As the temperature drops further towards $T = 0$, more and more of the spins line up in the same direction, and eventually as $T \rightarrow 0$ we get $|m| = 1$.

The study of phase transitions is an entire subject in itself and we refer the interested reader to other sources for more details of this interesting field. For our purposes the brief summary given above will be enough.

3.7.2 Critical fluctuations and critical slowing down

We are interested in the behaviour of the Ising model in the region close to T_c. This region is called the **critical region**, and the processes typical of the critical region are called **critical phenomena**. As we mentioned, the system tends to form into large clusters of predominantly up- or down-pointing spins as we approach the critical temperature from above. These clusters contribute significantly to both the magnetization and the energy of the system, so that, as they flip from one orientation to another, they produce large fluctuations in m and E, often called **critical fluctuations**. As the typical size ξ of the clusters diverges as $T \rightarrow T_c$, the size of the fluctuations does too. And since fluctuations in m and E are related to the magnetic susceptibility and the specific heat through Equations (3.15) and (3.16), we expect to get divergences in these quantities at T_c also. This is what we see in Figure 3.10. These divergences are some of the most interesting of critical phenomena, and a lot of effort, particularly using Monte Carlo methods, has been devoted to investigating their exact nature. Many Monte Carlo studies of many different models have focused exclusively on the critical region to the exclusion of all else. Unfortunately, it is in precisely this region that our Metropolis algorithm is least accurate.

There are two reasons for this. The first has to do with the critical fluctuations. The statistical errors in the measured values of quantities like the magnetization and the internal energy are proportional to the size of these critical fluctuations (see Section 3.4) and so grow as we approach T_c. In a finite-sized system like the ones we study in our simulations, the size of the fluctuations never actually diverges—that can only happen in the thermodynamic limit—but they can become very large, and this makes for large statistical errors in the measured quantities.

What can we do about this? Well, recall that the error on, for example, the magnetization m indeed increases with the size of the magnetization fluctuations, but it also decreases with the number of independent measurements of m that we make during our simulation (see Equation (3.37)). Thus, in order to reduce the error bars on measurements close to T_c, we need to run our program for longer, so that we get a larger number of measurements. This however, is where the other problem with the Metropolis algorithm comes in. As we saw in Figure 3.8, the correlation time τ of the simulation is also large in the region around T_c. In fact, like the susceptibility and the specific heat, the correlation time actually diverges at T_c in the thermodynamic limit. For the finite-sized systems of our Monte Carlo simulations τ does not diverge, but it can still become very large in the critical region, and a large correlation time means that the number of independent measurements we can extract from a simulation of a certain length is small (see Equation (3.19)). This effect on its own would increase the size of the errors on measurements from our simulation, even without the large critical fluctuations. The combination of both effects is particularly unfortunate, because it means that in order to increase the number of independent measurements we make during our simulation, we have to perform a much longer run of the program; the computer time necessary to reduce the error bars to a size comparable with those away from T_c increases very rapidly as we approach the phase transition.

The critical fluctuations which increase the size of our error bars are an innate physical feature of the Ising model. Any Monte Carlo algorithm which correctly samples the Boltzmann distribution will also give critical fluctuations. There is nothing we can do to change our algorithm which will reduce this source of error. However, the same is not true of the increase in correlation time. This effect, known as **critical slowing down**, is a property of the Monte Carlo algorithm we have used to perform the simulation, but not of the Ising model in general. Different algorithms can have different values of the correlation time at any given temperature, and the degree to which the correlation time grows as we approach T_c, if it grows at all, depends on the precise details of the algorithm. Therefore, if we are particularly interested in the behaviour of a model in the critical region, it may be possible to construct an algorithm which suffers less from critical slowing down than does the Metropolis algorithm, or even eliminates it completely, allowing us to achieve much greater accuracy for our measurements. In the next chapter we look at a number of other algorithms which do just this and which allow us to study the critical region of the Ising model more accurately.

Problems

3.1 Derive the appropriate generalization of Equation (3.10) for a simulation of an Ising model with non-zero external magnetic field B.

3.2 Suppose we have a set of n measurements $x_1 \ldots x_n$ of a real quantity x. Find an approximate expression for the error on our best estimate of the mean of their squares. Take the numbers below and estimate this error.

$$
\begin{array}{cccc}
20.27 & 19.61 & 20.06 & 20.73 \\
20.09 & 20.68 & 19.37 & 20.40 \\
19.95 & 20.55 & 19.64 & 19.94
\end{array}
$$

Now estimate the same quantity for the same set of numbers using the jackknife method of Section 3.4.4.

3.3 In this chapter we described methods for calculating a variety of quantities from Monte Carlo data, including internal energies, specific heats and entropies. Suggest a way in which we might measure the partition function using data from a Monte Carlo simulation.

3.4 Write a computer program to carry out a Metropolis Monte Carlo simulation of the one-dimensional Ising model in zero field. Use it to calculate the internal energy of the model for a variety of temperatures and check the results against the analytic solution from Problem 1.4.

4

Other algorithms for the Ising model

In the last chapter we saw that the Metropolis algorithm with single-spin-flip dynamics is an excellent Monte Carlo algorithm for simulating the Ising model when we are interested in temperatures well away from the critical temperature T_c. However, as we approach the critical temperature, the combination of large critical fluctuations and long correlation time makes the errors on measured quantities grow enormously. As we pointed out, there is little to be done about the critical fluctuations, since these are an intrinsic property of the model near its phase transition (and are, what's more, precisely the kind of interesting physical effect that we *want* to study with our simulations). On the other hand, the increase in the correlation time close to the phase transition is a function of the particular algorithm we are using—the Metropolis algorithm in this case—and it turns out that by using different algorithms we can greatly reduce this undesirable effect. In the first part of this chapter we will study one of the most widely used and successful such algorithms, the Wolff algorithm. Before introducing the algorithm however, we need to define a few terms.

4.1   Critical exponents and their measurement

As discussed in Section 3.7.1, the spins of an Ising model in equilibrium group themselves into clusters of a typical size ξ, called the correlation length, and this correlation length diverges as the temperature approaches T_c. Let us define a dimensionless parameter t, called the **reduced temperature**, which measures how far away we are from T_c:

$$t = \frac{T - T_c}{T_c}. \tag{4.1}$$

When $t = 0$, we are at the critical temperature. The divergence of the correlation length near the phase transition then goes like

$$\xi \sim |t|^{-\nu}. \tag{4.2}$$

The positive quantity ν is called a **critical exponent**. We take the absolute value $|t|$ of the reduced temperature so that the same expression can be used for temperatures both above and below T_c.[1]

It is not obvious that the divergence of ξ has to take the form of Equation (4.2), and indeed for some models it does not; there are models in which the correlation length diverges in other ways near T_c—logarithmically, for instance, or exponentially. However, it turns out that Equation (4.2) is the correct form for the Ising model which we are studying here. And what's more, it is believed that the value of the critical exponent ν is a property of the Ising model itself, and is independent of such things as the value of the coupling J, or the type of lattice we are studying the model on (square or triangular for example). This property is known as **universality** and is one of the most important results to come out of the study of the renormalization group, though to prove it would take us a long way off our path here. In Monte Carlo calculations the value of ν is also independent of the algorithm we use to perform the simulation. (The value of ν does depend on the dimensionality of the lattice we are using. Ising models in two and three dimensions, for example, have different values for ν.)

As we also discussed at the end of the last chapter, there are, along with the divergence of the correlation length, accompanying divergences of the magnetic susceptibility and the specific heat. Two other universal critical exponents govern these divergences:

$$\chi \sim |t|^{-\gamma}, \tag{4.3}$$
$$c \sim |t|^{-\alpha}. \tag{4.4}$$

To describe the divergence of the correlation time τ we define a slightly different sort of exponent, which we denote z:

$$\tau \sim |t|^{-z\nu}, \tag{4.5}$$

where τ is measured in Monte Carlo steps per lattice site. The exponent z is often called the **dynamic exponent**. (It should not be confused with the lattice coordination number introduced in the last chapter, which was also

---

[1] We could have different values for ν above and below the transition temperature, but as it turns out we don't. Although it is beyond the scope of this book, it is relatively straightforward to show using renormalization group arguments that this and the other critical exponents defined here must take the same values above and below T_c (see Binney *et al.* 1992). Note however that the constant of proportionality in Equation (4.2) need not be the same above and below the transition.

denoted z.) Its definition differs from that of the other exponents by the inclusion of the ν, which is the same ν as appeared in Equation (4.2); the exponent is really $z\nu$, and not just z. We're not quite sure why it is defined this way, but it's the way everyone does it, so we just have to get used to it. It is convenient in one respect, as we will see in a moment, because of the way in which z is measured.

The dynamic exponent gives us a way to quantify the critical slowing down effect. As we mentioned, the amount of critical slowing down we see in our simulations depends on what algorithm we use. This gives rise to different values of z for different algorithms—z is not a universal exponent in the way that ν, α and γ are.[2] A large value of z means that τ becomes large very quickly as we approach the phase transition, making our simulation much slower and hence less accurate. A small value of z implies relatively little critical slowing down and a faster algorithm near T_c. If $z = 0$, then there is no critical slowing down at all, and the algorithm can be used right up to the critical temperature without τ becoming large.[3]

The measurement of critical exponents is far from being a trivial exercise. Quite a lot of effort has in fact been devoted to their measurement, since their values have intrinsic interest to those concerned with the physics of phase transitions. At present we are interested in the value of the dynamic exponent z primarily as a measure of the critical slowing down in our algorithms. In Chapter 8 we will discuss in some detail the techniques which have been developed for measuring critical exponents, but for the moment we will just run briefly though one simple technique—a type of "finite size scaling"—to give an idea of how we gauge the efficiency of these algorithms.

Combining Equations (4.2) and (4.5), we can write

$$\tau \sim \xi^z. \tag{4.6}$$

This equation tells us how the correlation time gets longer as the correlation length diverges near the critical point. However, in a system of finite size, which includes all the systems in our Monte Carlo simulations, the correlation length can never really diverge. Recall that the correlation length is the typical dimension of the clusters of correlated spins in the system. Once this size reaches the dimension, call it L, of the system being simulated it can get no bigger. A volume of L^d, where d is the dimensionality of our system, is the largest cluster of spins we can have. This means that in fact

---

[2] The exponent z is still independent of the shape of the lattice, the spin–spin interaction J and so forth. Only changes in the dynamics affect its value. (Hence the name "dynamic exponent".)

[3] In fact, it is conventional to denote a logarithmic divergence $\tau \sim -\log|t|$ of the correlation time by $z = 0$, so a zero dynamic exponent does not necessarily mean that there is no critical slowing down. However, a logarithmic divergence is far less severe than a power-law one, so $z = 0$ is still a good thing.

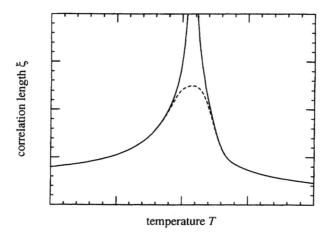

correlation length ξ

temperature T

FIGURE 4.1   Schematic depiction of the cutoff in the divergence of the
correlation length ξ near the critical temperature for a system of finite
size. The solid line indicates the way ξ would diverge in an infinite
system and the dashed one shows how it would be cut off around
$\xi = L$ for a finite system of finite dimension L.

the divergence of the correlation length, and as a result that of the correlation time, is "cut off" in the region for which $\xi > L$. This effect is pictured
schematically in Figure 4.1. Thus, for all temperatures sufficiently close to
the critical one, and particularly when we are at the critical temperature,
Equation (4.6) becomes

$$\tau \sim L^z. \tag{4.7}$$

Now suppose we *know* what the critical temperature T_c of our model is,
perhaps because, as in the 2D Ising model, an exact solution is available.
(More generally, we don't know the critical temperature with any accuracy,
in which case this method won't work, and we have to resort to some of the
more sophisticated ones described in Chapter 8.) In that case, we can use
Equation (4.7) to measure z. We simply perform a sequence of simulations
at $T = T_c$ for systems of a variety of different sizes L, and then plot τ
against L on logarithmic scales. The slope of the resulting plot should be the
exponent z. We have done exactly this in Figure 4.2 for the 2D Ising model
simulated using the Metropolis algorithm. The line is the best fit through
the points, given the errors, and its slope gives us a figure of $z = 2.09 \pm 0.06$
for the dynamic exponent. This calculation was rather a rough one, done
on one of the authors' home computers. Much more thorough Monte Carlo
calculations of z have been performed using this and a number of other
methods by a variety of people. At the time of the writing of this book the

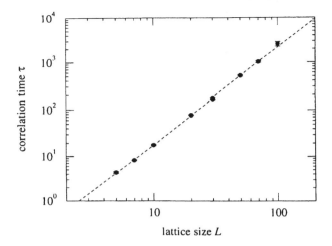

FIGURE 4.2   The correlation time τ for the 2D Ising model simulated using the Metropolis algorithm. The measurements were made at the critical temperature $T_c \simeq 2.269J$ for systems of a variety of different sizes $L \times L$, and then plotted on logarithmic scales against L. The slope $z = 2.09 \pm 0.06$ is an estimate of the dynamic exponent.

best figure available for this exponent was $z = 2.1665 \pm 0.0012$, obtained by Nightingale and Blöte (1996).

The value $z = 2.17$ is a fairly high one amongst Monte Carlo algorithms for the 2D Ising model, and indicates that the Metropolis algorithm is by no means the best algorithm for investigating the behaviour of the model near its phase transition. In the next section we will introduce a new algorithm which has a much lower dynamic exponent, albeit at the expense of some increase in complexity.

4.2 The Wolff algorithm

In the previous section we saw that the dynamic exponent z of the two-dimensional Ising model simulated using the Metropolis algorithm has a value around 2. Equation (4.7) shows us that when we are at the critical temperature, the correlation time increases with system size as L^z. The CPU time taken to perform a certain number of Monte Carlo steps per site increases like the number of sites, or in other words as L^d, where d is the dimensionality of the lattice. Thus the CPU time τ_{CPU} taken to simulate one correlation time increases with system size as

$$\tau_{\mathrm{CPU}} \sim L^{d+z}, \tag{4.8}$$

which for the 2D Ising model is about L^4. This makes measurements on large systems extremely difficult in the critical region. (The simulations which went into Figure 4.2, for instance, took more than two weeks of CPU time, with the largest system size ($L = 100$) requiring 150 billion Monte Carlo steps to get a reasonably accurate value of τ.)

The fundamental reason for the large value of z in the Metropolis algorithm is the divergence of the correlation length and the critical fluctuations present near the phase transition. When the correlation length becomes large close to T_c, large regions form in which all the spins are pointing in the same direction. These regions are often called **domains**.[4] It is quite difficult for the algorithm to flip over one of these large domains because it has to do it spin by spin, and each move has quite a high probability of being rejected because of the ferromagnetic interactions between neighbouring spins. The chances of flipping over a spin in the middle of a domain are particularly low, because it is surrounded by four others pointing in the same direction. In two dimensions, flipping such a spin costs us $8J$ in energy and, using the value of $T_c \simeq 2.269J$ given in Equation (3.53), the probability of accepting such a move, when we are close to the critical temperature, is

$$A(\mu \rightarrow \nu) \simeq e^{-8J/T_c} = 0.0294 \ldots \qquad (4.9)$$

or about three per cent, regardless of the value of J. The chance of flipping a spin on the edge of a domain is higher because it has a lower energy cost, and in fact it turns out that this process is the dominant one when it comes to flipping over entire domains. It can be very slow however, especially with the big domains that form near the critical temperature.

A solution to these problems was proposed in 1989 by Ulli Wolff, based on previous work by Swendsen and Wang (1987). The algorithm he suggested is now known as the **Wolff algorithm**. The basic idea is to look for clusters of similarly oriented spins and then flip them in their entirety all in one go, rather than trying to turn them over spin by painful spin. Algorithms of this type are referred to as **cluster-flipping algorithms**, or sometimes just **cluster algorithms**, and in recent years they have become popular for all sorts of problems, since it turns out that, at least in the case of the Ising model, they almost entirely remove the problem of critical slowing down.

How then do we go about finding these clusters of spins which we are going to flip? The simplest strategy which suggests itself, and the one which the Wolff algorithm uses, is just to pick a spin at random from the lattice and then look at its neighbours to see if any of them are pointing in the same

---

[4]Note that these domains are not quite the same thing as the clusters discussed in Section 3.7.1. By convention, the word "domain" refers to a group of adjacent spins which are pointing in the same direction, whereas the word "cluster" refers to a group of spins whose values are correlated with one another. We will shortly give a more precise definition of a cluster.

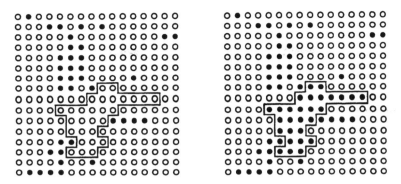

FIGURE 4.3 Flipping a cluster in a simulation of the 2D Ising model. The solid and open circles represent the up- and down-spins in the model.

direction. Then we can look at the neighbours of those neighbours, and so on, iteratively, until we have built up an entire cluster of spins. Clearly, however, we don't just want to flip over all of the spins which are pointing in the same direction as our first "seed" spin. How many we flip should depend on the temperature. We know for example that at high temperatures in the Ising model the spins tend to be uncorrelated with their neighbours, implying that they flip over singly, or in very small clusters. As we approach T_c, we know that the sizes of the clusters become much larger, and then, as we pass below T_c, ferromagnetism sets in, and most of the spins line up in the same direction, forming one big cluster that spans the entire lattice. In other words, the size of the clusters we flip should increase with decreasing temperature. We can achieve this if we build our clusters up from a seed, as described above, *but* we don't add every similarly oriented neighbour of the seed spin to the cluster. Instead, we have some probability P_{add} of adding a spin, which goes up as the temperature falls. Then we look at the neighbours of the spins we have added to the cluster, and add them with probability P_{add}, and so forth. Then, when we finally run out of similarly oriented spins to add to the cluster, we stop and flip the cluster over with some acceptance ratio which depends on the energy cost of flipping it. The question then is, given a particular choice of P_{add}, what is the correct acceptance ratio to make the algorithm satisfy detailed balance, and what is the best choice of P_{add} to make the average acceptance ratio as close to 1 as possible?

4.2.1 Acceptance ratio for a cluster algorithm

Let us work out what the acceptance ratio should be for this cluster-flipping algorithm so that detailed balance is obeyed. Consider two states of the system, μ and ν, illustrated in Figure 4.3. They differ from one another by the flipping of a single cluster of similarly oriented spins. The crucial thing to notice is the way the spins are oriented around the edge of the cluster (which is indicated by the line in the figure). Notice that in each of the two states, some of the spins just outside the cluster are pointing the same way as the spins in the cluster. The bonds between these spins and the ones in the cluster have to be "broken" when the cluster is flipped. Inevitably, the bonds which are *not* broken in going from μ to ν must be broken if we flip back again from ν to μ.

Consider now a move which will take us from μ to ν. There are in fact many such moves—we could choose any of the spins in the cluster as our seed spin, and then we could add the rest of the spins to it in a variety of orders. For the moment, however, let us just consider one particular move, starting with a particular seed spin and then adding the others to it in a particular order. Consider also the reverse move, which takes us back to μ from ν, starting with exactly the same seed spin, and adding the others to it in exactly the same way as in the forward move. The probability of choosing the seed is exactly the same in the two directions, as is the probability of adding each spin to the cluster. The only thing that changes between the two is the probability of "breaking" bonds around the edge of the cluster, because the bonds which have to be broken are different in the two cases. Suppose that, for the forward move, there are m bonds which have to be broken in order to flip the cluster. These broken bonds represent pairs of similarly oriented spins which were not added to the cluster by the algorithm. The probability of not adding such a spin is $1 - P_{\text{add}}$. Thus the probability of not adding all of them, which is proportional to the selection probability $g(\mu \to \nu)$ for the forward move, is $(1 - P_{\text{add}})^m$. If there are n bonds which need to be broken in the reverse move then the probability of doing it will be $(1 - P_{\text{add}})^n$. The condition of detailed balance, Equation (2.14), along with Equation (2.17), then tells us that

$$\frac{g(\mu \to \nu)A(\mu \to \nu)}{g(\nu \to \mu)A(\nu \to \mu)} = (1 - P_{\text{add}})^{m-n}\frac{A(\mu \to \nu)}{A(\nu \to \mu)} = e^{-\beta(E_\nu - E_\mu)}, \qquad (4.10)$$

where $A(\mu \to \nu)$ and $A(\nu \to \mu)$ are the acceptance ratios for the moves in the two directions. The change in energy $E_\nu - E_\mu$ between the two states also depends on the bonds which are broken. For each of the m bonds which are broken in going from μ to ν, the energy changes by $+2J$. For each of the n bonds which are made, the energy changes by $-2J$. Thus

$$E_\nu - E_\mu = 2J(m - n). \qquad (4.11)$$

Substituting into Equation (4.10) and rearranging we derive the following condition on the acceptance ratios:

$$\frac{A(\mu \to \nu)}{A(\nu \to \mu)} = [e^{2\beta J}(1 - P_{\text{add}})]^{n-m}. \qquad (4.12)$$

But now we notice a delightful fact. If we choose

$$P_{\text{add}} = 1 - e^{-2\beta J}, \tag{4.13}$$

then the right-hand side of Equation (4.12) is just 1, independent of any properties of the states μ and ν, or the temperature, or anything else at all. With this choice, we can make the acceptance ratios for both forward and backward moves unity, which is the best possible value they could take. Every move we propose is accepted and the algorithm still satisfies detailed balance. The choice (4.13) defines the Wolff cluster algorithm for the Ising model, whose precise statement goes like this:

1. Choose a seed spin at random from the lattice.

2. Look in turn at each of the neighbours of that spin. If they are pointing in the same direction as the seed spin, add them to the cluster with probability $P_{\text{add}} = 1 - e^{-2\beta J}$.

3. For each spin that was added in the last step, examine each of *its* neighbours to find the ones which are pointing in the same direction and add each of them to the cluster with the same probability P_{add}. (Notice that as the cluster becomes larger we may find that some of the neighbours are already members, in which case obviously we don't have to consider adding them again. Also, some of the spins may have been considered for addition before, as neighbours of other spins in the cluster, but rejected. In this case, they get another chance to be added to the cluster on this step.) This step is repeated as many times as necessary until there are no spins left in the cluster whose neighbours have not been considered for inclusion in the cluster.

4. Flip the cluster.

The implementation of this algorithm in an actual computer program is discussed in detail in Section 13.2.5.

As well as satisfying detailed balance, this algorithm clearly also satisfies the criterion of ergodicity (see Section 2.2.2). To see this, we need only note that there is a finite chance at any move that any spin will be chosen as the sole member of a cluster of one, which is then flipped. Clearly the appropriate succession of such moves will get us from any state to any other in a finite time, as ergodicity requires. Along with the condition of detailed balance, this then guarantees that our algorithm will generate a series of states of the lattice which (after equilibration) will appear with their correct Boltzmann probabilities.

Notice that the probability P_{add}, Equation (4.13), does indeed increase with decreasing temperature, from zero at $T = \infty$ to one at $T = 0$, just as we said it should, so that the sizes of the clusters we grow will increase

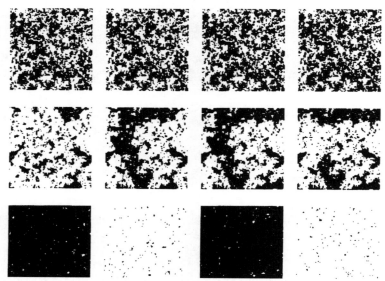

FIGURE 4.4   Consecutive states of a 100 × 100 Ising model in equilibrium simulated using the Wolff algorithm. The top row of four states
are at a temperature $T = 2.8J$, which is well above the critical temperature, the middle row is close to the critical temperature at $T = 2.3J$,
and the bottom row is well below the critical temperature at $T = 1.8J$.

as the temperature gets lower. The flipping of these clusters is a much less
laborious task than in the case of the Metropolis algorithm—as we shall see
it takes a time proportional to the size of the cluster to grow it and then
turn it over—so we have every hope that the algorithm will have a lower
dynamic exponent and less critical slowing down. In Section 4.3.1 we show
that this is indeed the case. First we look in a little more detail at the way
the Wolff algorithm works.

4.3   Properties of the Wolff algorithm

In this section we look at how the Wolff algorithm actually performs in practice. As we will see, it performs extremely well when we are near the critical
temperature, but is actually a little slower than the Metropolis algorithm at
very high or low temperatures.

Figure 4.4 shows a series of states generated by the algorithm at each of
three temperatures, one above T_c, one close to T_c, and one below it. Up-
and down-spins are represented by black and white dots. Consider first the
middle set of states, the ones near T_c. If you examine the four frames, it

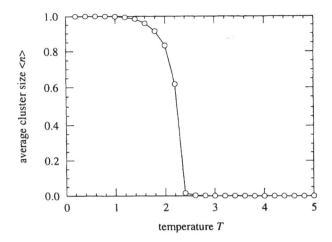

FIGURE 4.5   The average cluster size in the Wolff algorithm as a fraction of the size of the lattice measured as function of temperature. The error bars on the measurements are not shown, because they are smaller than the points. The lines are just a guide to the eye.

is not difficult to make out which cluster flipped at each step. Clearly the algorithm is doing its job, flipping large areas of spins when we are in the critical region. In the $T > T_c$ case, it is much harder to make out the changes between one frame and the next. The reason for this is that in this temperature region P_{add} is quite small (see Equation (4.13)) and this in turn makes the clusters small, so it is hard to see when they flip over. Of course, this is exactly what the algorithm is supposed to do, since the correlation length here is small, and we don't expect to get large regions of spins flipping over together. In Figure 4.5, we have plotted the mean size of the clusters flipped by the Wolff algorithm over a range of temperatures, and, as we can see, they do indeed become small at large temperatures.

When the temperature gets sufficiently high (around $T = 10J$), the mean size of the clusters becomes hardly greater than one. In other words, the single seed spin for each cluster is being flipped over with probability one at each step, but none of its neighbours are. This is exactly what the Metropolis algorithm does in this temperature regime also. When T is large, the Metropolis acceptance ratio, Equation (3.7), is 1, or very close to it, for any transition between two states μ and ν. Thus in the limit of high temperatures, the Wolff algorithm and the Metropolis algorithm become the same thing. But notice that the Wolff algorithm will actually be the slower of the two in this case, because for each seed spin it has to go through the business of testing each of the neighbours for possible inclusion in the cluster, whereas the Metropolis algorithm only has to decide whether to flip a single spin or not, a comparatively simple computational task. (To convince yourself of this, compare the programs for the two algorithms given in Appendix B. The Wolff algorithm is longer and more complex and will take more computer time per step than the Metropolis algorithm, even for a cluster with only one spin.) Thus, even if the Wolff algorithm is a good thing near the phase transition (and we will show that it is), there comes a point as the temperature increases where the Metropolis algorithm becomes better.

Now let us turn to the simulation at low temperature, the bottom row in Figure 4.4. The action of the algorithm is dramatically obvious in this case—almost every spin on the lattice is being flipped at each step. The reason for this is clear. When we are well below the critical temperature the Ising model develops a finite magnetization in one direction or the other, and the majority of the spins on the lattice line up in this direction, forming a "backbone" of similarly oriented spins which spans the entire lattice. When we choose our seed spin for the Wolff algorithm, it is likely that we will land on one of the spins comprising this backbone. Furthermore, the probability P_{add}, Equation (4.13), is large when T is small, so the neighbours of the seed spin are not only very likely to be aligned with it, but are also very likely to be added to the growing cluster. The result is that the cluster grows (usually) to fill almost the entire backbone of spontaneously magnetized spins, and then they are all flipped over in one step. Such a lattice-filling cluster is said to be a **percolating cluster**.

On the face of it, this seems like a very inefficient way to generate states for the Ising model. After all, we know that what should really be happening in the Ising model at low temperature is that most of the spins should be lined up with one another, except for a few "excitation" spins, which are pointing the other way (see the Figure 4.4). Every so often, one of these excitation spins flips back over to join the majority pointing the other way, or perhaps one of the backbone spins gets flipped by a particularly enthusiastic thermal excitation and becomes a new excitation spin. This of course is exactly what the Metropolis algorithm does in this regime. The Wolff algorithm on the other hand is removing the single-spin excitations by the seemingly extravagant measure of *flipping all the other spins on the entire lattice* to point the same way as the single lonely excitation. It's like working out which string of your guitar is out of tune and then tuning the other five to that one. In fact, however, it's actually not such a stupid thing to do. First, let us point out that, since the Wolff algorithm never flips spins which are pointing in the opposite direction to the seed spin, all the excitation spins on the entire lattice end up pointing the same way as the backbone when the algorithm flips over a percolating cluster. Therefore, the Wolff algorithm gets rid of *all* the excitation spins on the lattice in a single step. Second, since we know that the Wolff algorithm generates states with the correct

Boltzmann probabilities, it must presumably create some new excitation spins at the same time as it gets rid of the old ones. And indeed it does do this, as is clear from the frames in the figure. Each spin in the backbone gets a number of different chances to be added to the cluster; for most of the spins this number is just the lattice coordination number z, which is four in the case of the square lattice. Thus the chance that a spin will *not* be added to the cluster is $(1 - P_{\text{add}})^4$. This is a small number, but still non-zero. (It is 0.012 at the temperature $T = 1.8J$ used in the figure.) Thus there will be a small number of spins in the backbone which get left out of the percolating cluster and are not flipped along with the others. These spins become the new excitations on the lattice.

The net result of all this is that after only one Wolff Monte Carlo step, all the old excitations have vanished and a new set have appeared. This behaviour is clear in Figure 4.4. At low temperatures, the Wolff algorithm generates a complete new configuration of the model at every Monte Carlo step, though the payoff is that it has to flip very nearly every spin on every step. In the Metropolis algorithm at low temperatures it turns out that you have to do about one Monte Carlo step per site (each one possibly flipping one spin) to generate a new independent configuration of the lattice. To see this, we need only consider again the excitation spins. The average acceptance ratio for flipping these over will be close to 1 in the Metropolis algorithm, because the energy of the system is usually lowered by flipping them. (Only on the rare occasions when several of them happen to be close together might this not be true.) Thus, it will on average take N steps, where N is the number of sites on the lattice, before we choose any particular such spin and flip it—in other words, one Monte Carlo step per site. But we know that the algorithm correctly generates states with their Boltzmann probability, so, in the same N steps that it takes to find all the excitation spins and flip them over to join the backbone, the algorithm must also choose a roughly equal number of new spins to excite out of the backbone, just as the Wolff algorithm also did. Thus, it takes one sweep of the lattice to replace all the excitation spins with a new set, generating an independent state of the lattice. (This is in fact the best possible performance that any Monte Carlo algorithm can have, since one has to allow at least one sweep of the lattice to generate a new configuration, so that each spin gets the chance to change its value.)

Once more then, we find that the Wolff algorithm and the Metropolis algorithm are roughly comparable, this time at low temperatures. Both need to consider about N spins for flipping in order to generate an independent state of the lattice. Again, however, the additional complexity of the Wolff algorithm is its downfall, and the Metropolis algorithm has a slight edge in speed in this regime because of its extreme simplicity.

So, if the Metropolis algorithm beats the Wolff algorithm (albeit only by a slim margin) at both high and low temperatures, that leaves only the intermediate regime close to T_c in which the Wolff algorithm might be worthwhile. This of course is the regime in which we designed the algorithm to work well, so we have every hope that it will beat the Metropolis algorithm there, and indeed it does, very handily, as we can show by comparing the correlation times of the two algorithms.

4.3.1   The correlation time and the dynamic exponent

Before we measure the correlation time of the Wolff algorithm, we need to consider exactly how it should be defined. If we are going to compare the correlation times of the Wolff algorithm and the Metropolis algorithm near to the phase transition as a way of deciding which is the better algorithm in this region, it clearly would not be fair to measure it for both algorithms in terms of number of Monte Carlo steps (or steps per lattice site). A single Monte Carlo step in the Wolff algorithm is a very complicated procedure, flipping maybe hundreds of spins and potentially taking quite a lot of CPU time, whereas the Metropolis Monte Carlo step is a very simple, quick thing—we can do a million of them in a second on a good computer, though each one only flips at most one spin.

The time taken to complete one step of the Wolff algorithm is proportional to the number of spins n in the cluster.[5] Such a cluster covers a fraction n/L^d of the entire lattice, and thus it will on average take $\langle n \rangle / L^d$ steps to flip each spin once, where $\langle n \rangle$ is the mean cluster size in equilibrium. This is equivalent to one sweep of the lattice in the Metropolis algorithm, so the correct way to define the correlation time is to write $\tau \propto \tau_{\text{steps}} \langle n \rangle / L^d$, where τ_{steps} is the correlation time measured in steps (i.e., clusters flipped) in the Wolff algorithm. The conventional choice for the constant of proportionality is 1. This makes the correlation times for the Wolff and Metropolis algorithms equal in the limits of low and high temperature, for the reasons discussed in the last section. This is not quite fair, since, as we also pointed out in the last section, the Metropolis algorithm is slightly faster in these regimes because it is simpler and each step doesn't demand so much of the computer. However, this difference is slight compared with the enormous difference in performance between the two algorithms near the phase transition which we will witness in a moment, so for all practical purposes we can write

$$\tau = \tau_{\text{steps}} \frac{\langle n \rangle}{L^d}. \tag{4.14}$$

---

[5]This is fairly easy to see: for each spin we have to look at its neighbours to see if they should be included in the cluster, and then when the cluster is complete we have to flip all the spins. Since the same operations are performed for each spin in the cluster, the time taken to do one Monte Carlo step should scale with cluster size.

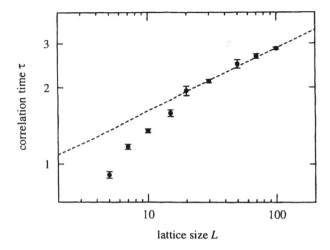

FIGURE 4.6   The correlation time τ for the 2D Ising model simulated using the Wolff algorithm. The measurements deviate from a straight line for small system sizes L, but a fit to the larger sizes, indicated by the dashed line, gives a reasonable figure of $z = 0.25 \pm 0.02$ for the dynamic exponent of the algorithm.

Using this definition, we can compare the performance of the two algorithms in the region close to the critical temperature, and we find that the Wolff algorithm does indeed dramatically outperform the Metropolis algorithm. For instance, in a 100×100 two-dimensional system right at the critical temperature, we measure the correlation time of the Wolff algorithm to be $\tau = 2.80 \pm 0.03$ spin-flips per site. The Metropolis algorithm by contrast has $\tau = 2570 \pm 330$. A factor of a thousand certainly outweighs any difference in the relative complexity of the two algorithms. It is this impressive performance on the part of the Wolff algorithm which makes it a worthwhile algorithm to use if we want to study the behaviour of the model close to T_c.

In Figure 4.6 we have plotted on logarithmic scales the correlation time of the Wolff algorithm for the two-dimensional Ising model at the critical temperature, over a range of different system sizes, just as we did for the Metropolis algorithm in Figure 4.2. Again, the slope of the line gives us an estimate of the dynamic exponent. Our best fit, given the errors on the data points, is $z = 0.25 \pm 0.02$. Again, this was something of a rough calculation, although our result is competitive with other more thorough ones. The best available figure at the time of writing was that of Coddington and Baillie (1992) who measured $z = 0.25 \pm 0.01$. This figure is clearly much lower than the $z = 2.17$ of the Metropolis algorithm, and gives us a quantitative measure of how much better the Wolff algorithm really is.

4.3.2   The dynamic exponent and the susceptibility

In most studies of the Wolff algorithm for the Ising model one does not
actually make use of Equation (4.14) to calculate τ. If we measure time
in Monte Carlo steps (that is, simple cluster flips), we can define a corresponding dynamic exponent z_{steps} in terms of the correlation time τ_{steps} of
Equation (4.14) thus:

$$\tau_{steps} \sim \xi^{z_{steps}}. \tag{4.15}$$

We can measure this exponent in exactly the same way as we did before,
and it turns out that it is related to the real dynamic exponent z for the
algorithm by

$$z = z_{steps} + \frac{\gamma}{\nu} - d, \tag{4.16}$$

where γ and ν are the critical exponents governing the divergences of the
magnetic susceptibility and the correlation length (see Equations (4.2) and
(4.3)) and d is the dimensionality of the model (which is 2 in the cases we
have been looking at). If we know the values of ν and γ, as we do in the
case of the 2D Ising model, then we can use Equation (4.16) to calculate
z without ever having to measure the mean cluster size in the algorithm,
which eliminates one source of error in the measurement, thereby making
the value of z more accurate.[6]

   The first step in demonstrating Equation (4.16) is to prove another useful
result, about the magnetic susceptibility χ. It turns out that, for temperatures $T \geq T_c$, the susceptibility is related to the mean size $\langle n \rangle$ of the clusters
flipped by the Wolff algorithm thus:

$$\chi = \beta \langle n \rangle. \tag{4.17}$$

In many simulations of the Ising model using the Wolff algorithm, the susceptibility is measured using the mean cluster size in this way.

   The demonstration of Equation (4.17) goes like this. Instead of carrying
out the Wolff algorithm in the way described in this chapter, imagine instead
doing it a slightly different way. Imagine that at each step we look at the
whole lattice and for every pair of neighbouring spins which are pointing
in the same direction, we make a "link" between them with probability
$P_{add} = 1 - e^{-2\beta J}$. When we are done, we will have divided the whole
lattice into many different clusters of spins, as shown in Figure 4.7, each of
which will be a correct Wolff cluster (since we have used the correct Wolff
probability P_{add} to make the links). Now we choose a single seed spin from
the lattice at random, and flip the cluster to which it belongs. Then we
throw away all the links we have made and start again. The only difference

---

[6] In cases such as the 3D Ising model, for which we don't know the exact values of the
critical exponents, it may still be better to make use of Equation (4.14) and measure z
directly, as we did in Figure 4.6.

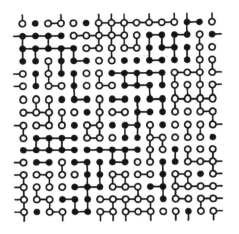

FIGURE 4.7   The whole lattice
can be divided into clusters simply
by putting "links" down with
probability P_{add} between any two
neighbouring spins which are
pointing in the same direction.

between this algorithm and the Wolff algorithm as we described it is that
here we make the clusters first and then choose the seed spin afterwards,
rather than the other way around. This would not be a very efficient way
of implementing the Wolff algorithm, since it requires us to create clusters
all over the lattice, almost all of which never get flipped, but it is a useful
device for proving Equation (4.17). Why? Well, we can write the total
magnetization M of the lattice in a particular state as a sum over all the
clusters on the lattice thus:

$$M = \sum_i S_i n_i. \tag{4.18}$$

Here i labels the different clusters, n_i is their size (a positive integer), and
$S_i = \pm 1$ depending on whether the ith cluster is pointing up or down, thereby
making either a positive or negative contribution to M. In order to calculate
the mean square magnetization we now take the square of this expression
and average it over a large number of spin configurations:

$$\langle M^2 \rangle = \left\langle \sum_i S_i n_i \sum_j S_j n_j \right\rangle = \left\langle \sum_{i \neq j} S_i S_j n_i n_j \right\rangle + \left\langle \sum_i S_i^2 n_i^2 \right\rangle. \tag{4.19}$$

Now, since the two values ± 1 which the variables S_i can take are equally
likely on average, the first term in this expression is an average over a large
number of quantities which are randomly either positive or negative and
which will therefore tend to average out to zero. The second term on the
other hand is an average over only positive quantities since $S_i^2 = +1$ for all

i. This term therefore is definitely not zero[7] and

$$\langle M^2 \rangle = \left\langle \sum_i n_i^2 \right\rangle. \tag{4.20}$$

And for the magnetization per spin we have

$$\langle m^2 \rangle = \frac{1}{N^2} \left\langle \sum_i n_i^2 \right\rangle. \tag{4.21}$$

Now consider the average $\langle n \rangle$ of the size of the clusters which get flipped in the Wolff algorithm. This is not quite the same thing as the average size of the clusters over the entire lattice, because when we choose our seed spin for the Wolff algorithm it is chosen at random from the entire lattice, which means that the probability p_i of it falling in a particular cluster i is proportional to the size of that cluster:

$$p_i = \frac{n_i}{N}. \tag{4.22}$$

The average cluster size in the Wolff algorithm is then given by the average of the probability of a cluster being chosen times the size of that cluster:

$$\langle n \rangle = \left\langle \sum_i p_i n_i \right\rangle = \frac{1}{N} \left\langle \sum_i n_i^2 \right\rangle = N \langle m^2 \rangle. \tag{4.23}$$

Now if we employ Equation (1.36), and recall that $\langle m \rangle = 0$ for $T \geq T_c$, we get

$$\chi = \beta \langle n \rangle \tag{4.24}$$

as promised. Thus, we can measure the magnetic susceptibility of the Ising model simply by counting the spins in our Wolff clusters and taking the average.[8]

The reason this result does not extend below the critical temperature is that for $T < T_c$ the average magnetization $\langle m \rangle$ is non-zero and must be included in Equation (1.36), but there is no simple expression for $\langle m \rangle$ in terms of the sizes of the clusters flipped. For $T \geq T_c$ however, we can use (4.24) to rewrite Equation (4.14) thus:

$$\tau = \tau_{\text{steps}} \frac{\chi}{\beta L^d}. \tag{4.25}$$

---

[7]Strictly, the second term scales like the number of spin configurations in the average and the first scales like its square root, so the second term dominates for a large number of configurations.

[8]In fact, measuring the susceptibility in this way is not only simpler than a direct measurement, it is also superior, giving, as it turns out, smaller statistical errors. For this reason the expression (4.24) is sometimes referred to as an **improved estimator** for the susceptibility (Sweeny 1983, Wolff 1989).

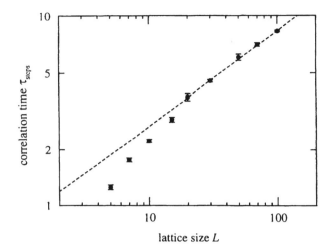

FIGURE 4.8   The correlation time τ_{steps} of the 2D Ising model simulated using the Wolff algorithm, measured in units of Monte Carlo
steps (i.e., cluster flips). The fit gives us a value of $z_{\text{steps}} = 0.50 \pm 0.01$
for the corresponding dynamic exponent.

Now if we use Equations (4.2), (4.3), (4.6) and (4.15), this implies that
slightly above T_c we must have

$$\xi^z \sim \xi^{z_{\text{steps}}} \xi^{\gamma/\nu} L^{-d}. \tag{4.26}$$

Then, as we did in Section 4.1, we note that, close to T_c, the correlation
length is just equal to the dimension L of the system, so we replace ξ every-
where by L, and we get

$$L^z \sim L^{z_{\text{steps}}} L^{\gamma/\nu} L^{-d}, \tag{4.27}$$

which then implies that

$$z = z_{\text{steps}} + \frac{\gamma}{\nu} - d \tag{4.28}$$

as we suggested earlier on.

In Figure 4.8 we have replotted the results from our Wolff algorithm
simulations of the 2D Ising model using the correlation time measured in
Monte Carlo steps (i.e., cluster flips). The best fit to the data gives a figure
of $z_{\text{steps}} = 0.50 \pm 0.01$. Using the accepted values $\nu = 1$ and $\gamma = \frac{7}{4}$ for the
2D Ising model (Onsager 1944), and setting $d = 2$, Equation (4.28) then
gives us $z = 0.25 \pm 0.01$, which is as good as the best published result for
this exponent.

4.4   Further algorithms for the Ising model

We have looked at two algorithms for simulating the Ising model—the Metropolis algorithm and the Wolff algorithm—and for all practical purposes, these two are all we need for simulating the model in equilibrium. When we are interested in the behaviour of the model well away from the critical temperature, the Metropolis algorithm provides a simple and efficient way of getting results which is not bettered by any other algorithm. Close to T_c the Wolff algorithm is a better choice than the Metropolis algorithm; although it is more complex than the Metropolis algorithm, the Wolff algorithm has a very small dynamic exponent, which means that the time taken to perform a simulation scales roughly like the size of the system, which is the best that we can hope for in any algorithm. The combination of these two algorithms allows us to study large systems at any temperature—calculations have been performed on systems with $L = 200$ or greater in both two and three dimensions.

However, many other algorithms have been suggested for the simulation of the Ising model. Although the algorithms studied so far pretty much cover the ground, it is still useful to know about some of these others, since many of them are in use and have been employed in published studies. In this section we talk about a few of these alternative algorithms.

4.4.1   The Swendsen–Wang algorithm

After the Metropolis and Wolff algorithms, probably the most important other algorithm is that of Swendsen and Wang (1987), which, like the Wolff algorithm, is a cluster-flipping algorithm. In fact this algorithm is very similar to the Wolff algorithm and Wolff took the idea for his algorithm directly from it. (Swendsen and Wang in turn got the idea from the work of Fortuin and Kasteleyn (1972) and Coniglio and Klein (1980).) Actually, we have seen the central idea behind the Swendsen–Wang algorithm already. In Section 4.3.2 we considered an alternative implementation of the Wolff algorithm in which the entire lattice of spins is divided up into clusters by making "links" with probability $P_{\text{add}} = 1 - e^{-2\beta J}$ between similarly oriented neighbouring spins—any group of spins which is connected together by links is a cluster. We then imagined choosing a single spin from the lattice and flipping over the whole of the cluster to which it belongs. This procedure is clearly equivalent to the Wolff algorithm, although in practice it would be a very inefficient way of carrying it out.

The Swendsen–Wang algorithm divides the entire lattice into clusters in exactly the same way, with this same probability P_{add} of making a link. But then, instead of flipping just one cluster, all the clusters are flipped with probability $\frac{1}{2}$. That is, for each cluster in turn, we decide independently with probability $\frac{1}{2}$ whether to flip it or not. We notice the following facts about this algorithm:

1. The algorithm satisfies the requirement of ergodicity (Section 2.2.2). To see this, we note that there is always a finite chance that *no* links will be made on the lattice at all (in fact this is what usually happens when T becomes very large), in which case the subsequent randomizing of the clusters is just equivalent to choosing a new state at random for the entire lattice. Thus, it is in theory possible to go from any state to any other in one step.

2. The algorithm also satisfies the condition of detailed balance. The proof of this fact is exactly the same as it was for the Wolff algorithm. If the number of links broken and made in performing a move are m and n respectively (and the reverse for the reverse move), then the energy change for the move is $2J(m - n)$ (or $2J(n - m)$ for the reverse move). The selection probabilities for choosing a particular set of links differ between forward and reverse moves only at the places where bonds are made or broken, so the ratio of the two selection probabilities is $(1 - P_{\mathrm{add}})^{m-n}$, just as it was before. By choosing $P_{\mathrm{add}} = 1 - e^{-2\beta J}$, we then ensure, just as before, that the acceptance probability is independent of m and n and everything else, so any choice which makes it the same in each direction, such as flipping all clusters with probability $\frac{1}{2}$, will make the algorithm correct. Notice however that many other choices would also work. It doesn't matter how we choose to flip the clusters, though the choice made here is good because it minimizes the correlation between the direction of a cluster before and after a move, the new direction being chosen completely at random, regardless of the old one.

3. The algorithm updates the entire lattice on each move. In measuring correlation times for this algorithm, one should therefore measure them simply in numbers of Monte Carlo steps, and not steps per site as with the Metropolis algorithm. (In fact, on average, only half the spins get flipped on each move, but the number flipped scales like the size of the system, which is the important point.)

4. The Swendsen–Wang algorithm is essentially the same as the Wolff algorithm for low temperatures. Well below T_c, one of the clusters chosen by the algorithm will be the big percolating cluster, and the rest will correspond to the "excitations" discussed in Section 4.3, which will be very small. Ignoring these very small clusters then, the Swendsen–Wang algorithm will tend to turn over the percolating backbone of the lattice on average every two steps (rather than every step as in the Wolff algorithm—see Figure 4.4), but otherwise the two will behave almost identically. Thus, as with the Wolff algorithm, we can expect

the performance of the Swendsen–Wang algorithm to be similar to that of the Metropolis algorithm at low T, though probably a little slower on average due to the complexity of the algorithm.

5. At high temperatures, the Swendsen–Wang algorithm tends to divide the lattice into very small clusters because P_{add} becomes small. As $T \to \infty$ the clusters will just be one spin each, and the algorithm will just change all the spins to new random values on each move. This is also what the Metropolis algorithm does at high temperatures in one sweep of the lattice, though again the Metropolis algorithm can be expected to be a little more efficient in this regime, since it is a simpler algorithm which takes few operations on the computer to flip each spin.

The combination of the last two points here implies that the only regime in which the Swendsen–Wang algorithm can be expected to outperform the Metropolis algorithm is the one close to the critical temperature. The best measurement of the dynamic exponent of the algorithm is that of Coddington and Baillie (1992), who found $z = 0.25 \pm 0.01$ in two dimensions, which is clearly much better than the Metropolis algorithm, and is in fact exactly the same as the result for the Wolff algorithm. So the Swendsen–Wang algorithm is a pretty good algorithm for investigating the 2D Ising model close to its critical point. However, the Wolff algorithm is always at least a factor of two faster, since in the Wolff algorithm every spin in every cluster generated is flipped, whereas the spins in only half of the clusters generated are flipped in the Swendsen–Wang algorithm. In addition, as Table 4.1 shows, for higher dimensions the Swendsen–Wang algorithm has a significantly higher dynamic exponent than the Wolff algorithm, making it slower close to T_c. The reason for this is that close to T_c the properties of the Ising model are dominated by the fluctuations of large clusters of spins. As the arguments of Section 4.3.2 showed, the Wolff algorithm preferentially flips larger clusters because the chance of the seed spin belonging to any particular cluster is proportional to the size of that cluster. The Swendsen–Wang algorithm on the other hand treats all clusters equally, regardless of their size, and therefore wastes a considerable amount of effort on small clusters which make vanishingly little contribution to the macroscopic properties of the system for large system sizes. This, coupled with the fact that the Swendsen–Wang algorithm is slightly more complicated to program than the Wolff algorithm, makes the Wolff algorithm the algorithm of choice for most people.[9]

---

[9]There is one important exception to this rule; as discussed in Section 14.2.2, the Swendsen–Wang algorithm can be implemented more efficiently on a parallel computer than can the Wolff algorithm.

dimension d	Metropolis	Wolff	Swendsen–Wang
2	2.167 ± 0.001	0.25 ± 0.01	0.25 ± 0.01
3	2.02 ± 0.02	0.33 ± 0.01	0.54 ± 0.02
4	–	0.25 ± 0.01	0.86 ± 0.02

TABLE 4.1 Comparison of the values of the dynamic exponent z for the Metropolis, Wolff and Swendsen–Wang algorithms in various dimensions. The values are taken from Coddington and Baillie (1992), Matz *et al.* (1994), and Nightingale and Blöte (1996). To our knowledge, the dynamic exponent of the Metropolis algorithm has not been measured in four dimensions.

4.4.2 Niedermayer's algorithm

Another variation on the general cluster algorithm theme was proposed by Ferenc Niedermayer in 1988. His suggestion is really just an extension of the ideas used in the Wolff and Swendsen–Wang algorithms. Niedermayer's methods are very general and can be applied to all sorts of models, such as the glassy spin models that we will study in Chapter 6. Here we will just consider their application to the ordinary Ising model.

Niedermayer pointed out that it is not necessary to constrain the "links" with which we make clusters to be only between spins which are pointing in the same direction. In general, we can define two different probabilities for putting links between sites—one for parallel spins and one for anti-parallel ones. The way Niedermayer expressed it, he considered the energy contribution E_{ij} that a pair of spins i and j makes to the Hamiltonian. In the case of the Ising model, for example,

$$E_{ij} = -Js_i s_j. \tag{4.29}$$

He then wrote the probability for making a link between two neighbouring spins as a function of this energy $P_{\text{add}}(E_{ij})$. In the Ising model E_{ij} can only take two values $\pm J$, so the function $P_{\text{add}}(E_{ij})$ only needs to be defined at these points, but for some of the more general models Niedermayer considered it needs to be defined elsewhere as well. Clearly, if for the Ising model we make $P_{\text{add}}(-J) = 1 - \mathrm{e}^{-2\beta J}$ and $P_{\text{add}}(J) = 0$, then we recover the Wolff algorithm or the Swendsen–Wang algorithm, depending on whether we flip only a single cluster on each move, or many clusters over the entire lattice—Niedermayer's formalism is applicable in either case.

To be concrete about things, let us look at the case of the single-cluster, Wolff-type version of Niedermayer's algorithm. First, it is clear that for any choice of P_{add} (except the very stupid choice $P_{\text{add}}(E) = 1$ for all E), the algorithm will satisfy the condition of ergodicity. Just as in the Wolff algorithm, there is a finite probability that any spin on the lattice will find itself the sole member of a cluster of one. Flipping a succession of such clusters will clearly get us from any state to any other in a finite number of moves. Second, let us apply the condition of detailed balance to the algorithm. Consider, as we did in the case of the Wolff algorithm, two states of our system which differ by the flipping of a single cluster. (You can look again at Figure 4.3 if you like, but bear in mind that, since we are now allowing links between anti-parallel spins, not all the spins in the cluster need be pointing in the same direction.) As before, the probability of forming the cluster itself is exactly the same in the forward and reverse directions, except for the contributions which come from the borders. At the borders, there are some pairs of spins which are parallel and some which are anti-parallel. Suppose that in the forward direction there are m pairs of parallel spins at the border—bonds which will be broken in flipping the cluster—and n pairs which are anti-parallel—bonds which will be made. By definition, no links are made between the spins of any of these border pairs, and the probability of that happening is $[1 - P_{\text{add}}(-J)]^m [1 - P_{\text{add}}(J)]^n$. In the reverse direction the corresponding probability is $[1 - P_{\text{add}}(-J)]^n [1 - P_{\text{add}}(J)]^m$. Just as in the Wolff case, the energy cost of flipping the cluster from state μ to state ν is

$$E_\nu - E_\mu = 2J(m - n). \tag{4.30}$$

Thus, the appropriate generalization of the acceptance ratio relation, Equation (4.12), is

$$\frac{A(\mu \to \nu)}{A(\nu \to \mu)} = \left[e^{2\beta J} \frac{1 - P_{\text{add}}(-J)}{1 - P_{\text{add}}(J)} \right]^{n-m} \tag{4.31}$$

Any choice of acceptance ratios $A(\mu \to \nu)$ and $A(\nu \to \mu)$ which satisfies this relation will satisfy detailed balance. For the Wolff choice of P_{add} we get acceptance ratios which are always unity, but Niedermayer pointed out that there are other ways to achieve this. In fact, all we need to do is choose P_{add} to satisfy

$$\frac{1 - P_{\text{add}}(-E)}{1 - P_{\text{add}}(E)} = e^{-2\beta E} \tag{4.32}$$

and we will get acceptance ratios which are always one. Niedermayer's solution to this equation was $P_{\text{add}}(E) = 1 - \exp[\beta(E - E_0)]$ where E_0 is a free parameter whose value we can choose as we like. Notice however that since $P_{\text{add}}(E)$ is supposed to be a probability, it is not allowed to be less than zero. Thus the best expression we can write for the probability $P_{\text{add}}(E_{ij})$ of adding a link between sites i and j is

$$P_{\text{add}}(E_{ij}) = \begin{cases} 1 - e^{\beta(E_{ij} - E_0)} & \text{if } E_{ij} \leq E_0 \\ 0 & \text{otherwise.} \end{cases} \tag{4.33}$$

And this defines Niedermayer's algorithm. Notice the following things about this algorithm:

1. As long as all E_{ij} on the lattice are less than or equal to E_0, the right-hand side of Equation (4.31) is always unity for the choice of P_{add} given in (4.33), so the two acceptance ratios can be chosen to be one for every move. Since we are at liberty to choose E_0 however we like, we can always satisfy this condition by making it greater than or equal to the largest value that E_{ij} can take, which is J in the case of the Ising model. This gives us a whole spectrum of Wolff-type algorithms for various values $E_0 \geq J$, which all have acceptance ratios of one. The Wolff algorithm itself is equivalent to $E_0 = J$. If we increase E_0 above this value, the probabilities $P_{add}(E_{ij})$ tend closer and closer to one, making the clusters formed larger and larger. This gives us a way of controlling the sizes of the clusters formed in our algorithm, all the way up to clusters which encompass (almost) every spin on the lattice at every move.

2. If we choose E_0 to be less than the largest possible value of E_{ij} then the right-hand side of Equation (4.31) is no longer equal to one, and we can no longer choose the acceptance ratios to be unity. For the Ising model, if we choose $-J \leq E_0 < J$ then we have

$$\frac{A(\mu \to \nu)}{A(\nu \to \mu)} = \left[e^{2\beta J}e^{-\beta(J+E_0)}\right]^{n-m} = \left[e^{\beta(J-E_0)}\right]^{n-m}. \qquad (4.34)$$

   Just as with the Metropolis algorithm, the optimal choice of acceptance ratios is then to make the larger of the two equal to one, and choose the smaller to satisfy this equation. If we do this, we again achieve detailed balance, and we now have an algorithm which, as E_0 is made smaller and smaller, produces smaller and smaller clusters, although it does so at the expense of an exponentially decreasing acceptance ratio for cluster moves which increase the energy of the system.

3. If E_0 is chosen to be smaller than the smallest possible value of E_{ij}, which is $-J$, then $P_{add}(E_{ij}) = 0$ for all pairs of spins i, j, so every cluster formed has only one spin in it and the acceptance ratios are given by

$$\frac{A(\mu \to \nu)}{A(\nu \to \mu)} = \left[e^{2\beta J}\right]^{n-m}. \qquad (4.35)$$

   Bearing in mind that m is the number of neighbours of this single spin which are pointing in the same direction as it, and n is the number which point in the opposite direction, we can see that this is exactly the same as Equation (3.7) which defines the Metropolis algorithm.

Thus we see that by varying the parameter E_0, Niedermayer's cluster algorithm includes as special cases both the Metropolis algorithm and the Wolff algorithm, and interpolates smoothly from one to the other and beyond, varying the average cluster size from one spin all the way up to the entire lattice. The trouble with the algorithm is that no one really knows what value one should choose for E_0. Niedermayer himself conjectured that the Wolff choice $E_0 = J$ might not give the optimal correlation times and that shorter ones could be achieved by making other choices. He gave preliminary evidence in his 1988 paper that, in some cases at least, a smaller value of E_0 (i.e., a value which produces smaller clusters on average, at the expense of a lowered acceptance probability) gives a shorter correlation time. However, to our knowledge no one has performed an extensive study of the dynamic exponent of the algorithm as a function of E_0 so, for the moment at least, the algorithm remains an interesting extension of the Wolff idea which has yet to find use in large-scale simulations.

4.4.3 Multigrid methods

Also worthy of mention is a class of methods developed by Kandel *et al.* (1989, Kandel 1991), which are referred to as **multigrid methods**. These methods are also aimed at reducing critical slowing down and accelerating simulations at or close to the critical temperature. Multigrid methods have not been used to a very great extent in large-scale Monte Carlo simulations because they are considerably more complex to program than the cluster algorithms discussed above. However, they may yet prove useful in some contexts because they appear to be faster than cluster algorithms for simulating very large systems.

The fundamental idea behind these methods is the observation that, with the divergence of the correlation length at the critical temperature, we expect to see fluctuating clusters of spins of all sizes up to the size of the entire lattice. The multigrid methods therefore split the CPU time of the simulation up, spending varying amounts of time flipping blocks of spins of various sizes. This idea certainly has something in common with the ideas behind the Wolff and Swendsen–Wang algorithms, but the multigrid methods are more deliberate about flipping blocks of certain sizes, rather than allowing the sizes to be determined by the temperature and configuration of the lattice. Kandel and his co-workers gave a number of different, similar algorithms, which all fall under the umbrella of multigrid methods. Here, we describe one example, which is probably the simplest and most efficient such algorithm for the Ising model. The algorithm works by grouping the spins on the lattice into blocks and then treating the blocks as single spins and flipping them using a Metropolis algorithm. In detail, what one does is this.

In the Swendsen–Wang algorithm we made "links" between similarly oriented spins, which effectively tied those spins together into a cluster, so that they flipped as one. Any two spins which were not linked were free to flip separately—there was not even a ferromagnetic interaction between them to encourage them to point in the same direction. In the multigrid method, pairs of adjacent spins can be in three different configurations: they can be linked as in the Swendsen–Wang case, so that they must flip together, they can have no connection between them at all, so that they can flip however they like, or they can have a normal Ising interaction between them of strength J, which encourages them energetically to point in the same direction, but does not force them to as our links do. By choosing one of these three states for every pair of spins, the lattice is divided up into clusters of linked spins which either have interactions between them, or which are free to flip however they like. The algorithm is contrived so that only clusters of one or two spins are created. No clusters larger than two spins appear. The procedure for dividing up the lattice goes like this.

1. We take a spin on the lattice, and examine each of its neighbours in turn. If a neighbour is pointing in the opposite direction to the spin, then we leave it alone. In Kandel's terms, we "delete" the bond between the two spins, so that they are free to assume the same or different directions with no energy cost. If a neighbour is pointing in the same direction as our spin, then we make a link between the two with the same probability $P_{\mathrm{add}} = 1 - e^{-2\beta J}$ as we used in the Wolff and Swendsen–Wang algorithms. Kandel calls this "freezing" the bond between the spins.

2. Since we only want to create clusters of at most two spins, we stop looking at neighbours once we have created a link to any one of them. In fact, what we do is keep the normal Ising interactions between the spin and all the remaining neighbours that we have not yet looked at. We do this regardless of whether they are pointing in the same direction as our first spin or not. Kandel describes this as "leaving the bond active".

3. Now we move on to another spin and do the same thing, and in this way cover the entire lattice. Notice that, if we come to a spin which is adjacent to one we have considered before, then one or more of the spin's bonds will already have been frozen, deleted or marked as active. In this case we leave those bonds as they are, and only go to work on the others which have not yet been considered. Notice also that if we come to a spin and it has already been linked ("frozen") to another spin, then we know immediately that we need to leave the interactions on all the remaining bonds to that spin untouched.

In this way, we decide the fate of every spin on the lattice, dividing them up into clusters of one or two, joined by bonds which may or may not have interactions associated with them. Then we treat those clusters as single spins, and we carry out the Metropolis algorithm on them, for a few sweeps of the lattice.

But this is not the end. Now we do the whole procedure again, treating the clusters as spins, and joining them into bigger clusters of either one or two elements each, using exactly the same rules as before. (Note that the lattice of clusters is not a regular lattice, as the original system was, but this does not stop us from carrying out the procedure just as before.) Then we do a few Metropolis sweeps of this coarser lattice too. And we keep repeating the whole thing until the size of the blocks reaches the size of the whole lattice. In this way, we get to flip blocks of spins of all sizes from single spins right up to the size of the entire system. Then we start taking the blocks apart again into the blocks that made them up, and so forth until we get back to the lattice of single spins. In fact, Kandel and co-workers used a scheme where at each level in the blocking procedure they either went towards bigger blocks ("coarsening") or smaller ones ("uncoarsening") according to the following rule. At any particular level of the procedure we look back and see what we did the previous times we got to this level. If we coarsened the lattice the previous two times we got to this point, then on the third time only, we uncoarsen. This choice has the effect of biasing the algorithm towards working more at the long length-scales (bigger, coarser blocks).

Well, perhaps you can see why the complexity of this algorithm has put people off using it. The proof that the algorithm satisfies detailed balance is quite involved, and, since you're probably not dying to hear about it right now, we'll refer you to the original paper for the details. In the same paper it is demonstrated that the dynamic exponent for the algorithm is in the region of 0.2 for the two-dimensional Ising model—a value similar to, though not markedly better than, the Wolff algorithm. Lest you dismiss the multigrid method out of hand, however, let us just point out that the simulations do indicate that its performance is superior to cluster algorithms for large systems. These days, with increasing computer power, people are pushing simulations towards larger and larger lattices, and there may well come a point at which using a multigrid method could win us a factor of ten or more in the speed of our simulation.

4.4.4 The invaded cluster algorithm

Finally, in our round-up of Monte Carlo algorithms for the Ising model, we come to an unusual algorithm proposed by Jonathan Machta and co-workers called the **invaded cluster algorithm** (Machta *et al.* 1995, 1996). The thing which sets this algorithm apart from the others we have looked at so far is that it is not a general purpose algorithm for simulating the Ising model at any temperature. In fact, the invaded cluster algorithm can only be used to simulate the model at the critical point; it does not work at any other temperature. What's more, for a system at the critical point in the thermodynamic limit the algorithm is equivalent to the Swendsen–Wang cluster algorithm of Section 4.4.1. So what's the point of the algorithm? Well, there are two points. First, the invaded cluster algorithm can find the critical point all on its own—we don't need to know what the critical temperature is beforehand in order to use the algorithm. Starting with a system at any temperature (for example $T = 0$ at which all the spins are pointing in the same direction) the algorithm will adjust its simulation temperature until it finds the critical value T_c, and then it will stay there. This makes the algorithm very useful for actually measuring the critical temperature, something which otherwise demands quite sophisticated techniques such as the finite-size scaling or Monte Carlo RG techniques of Chapter 8. Second, the invaded cluster algorithm equilibrates extremely fast. Although the algorithm is equivalent to the Swendsen–Wang algorithm once it reaches equilibrium at T_c, its behaviour while getting there is very different, and in a direct comparison of equilibration times between the two, say for systems starting at $T = 0$, there is really no contest. For large system sizes, the invaded cluster algorithm can reach the critical point as much as a hundred times faster than the Swendsen–Wang algorithm. (A comparison with the Wolff algorithm yields similar results—the performance of the Wolff and Swendsen–Wang algorithms is comparable.)

So, how does the invaded cluster algorithm work? Basically, it is just a variation of the Swendsen–Wang algorithm in which the temperature is continually adjusted to look for the critical point. From the discussions of Section 4.4.1 we know that below the critical temperature the Swendsen–Wang algorithm generates a percolating cluster of either up- or down-pointing spins which stretches across the entire lattice. The algorithm looks for the temperature at which this percolating cluster first appears, and takes that to be the critical temperature. Here's how it goes:

1. First we choose some starting configuration of the spins. It doesn't matter what choice we make, but it could, for example, be the $T = 0$ state.

2. We perform a single step of the Swendsen–Wang algorithm in which we consider in turn every pair of neighbouring spins which are pointing in the same direction and make a bond between them with probability $P_{\text{add}} = 1 - e^{-2\beta J}$. But the trick is, we adjust the temperature (or the inverse temperature $\beta = (kT)^{-1}$) to find the point at which one of the clusters produced in this way just starts to percolate. For example, if we start off with all the spins pointing the same way, then every

single pair of neighbouring spins on the lattice are aligned, and are therefore candidates for linking with probability P_{add}. In this case, it is known that for the square lattice in two dimensions for instance, percolation will set in when the probability of making a link between two spins is $P_{add} = \frac{1}{2}$. Solving for the temperature, this is equivalent to $\beta J = \frac{1}{2} \log 2 = 0.346 \ldots$, so this is the temperature at which the first step should be conducted.

3. Once the links are made, we flip each cluster separately with probability $\frac{1}{2}$, just as we do in the normal Swendsen–Wang algorithm. Then the whole procedure is repeated from step 2 again.

So what's the point here? Why does the algorithm work? Well, consider what happens if the system is below the critical temperature. In that case, the spins will be more likely to be aligned with their neighbours than they are at the critical temperature, and therefore there will be more neighbouring pairs of aligned spins which are candidates for making links. Thus, in order to make enough links to get one cluster to percolate across the entire lattice we don't need as high a linking probability P_{add} as we would at T_c. But a lower value of P_{add} corresponds to a higher value $T > T_c$ of the corresponding temperature. In other words, when the system is below the critical temperature, the algorithm automatically chooses a temperature $T > T_c$ for its Swendsen–Wang procedure. Conversely, if the system is at a temperature above the critical point, then neighbouring spins will be less likely to be aligned with one another than they are at T_c. As a result, there will be fewer places that we can put links on the lattice, and P_{add} will need to be higher than it would be at T_c in order to make one of the clusters percolate and fill the entire lattice.[10] The higher value of P_{add} corresponds to a lower value of the temperature $T < T_c$, and so, for a state of the system above the critical temperature, the algorithm will automatically choose a temperature $T < T_c$ for its Swendsen–Wang procedure.

The algorithm therefore has a kind of negative feedback built into it, which always drives the system towards the critical point, and when it finally reaches T_c it will just stay there, performing Swendsen–Wang Monte Carlo steps at the critical temperature for the rest of the simulation. We do not need to know what the critical temperature is for the algorithm to work. It finds T_c all on its own, and for that reason the algorithm is a good way of measuring T_c. Notice also that when the temperature of the system is lower than the critical temperature, the algorithm performs Monte Carlo steps with $T > T_c$, and *vice versa*. Thus, it seems plausible that the algorithm would drive itself towards the critical point quicker than simply performing

---

[10]In fact, for sufficiently high temperatures it may not be possible to produce a percolating cluster at all. In this case the algorithm will perform a Monte Carlo step at $T = 0$.

a string of Swendsen Wang Monte Carlo steps exactly at T_c. As discussed below, this is indeed the case.

Before we look at the results from the invaded cluster algorithm, let us discuss briefly how it is implemented. There are a couple of questions that need to be answered. What is the most efficient way of varying the temperature T to find the value of P_{add} at which a percolating cluster first appears on the lattice? And how do we identify the percolating cluster? There is a very elegant answer to the first question: at the beginning of each step of the algorithm, we go over the entire lattice and for every pair of neighbouring spins which are pointing in the same direction, we generate a random real number between zero and one. We can imagine writing down these numbers on the bonds that they belong to. Then we go through those random numbers in order, starting with the smallest of them and working up, making links between the corresponding pairs of spins one by one. At some point, we will have added enough links to make one of the clusters on the lattice percolate.[11] At that point, we stop, and work out what fraction of all the possible links on the lattice we have made. This fraction is just equal to the probability P_{add} we would have to use to produce the percolating cluster. Rearranging our expression for P_{add}, Equation (4.13), we can then calculate the corresponding temperature of the Monte Carlo step:

$$T = -\frac{2J}{\log(1 - P_{\text{add}})}. \tag{4.36}$$

Our other question—how we know when we have a percolating cluster does not have such an elegant solution. Machta *et al.* (1995), who invented the algorithm, suggest two ways of resolving the problem. One is to measure the dimensions of each of the clusters along the axes of the lattice. When a cluster has a length along one of the axes which is equal to the dimension L of the lattice, it is declared to be percolating and no more links are added to the system. The other suggestion is that, given that the system has periodic boundary conditions, you wait until one of the clusters has gone off one side of the lattice and come back on the other and joined up with itself once more.[12] Machta *et al.* find that these two criteria yield very similar results when used in actual simulations, implying that the algorithm is not particularly sensitive to the exact method used.

The results for the invaded cluster algorithm are impressive. Machta *et al.* found equilibration times 20 or more times faster than the for the Swendsen Wang algorithm at T_c for the two- and three-dimensional Ising systems they

---

[11] Again, this may not be true at high temperatures, where it is sometimes not possible to produce a percolating cluster for any value of P_{add}. In this case we stop when we have added links between all pairs of aligned spins.

[12] An efficient way of implementing this second criterion using a tree data structure is described by Barkema and Newman (1999).

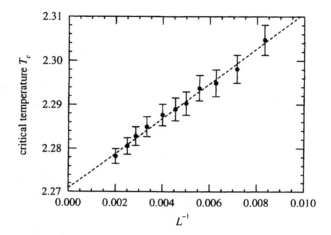

FIGURE 4.9 The critical temperature of the two-dimensional Ising model measured using the invaded cluster algorithm for systems of a variety of sizes from $L = 120$ up to $L = 500$. Here they are plotted against L^{-1} and the extrapolation to $L = \infty$ at the vertical axis gives an estimate of $T_c = 2.271 \pm 0.002$ for the critical temperature in the thermodynamic limit. The data are taken from Machta *et al.* (1995).

examined, and of course, the invaded cluster algorithm allowed them to measure the value of T_c, which is not directly possible with the normal Swendsen–Wang algorithm. Figure 4.9 shows some of their results for the critical temperature of two-dimensional Ising systems of different sizes. On the finite lattice, the average estimate of T_c which the algorithm makes is a little high.[13] However, if, as has been done here, one measures T_c for a variety of system sizes, one can extrapolate to the limit $L = \infty$ to get an estimate of the true critical temperature in the thermodynamic limit. In this case we do this by plotting the measured T_c as a function of L^{-1} and extrapolating to the $L^{-1} = 0$ axis. The result is $T_c = 2.271 \pm 0.002$, which compares favourably with the known exact result of $T_c = 2.269$, especially given the small amount of CPU taken by the simulation. (The runs were 10 000 Monte Carlo steps for each system size.)

The invaded cluster algorithm can also be used to measure other quantities at the critical temperature—magnetization for example, or internal energy. However, a word of caution is in order here. For lattices of finite size, which of course include all the lattices in our Monte Carlo simulations,

---

[13]This behaviour depends on the criterion used to judge when percolation takes place. For some criteria the estimate of T_c approaches the infinite system result from below rather than above.

the invaded cluster algorithm does not sample the Boltzmann distribution exactly. In particular, the fluctuations in quantities measured using the algorithm are different from those you would get in the Boltzmann distribution. To see this, consider what happens once the algorithm has equilibrated at the critical temperature. At this point, as we argued before, it should stop changing the temperature and just become equivalent to the Swendsen–Wang algorithm at T_c, which certainly samples the Boltzmann distribution correctly. However, in actual fact, because the lattice is finite, the innate randomness of the Monte Carlo method will give rise to variations in the temperature T of successive steps in the simulation. The negative feedback effect that we described above will ensure that T always remains close to T_c, but the size of fluctuations is very sensitive to small changes in temperature near to T_c and as a result the measured fluctuations are not a good approximation to those of the true Boltzmann distribution.[14] Thus the invaded cluster algorithm is not well suited to estimating, for instance, the specific heat c of the Ising model at T_c, which is determined by measuring fluctuations. On the other hand, one could use the algorithm to determine the value of T_c, and then use the normal Wolff or Swendsen–Wang algorithm to perform a simulation at that temperature to measure c. Determining T_c in this way could also be a useful preliminary to determining the dynamic exponent of another algorithm, as we did earlier in the chapter for both the Metropolis and Wolff algorithms. Those calculations demanded a knowledge of the value of T_c, which is known exactly for the 2D Ising model, but not for many other models (including the 3D Ising model).

4.5 Other spin models

The Ising model, which we have spent the last two chapters examining, is the best studied spin model in statistical mechanics. There are however many others, some of which are of great importance. In the next few chapters we will look at some variations on the Ising model theme, including the conserved-order-parameter Ising model, in which the total magnetization of the lattice is held constant, Ising spin glasses, in which the strengths of the bonds between spins vary randomly over the lattice, and the random-field Ising model, in which each spin is subjected to its own local magnetic field which has a randomly chosen value. These models are interesting because, for a variety of reasons, their simulation demands Monte Carlo algorithms which are fundamentally different from the ones we have seen so far.

---

[14] As the size of the system gets larger however, the algorithm samples the Boltzmann distribution more accurately. This is the justification for our earlier claim that, in the thermodynamic limit, the invaded cluster algorithm is identical to the Swendsen–Wang algorithm at T_c.

In this section, we extend the Ising model in a different way. We look at a few of the many models which consider more general types of spins on the vertices of the lattice, spins which can take more than two discrete values, or spins which take a continuum of values. For the most part, these models can be simulated with techniques similar to the ones we have already seen, which is why we are considering them now. First we will look at Potts models.

4.5.1   Potts models

Potts models are a class of models which are similar to the Ising model except that the spins s_i on each lattice site can take more than two different discrete values. Usually these values are represented by positive integers starting at 1, and a q-state Potts model is one in which each spin can have integer values $s_i = 1 \ldots q$. Any two neighbouring spins then contribute an amount $-J$ to the Hamiltonian if they have the same value, or zero otherwise. The Hamiltonian can thus be written:

$$H = -J \sum_{\langle ij \rangle} \delta_{s_i s_j}, \tag{4.37}$$

where δ_{ij} is the Kronecker δ-symbol, which is 1 when $i = j$ and zero otherwise.

For the case $q = 2$, the Potts model is equivalent to the Ising model, up to an additive constant in the Hamiltonian. To see this we note that Equation (4.37) can be rewritten as

$$H = -\tfrac{1}{2}J \sum_{\langle ij \rangle} 2\big(\delta_{s_i s_j} - \tfrac{1}{2}\big) - \sum_{\langle ij \rangle} \tfrac{1}{2}J. \tag{4.38}$$

But $2\big(\delta_{s_i s_j} - \tfrac{1}{2}\big)$ is $+1$ when $s_i = s_j$ and -1 when they are different, so this expression is indeed equivalent to the Ising Hamiltonian except for the constant term $-\sum_{\langle ij \rangle} \tfrac{1}{2}J$. (Note though that the interaction energy is changed by a factor of two $J \to \tfrac{1}{2}J$ from Equation (3.1).)

For higher values of q the Potts model behaves similarly in some ways to the Ising model. For $J > 0$ (the ferromagnetic case) it has q equivalent ground states in which all the spins have the same value, and as the temperature is increased it undergoes a phase transition to a state in which each of the q spin states occurs with equal frequency across the lattice. There are differences with the Ising model however. In particular, the entropy of a Potts model with $q > 2$ is higher than that of the Ising model at an equivalent temperature, because the density of states of the system as a function of energy is higher. For example, at temperatures just above $T = 0$, almost all of the spins will be in one state—say $q = 1$—but there will be a few isolated excitations on the lattice, just as there were in the Ising model. Unlike the

Ising model however, each excitation can take any of the possible spin values other than 1, which may be very many if q is large, and this gives rise to many more low-lying excitation states than in the Ising case.

Monte Carlo simulation of Potts models is quite similar to the simulation of the Ising model. The simplest thing one can do is apply the single-spin-flip Metropolis algorithm, which would go like this. First we pick a spin i from the lattice at random. It will have some value s_i. We choose at random a new value $s'_i \neq s_i$ from the $q - 1$ available possibilities. Then we calculate the change in energy ΔE that would result if we were to make this change to this spin, and either accept or reject the move, with acceptance probability

$$A = \begin{cases} e^{-\beta \Delta E} & \text{if } \Delta E > 0 \\ 1 & \text{otherwise.} \end{cases} \qquad (4.39)$$

(See Equation (3.7).) This algorithm satisfies the conditions of both ergodicity and detailed balance, for the same reasons that it did in the case of the Ising model. However, even for a single-spin-flip algorithm, the Metropolis algorithm is not a very good one to use for Potts models, especially when q becomes large. To see this, let us consider an extreme case: the $q = 100$ Potts model on a square lattice in two dimensions.

At high temperatures, the acceptance ratio (4.39) is always either 1 or close to it because β is small, so the algorithm is reasonably efficient. However, at lower temperatures this ceases to be the case. As the temperature decreases, more and more spins will take on the same values as their neighbours, forming ferromagnetic domains, just as in the Ising model. Consider the case of a spin which has four neighbours which all have different values. The four states of this spin in which it is aligned with one of its neighbours have lower energy, and therefore higher Boltzmann weight, than the other 96 possible states. If our spin is in one of the 96 higher energy states, how long will it take to find one of the four aligned states? Well, until we find one of them, all the states that the system passes through will have the same energy, so the Metropolis acceptance ratio will be 1. On average therefore, it will take about $100/4 = 25$ steps to find one of the four desirable states. As q increases this number will get larger; it will take longer and longer for the system to find the low-lying energy states, despite the fact that the acceptance ratio for all moves is 1, simply because there are so many states to sort through.

Conversely, if the spin is in one of the four low-lying states, then there is an energy cost associated with excitation to one of the other 96, which means that the acceptance ratio will be less than one, and possibly very small (if the temperature is low). Thus it could be that nearly 96 out of every 100 attempted moves is rejected, giving an overall acceptance ratio little better than 4%.

One way to get around these problems is to use the **heat-bath algo-**

rithm. The heat-bath algorithm is also a single-spin-flip algorithm, but one which is more efficient at finding the energetically desirable states of the spins. The algorithm goes like this. First we choose a spin k at random from the lattice. Then, regardless of its current value, we choose a new value s_k for the spin in proportion to the Boltzmann weights for the different values—the values are drawn from a so-called "heat bath". In other words, we give the spin a value n lying between 1 and q with a probability

$$p_n = \frac{e^{-\beta E_n}}{\sum_{m=1}^{q} e^{-\beta E_m}} \tag{4.40}$$

where E_n is the energy of the system when $s_k = n$. Note that we have normalized the probabilities to add up to one, so *some* value of the spin gets chosen on every step, even if it is just the same as the old value (though this will become increasingly unlikely as q becomes large). Clearly this algorithm satisfies ergodicity, since on a lattice of N spins the appropriate string of single-spin moves like this can take us from any state to any other in N moves or fewer. And since we are choosing states with probabilities proportional to their Boltzmann weights, it should not come as a great surprise that this algorithm also satisfies the condition of detailed balance, though just to be thorough let's prove it. The probability of making the transition from a state in which $s_k = n$ to one in which $s_k = n'$ is just $p_{n'}$, and the probability of going back again is p_n. (Note that these probabilities do not depend on the initial state, only on the final one, unlike the Metropolis algorithm.) Then, using Equation (4.40), the ratio of forward and backward transition rates is

$$\frac{P(n \to n')}{P(n' \to n)} = \frac{p_{n'}}{p_n} = \frac{e^{-\beta E_{n'}}}{\sum_{m=1}^{q} e^{-\beta E_m}} \times \frac{\sum_{m=1}^{q} e^{-\beta E_m}}{e^{-\beta E_n}} = e^{-\beta(E_{n'} - E_n)}, \tag{4.41}$$

exactly as detailed balance demands.

This algorithm is much more efficient than the Metropolis algorithm for large values of q, because it will choose the states with the highest Boltzmann weights most often, and can find them in one move, rather than wandering through large numbers of unfavourable states at random before finding them. In practice how is it implemented? First, we observe that once we have chosen which spin k we are going to change, we can split the Hamiltonian, Equation (4.37), into the terms which involve s_k, and those which don't:

$$H = -J \sum_{\langle ij \rangle \, (i,j \neq k)} \delta_{s_i s_j} - J \sum_{i \text{ n.n. to } k} \delta_{s_i s_k}. \tag{4.42}$$

The first sum here is the same for all q possible values of s_k, and therefore cancels out of the expression (4.40). The second sum has only z terms, where z is the lattice coordination number (which is four in the square

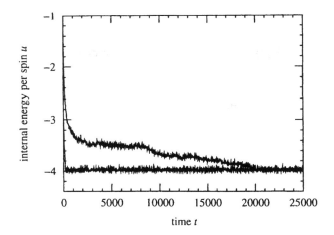

FIGURE 4.10   The internal energy of a $q = 10$ Potts model on a 20×20 square lattice at a temperature $T = 0.5J$, simulated using the Metropolis algorithm (upper line) and the heat-bath algorithm (lower line).

lattice case we have been considering). Just calculating these terms, rather than evaluating the entire Hamiltonian, makes evaluating p_n much quicker.

Second, we notice that there are at most z states of the spin s_k in which it is the same as at least one of its neighbours. These are the only states in which the spin makes a contribution to the Hamiltonian. In all the other states it makes no contribution to the Hamiltonian. Thus, evaluating the Boltzmann factors $e^{-\beta E_n}$ appearing in Equation (4.40) requires us to calculate the values of, at most, z exponentials, all the other terms being just equal to 1.

Finally, we note that there is only a small spectrum of possible values for each Boltzmann factor, since the second term in Equation (4.42) can only take values which are multiples of $-J$, from zero up to $-zJ$. Thus we can, as we did for the Ising model, calculate the values of all the exponentials we are going to need at the very beginning of our simulation, and thereafter simply look up those values whenever we need to know one of them. The calculation of exponentials being a costly process in terms of CPU time, this trick considerably improves the efficiency of the algorithm.

In Figure 4.10 we compare the internal energy of a $q = 10$ Potts model on a 20×20 square lattice at low temperature simulated using the heat-bath and Metropolis algorithms. In this case, the heat-bath algorithm takes about 200 sweeps of the lattice to come to equilibrium, by contrast with the Metropolis algorithm which takes about 20 000 sweeps. This is an impressive gain in speed, although as usual we need to be careful about our claims. The heat-bath algorithm is a more complex algorithm than the Metropolis algorithm, and each Monte Carlo step takes longer in the heat-bath case than in the Metropolis case. In this particular simulation, there was about a factor of three difference between the CPU time taken per step in the two algorithms. Even allowing for this however, the heat-bath algorithm is still more than thirty times faster at coming to equilibrium than the Metropolis algorithm. And this factor will increase as q gets larger, making the heat-bath algorithm very definitely the algorithm of choice for large q Potts models.

Before we move on, here is one more question about the heat-bath algorithm: what does the algorithm look like for the normal Ising model? The answer is easy enough to work out. In the Ising case the algorithm would involve choosing a spin at random and then, regardless of its current state, setting it either up or down with probabilities p_{up} and p_{down}. From Equation (4.40) these two probabilities are:

$$p_{up} = \frac{e^{-\beta E_{up}}}{e^{-\beta E_{up}} + e^{-\beta E_{down}}}, \qquad p_{down} = \frac{e^{-\beta E_{down}}}{e^{-\beta E_{up}} + e^{-\beta E_{down}}}. \qquad (4.43)$$

Now suppose that the initial state of the spin is down. What acceptance ratio A do these equations represent for the move that flips the spin to the up state? Well, clearly the acceptance ratio is simply $A = p_{up}$. If we multiply numerator and denominator by $e^{\frac{1}{2}\beta(E_{up}+E_{down})}$, we can write it in terms of the change in energy $\Delta E = E_{up} - E_{down}$:

$$A = \frac{e^{-\frac{1}{2}\beta\Delta E}}{e^{-\frac{1}{2}\beta\Delta E} + e^{\frac{1}{2}\beta\Delta E}}. \qquad (4.44)$$

It is easy to show that the same expression applies for the move which flips the spin from up to down. Thus for the Ising model, the heat-bath algorithm can be regarded as a single-spin-flip algorithm with an acceptance ratio which is a function of the energy change ΔE, just as it is in the Metropolis algorithm, although the exact functional dependence of A on ΔE is different from that of the Metropolis algorithm. In Figure 4.11 we show A as a function of ΔE for the two algorithms. For all values of ΔE, the heat-bath acceptance ratio is lower than the corresponding value for the Metropolis algorithm, Equation (3.7), so it is always more efficient to use the Metropolis algorithm. This should come as no surprise, since, as we pointed out in the last chapter, the Metropolis algorithm is the most efficient possible single-spin-flip algorithm for simulating the Ising model. However, it is worth being familiar with Equation (4.44), because, for reasons which are not completely clear to the authors, some people still use it for simulating the Ising model, despite its relative inefficiency.

We have shown then that if we are going to simulate a Potts model with a single-spin-flip algorithm, the algorithm that we choose should depend

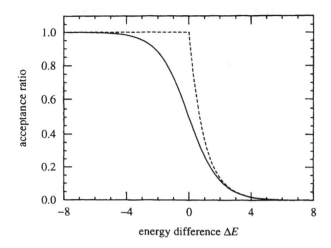

FIGURE 4.11   The acceptance ratio of the heat-bath algorithm for the
Ising model (solid line) as a function of the energy difference between
states ΔE. For comparison, the acceptance ratio of the Metropolis
algorithm is also shown (dashed line).

on the value of q. For small q (for example, the Ising model, which is
equivalent to $q = 2$) the Metropolis algorithm is the most efficient algorithm,
but as q gets larger there comes a point at which it is advantageous to
switch over to the heat-bath algorithm. Where exactly the cross-over occurs
is unfortunately dependent on what temperature we are interested in. At
the low temperatures used for the simulations in Figure 4.10, clearly the
cross-over was well below $q = 10$. However, at high temperatures where
the probabilities of occurrence of the states of any particular spin become
increasingly independent of the states of the spin's neighbours, there is less
and less difference between the performance of the two algorithms, and we
have to go to higher values of q to make the extra complexity of the heat-bath
algorithm worthwhile. In practice, for any serious study of a Potts model,
one should conduct short investigatory simulations with both algorithms to
decide which is more suited to the problem in hand.

4.5.2   Cluster algorithms for Potts models

In the last section we studied single-spin-flip algorithms for Potts models, and
demonstrated that both Metropolis and heat-bath algorithms are efficient
ways to perform simulations for certain ranges of temperature and q. Near
the phase transition, however, these algorithms suffer from exactly the same
problems as they did in the case of the Ising model: the correlation length
diverges and the time taken to flip over the large correlated clusters that

appear gets longer and longer as we approach T_c. As with the Ising model, we can define a dynamic exponent z, and show that the correlation time τ at the critical point, measured in sweeps of the lattice, increases as $\tau \sim L^z$ with the size L of the system simulated. If z takes a large value, this can place severe limitations on the size of the system we can work with. As it turns out z, is indeed quite large. For the case of the $q = 3$ Potts model, for example, Schülke and Zheng (1995) have measured a value of 2.198 ± 0.008 for the dynamic exponent of the Metropolis algorithm in two dimensions, which is comparable to the value for the Ising model, and it is believed that z has similar values for larger q also (Landau *et al.* 1988). The solution to this problem is exactly the same as it was in the Ising model case: to try and flip the large correlated clusters all in one go using a cluster-flipping algorithm. In fact, the Wolff algorithm which we described in Section 4.2 generalizes to the Potts case in a very straightforward manner. The correct generalization is as follows.

1. Choose a seed spin at random from the lattice.

2. Look in turn at each of the neighbours of that spin. If they have the same value as the seed spin, add them to the cluster with probability $P_{\text{add}} = 1 - e^{-\beta J}$. (Note that the 2 has vanished from the exponent. This is the same factor of two that appeared in the interaction constant when we compared the Ising model to the $q = 2$ Potts model.)

3. For each spin that was added in the last step, examine each of *its* neighbours to find which ones, if any, have the same value and add each of them to the cluster with the same probability P_{add}. (As with the Ising model, we notice that some of the neighbours may already be members of the cluster, in which case you don't have to consider adding them again. Also, some of the spins may have been considered for addition before, as neighbours of other spins in the cluster, but rejected. In this case, they get another chance to be added to the cluster on this step.) This step is repeated as many times as necessary until there are no spins left in the cluster whose neighbours have not been considered for inclusion in the cluster.

4. Choose at random a new value for the spins in the cluster, different from the present value, and set all the spins to that new value.

The proof that this algorithm satisfies detailed balance is exactly the same as it was for the Ising model. If we consider two states μ and ν of the system which differ by the changing of just one cluster, then the ratio $g(\mu \rightarrow \nu)/g(\nu \rightarrow \mu)$ of the selection probabilities for the moves between these states depends only on the number of bonds broken m and the number made n around the edges of the cluster. This gives us an equation of detailed balance which reads

$$\frac{g(\mu \to \nu)A(\mu \to \nu)}{g(\nu \to \mu)A(\nu \to \mu)} = (1 - P_{\text{add}})^{m-n}\frac{A(\mu \to \nu)}{A(\nu \to \mu)} = e^{-\beta(E_\nu - E_\mu)}, \qquad (4.45)$$

just as in the Ising case. The change in energy is also given by the same expression as before, except for a factor of two:

$$E_\nu - E_\mu = J(m - n), \qquad (4.46)$$

and so the ratio of the acceptance ratios $A(\mu \to \nu)$ and $A(\nu \to \mu)$ for the two moves is

$$\frac{A(\mu \to \nu)}{A(\nu \to \mu)} = [e^{\beta J}(1 - P_{\text{add}})]^{n-m}. \qquad (4.47)$$

For the choice of P_{add} given above, this is just equal to 1. This equation is satisfied by making the acceptance ratios also equal to 1, and so, with this choice, the algorithm satisfies detailed balance.

The algorithm also satisfies the condition of ergodicity for the same reason that it did for the Ising model. Any spin can find itself the sole member of a cluster of one, and flipping the appropriate succession of such single-spin clusters can take us from any state to any other on a finite lattice in a finite number of steps.

As in the case of the Ising model, the Wolff algorithm gives an impressive improvement in performance near to the critical temperature. Coddington and Baillie (1991) have measured a dynamic exponent of $z = 0.60 \pm 0.02$ for the algorithm in the $q = 3$ case in two dimensions—a considerable improvement over the $z = 2.2$ of the Metropolis algorithm. As before, the single-spin-flip algorithms come into their own well away from the critical point because critical slowing down ceases to be a problem and the relative simplicity of these algorithms over the Wolff algorithm tends to give them the edge. One's choice of algorithm should therefore (as always) depend on exactly which properties of the model one wants to investigate, but the Wolff algorithm is definitely a good choice for examining critical properties.

The Swendsen–Wang algorithm can also be generalized for use with Potts models in a straightforward fashion, as can all of the other algorithms described in Section 4.4. In general however, the three algorithms we have described here—Metropolis, heat-bath and Wolff—should be adequate for most purposes.

4.5.3 Continuous spin models

A different generalization of the Ising model gives us the **continuous spin models**, in which the spins on the lattice have a continuous range of values, rather than a discrete spectrum like the Ising and Potts models. The two most widely studied such models are the **XY model** and the **Heisenberg**

model. In the XY model the spins are two-component vectors of unit length, which can point in any direction on a two-dimensional plane. Thus the spins can be represented by their components s_x, s_y, with the constraint that $s^2 \equiv s_x^2 + s_y^2 = 1$, or they can be represented by a single angle variable θ, which records the direction that the spin points in. Note that although the spins are two-dimensional there is no reason why the lattice need be two-dimensional. For example, we can put XY spins on a cubic lattice in three dimensions if we want to.

The Heisenberg model follows the same idea, but the spins are three-dimensional unit vectors. (Again, the dimensionality of the spins and that of the lattice are independent. We can put Heisenberg spins on a two-dimensional lattice for instance.) Heisenberg spins can be represented either as three-component vectors with $s^2 \equiv s_x^2 + s_y^2 + s_z^2 = 1$, or by two angle variables, such as the θ and ϕ of spherical coordinates.

The Hamiltonian for each of these models is the obvious generalization of the Ising Hamiltonian:

$$H = -J \sum_{\langle ij \rangle} \mathbf{s}_i \cdot \mathbf{s}_j, \qquad (4.48)$$

where $\mathbf{s}_i$ and $\mathbf{s}_j$ are the vectors representing the spins on sites i and j. When $J > 0$ the spins can lower their energy by lining up with one another and the model is ferromagnetic, just as in the Ising case. When $J < 0$, the model is anti-ferromagnetic.

We can also define similar models with higher-dimensional spins—unit vectors in four or more dimensions—and there are many other generalizations of the idea to continuous-valued spins which have symmetries other than the simple spherical symmetry of these models. In this section we concentrate on the XY/Heisenberg class of models as examples of how Monte Carlo algorithms can be constructed for continuous spin models.

The first thing we notice about these continuous spin models is that they have a continuous spectrum of possible states, with energies which can take any real value between the ground state energy and the energy of the highest-lying state. The appropriate generalization of the Boltzmann distribution to such a model is that the probability of finding the system in a state with energy between E and $E + \mathrm{d}E$ is

$$p(E)\,\mathrm{d}E = \frac{\mathrm{e}^{-\beta E}\rho(E)\,\mathrm{d}E}{Z}, \qquad (4.49)$$

where $\rho(E)$ is the **density of states**, defined such that $\rho(E)\,\mathrm{d}E$ is the number of states in the interval E to $E + \mathrm{d}E$. You may well ask what exactly it means to talk about the number of states in an energy interval when we have a continuous spectrum of states. This was one of the toughest problems which plagued the nineteenth century study of statistical mechanics, and in fact it cannot be answered within the realm of the classical systems we are studying here. Only by studying quantum mechanical systems and then taking their classical limit can the question be answered properly. However, for the purposes of Monte Carlo simulation, we don't actually need to know what the density of states in the system is, just as we didn't need to know the complete spectrum of energies in the discrete case—the Monte Carlo method ensures that, provided we wait long enough for the system to equilibrate, all states will appear with their correct equilibrium probabilities in the simulation.[15]

The partition function Z for a system with a continuous energy spectrum is the obvious generalization of the discrete case:

$$Z = \int e^{-\beta E} \rho(E) \, dE, \qquad (4.50)$$

so that the probabilities in Equation (4.49) are properly normalized.

Note also that these continuous spin models have a continuum of ground states. For the ferromagnetic XY model, for example, the ground state is one in which all the spins are lined up pointing in the same direction. However, that direction can equally well be any of the allowed directions of the spins. The system therefore has one degree of freedom—rotation of all the spins by the same amount—which costs it nothing in energy, and this result extends to all the higher energy states of the lattice also. (Field theorists refer to this as a **Goldstone mode**.)

So, how do we go about simulating a continuous spin system? Let us take the XY model as our example. The simplest Monte Carlo algorithm for this model is the single-spin-flip Metropolis algorithm. The algorithm works in much the same way as it did for the Ising and Potts models. The only question we need to answer is what is the equivalent of "flipping" an Ising spin for a model in which the spins have continuous values? Clearly if we simply reversed the direction of a spin on each move we would not get an ergodic algorithm—each spin could then only point in the direction it started in or the opposite direction and there would be no way for it to reach any of the other possible directions that it is allowed to point in. So instead, rather than reversing the spin, we choose a new direction for it at random. Then the algorithm would look like this:

1. Choose a spin at random from the lattice, and choose a new direction for it by generating a random number between 0 and 2π to represent its new angle.[16]

---

[15] However, if the reader is interested in the derivation of the quantum mechanical density of states and its classical limit, we can recommend the excellent treatment by Pathria (1972).

[16] In the case of the Heisenberg model, whose spins are three-dimensional, choosing

2. Calculate the change in energy ΔE if we were to make this move.

3. Accept or reject the move with an acceptance ratio A thus:

$$A = \begin{cases} e^{-\beta \Delta E} & \text{if } \Delta E > 0 \\ 1 & \text{otherwise.} \end{cases} \qquad (4.51)$$

This algorithm now satisfies ergodicity, since we have a finite chance of accepting any move, and can therefore in theory get from any state of the lattice to any other in N moves or fewer, where N is the number of spins on the lattice. It also satisfies detailed balance, since the selection probabilities $g(\mu \to \nu)$ and $g(\nu \to \mu)$ for a move to go from state μ to state ν or back again are the same, and the ratio of the acceptance probabilities is just $\exp(-\beta \Delta E)$, as it should be.

This form of the Metropolis algorithm is the most widely used single-spin-flip algorithm for the XY model. However, we should point out that it is not the only form the algorithm can take, nor is it known whether this is the most efficient form. The thing is, there is quite a lot of flexibility about the choice of the new state for the spin we are proposing to change. The only restriction that the algorithm places on it is that the selection probabilities $g(\mu \to \nu)$ and $g(\nu \to \mu)$ for the forward and reverse moves be equal. This in turn only requires that the probability of picking a new direction θ' for the spin be a function solely of the angle $|\theta' - \theta|$ between the two states. Any function will do however, not just the uniform one we used above. If we choose a function which is biased towards small changes in the direction of our spin, then we will increase the acceptance ratio of the algorithm because the energy cost of any move will be small. However, the change of state will be small too, so the algorithm may take a lot of moves to reach equilibrium. (For example, many small changes in the direction of single spins would be necessary to cool the system from an initial $T = \infty$ state in which the directions of the spins are all random to a low-temperature state in which they all point in approximately the same direction.) Conversely, an algorithm in which we preferentially propose larger changes in direction $|\theta' - \theta|$ for our Monte Carlo moves would take fewer accepted moves to get from one state to another on average, but would tend to suffer from a lower mean acceptance ratio, because with larger changes in spin direction moves can have a larger energy cost.

As we mentioned, no one has, to our knowledge, investigated in any detail the question of how best to compromise between these two cases—all the large Metropolis studies of models like the XY and Heisenberg models have been carried out with the uniform choice of directions described above. It seems likely that the algorithm would benefit from a choice of smaller

---

a new direction at random isn't quite so simple. The problem of choosing a random spherically symmetric direction is discussed in Section 16.2.1.

angles at low temperatures, where the acceptance ratio for higher energy excitations becomes extremely small, and larger angles at high temperatures, where the important point is to cover the space of possible states quickly, the acceptance ratio being close to unity regardless of the moves proposed. A detailed investigation of this point would make an interesting study.[17]

Many continuous spin models exhibit a phase transition similar to that of the Ising model between a ferromagnetic and a paramagnetic phase, with the accompanying problem of critical slowing down. As with the Ising model, cluster algorithms can reduce these problems. A version of the Wolff algorithm which is suitable for the XY and Heisenberg models was given by Wolff in his original paper in 1989. The basic idea is that we choose at random both a seed spin from which to start building our cluster and a random direction, which we will denote by a unit vector $\hat{\mathbf{n}}$. Then we treat the components $\hat{\mathbf{n}} \cdot \mathbf{s}_i$ of the spins in that direction roughly in the same way as we did the spins in the Ising model. A neighbour of the seed spin whose component in this direction has the same sign as that of the seed spin can be added to the cluster by making a link between it and the seed spin with some probability P_{add}. If the components point in opposite directions then the spin is not added to the cluster. When the complete cluster has been built, it is "flipped" by reflecting all the spins in the plane perpendicular to $\hat{\mathbf{n}}$.

The only complicating factor is that, in order to satisfy detailed balance, the expression for P_{add} has to depend on the exact values of the spins which are joined by links thus:

$$P_{\text{add}}(\mathbf{s}_i, \mathbf{s}_j) = 1 - \exp[-2\beta(\hat{\mathbf{n}} \cdot \mathbf{s}_i)(\hat{\mathbf{n}} \cdot \mathbf{s}_j)]. \qquad (4.52)$$

Readers might like to demonstrate for themselves that, with this choice, the ratio of the selection probabilities $g(\mu \rightarrow \nu)$ and $g(\nu \rightarrow \mu)$ is equal to $e^{-\beta \Delta E}$, where ΔE is the change in energy in going from a state μ to a state ν by flipping a single cluster. Thus, detailed balance is obeyed, as in the Ising case, by an algorithm for which the acceptance probability for the cluster flip is 1. This algorithm has been used, for example, by Gottlob and Hasenbusch (1993) to perform extensive studies of the critical properties of the XY model in three dimensions.

Similar generalizations to continuous spins are possible for all the algorithms we discussed in Section 4.4. However, the Metropolis algorithm and the Wolff algorithm are probably adequate to cover most situations.

---

[17] In a system where the energy varies quadratically about its minimum as we change the values of the spins, one can achieve a roughly constant acceptance ratio by making changes of direction whose typical size varies as $\sqrt{T}$. This might also be a reasonable approach for systems such as the XY and Heisenberg models when we are at low temperatures because, to leading order, the energies of these systems are quadratic about their minima.

Problems

4.1 An Ising system can be considered as being composed of regions of predominantly up- or down-pointing spins separated by domain walls. Assuming that in equilibrium the walls roughly speaking perform a random walk about the system, show that the dynamic exponent of the model (in any number of dimensions) should be about $z = 2$.

4.2 Suggest a simple modification of the Wolff algorithm that would allow us to simulate Ising systems in a non-zero magnetic field B.

4.3 What is the equivalent of Equation (4.24) for the Swendsen–Wang algorithm?

4.4 Modify the Metropolis program given in Appendix B to carry out the heat-bath algorithm for the two-dimensional Ising model (see Equation (4.44)).

4.5 Write a program to implement the Wolff algorithm for a q-state Potts model. (Hint: you might want to use the Wolff algorithm program for the Ising model given in Appendix B as a starting point.)

5

The conserved-order-parameter Ising model

In the last two chapters we have looked in detail at the Ising model, which is primarily of interest as a model of a ferromagnet. In this chapter we look at a variation on the Ising model idea, the conserved-order-parameter Ising model. Although mathematically similar to the normal Ising model, the conserved-order-parameter Ising model is used to model very different systems. In particular, it can be used to study the properties of **lattice gases**.

A lattice gas is a simple model of a gas in which a large number of particles (representing atoms or molecules) move around on the vertices of a lattice. Confining the particles to a lattice in this way makes the model considerably simpler to deal with than a true gas model in which particles can take any position in space, although it also makes the model less realistic. However, lattice gases can give a good deal of insight into the general behaviour of real gases without necessarily being quantitatively accurate representations.

Many different types of lattice gases have been studied, including ones in which the particles possess inertia, or do not, ones in which particular types of particle collisions occur, and ones with a variety of different types of particle–particle interactions. In this chapter we will look at one of the simplest types which, as we will show, is equivalent to an Ising model. In our lattice gas the particles have no inertia and simply walk at random around the lattice under the influence of thermal excitations. The model is defined as follows. We start off with a lattice possessing N sites, of which a fraction ρ are occupied by particles. The particles satisfy the following rules:

1. The total number of particles is fixed, so that if a particle disappears from one site it must reappear at another. (Equivalently, we could say that the particle density ρ is constant.)

2. A lattice site can be occupied by at most one particle at any time. This **site exclusion** rule has a similar effect to the hard-sphere repulsion seen in real systems, which is the result primarily of Pauli exclusion.

3. If two particles occupy nearest-neighbour sites on the lattice, they feel an attraction with a fixed energy ϵ. This rule mimics the effect of the attractive forces between molecules.

Rules 2 and 3 are a fairly poor approximation to the forces between real molecules. For one thing, the model assumes that the repulsive and attractive forces between molecules act on the same length-scale, whereas in reality the attractive forces are normally of longer range. Nonetheless the model proves useful. This is in part because, as we mentioned above, we are often interested more in the qualitative behaviour of the system than in simulating it faithfully, but also because of the model's critical properties. As we will see, the model defined above possesses a phase transition, a critical point at which it switches from a homogeneous gas phase to a solid/vapour coexistence phase. In the vicinity of this phase transition some properties of the model become independent of the exact nature of the interactions between molecules and, simple though it is, our lattice gas is in this region capable of making quantitative predictions about real systems. This is another aspect of the phenomenon of universality which was introduced in Section 4.1.

The first step in studying our lattice gas model is to derive a Hamiltonian for it. To represent the state of the gas we define a set of variables σ_i, one on each lattice site, such that σ_i is 1 if site i is occupied by a particle and 0 if the site is vacant. We use the notation σ_i to distinguish these variables from the spins s_i of the previous chapters, which took values of ± 1.

Already, this notation ensures that Rule 2 above is obeyed. A site can only be occupied or not; we cannot have two or more particles on the same site. In order to ensure that Rule 1 is obeyed, we require that the total number of ones on the lattice, which is the number of occupied sites, should be a constant. We can express this mathematically as

$$\sum_i \sigma_i = \rho N. \tag{5.1}$$

It is Rule 3 above which specifies the Hamiltonian of our model. If two particles are located on nearest-neighbour sites, they contribute an energy $-\epsilon$ to the energy of the whole system. In terms of our particle variables we can write this as

$$H = -\epsilon \sum_{\langle ij \rangle} \sigma_i \sigma_j, \tag{5.2}$$

where $\langle ij \rangle$ denotes pairs of spins i and j which are nearest neighbours, just as in previous chapters. Now let us define a new set of variables

$$s_i = 2\sigma_i - 1. \tag{5.3}$$

These new variables are precisely the Ising spins of previous chapters. The variable s_i takes the value $+1$ on a site occupied by a particle, and -1 on an unoccupied site. Inverting (5.3) to get $\sigma_i = \frac{1}{2}(s_i + 1)$ and substituting into Equation (5.2) we then find

$$
H = -\frac{1}{4}\epsilon \sum_{\langle ij \rangle} (s_i + 1)(s_j + 1)
$$

$$
= -\frac{1}{4}\epsilon \sum_{\langle ij \rangle} s_i s_j - \frac{1}{2}z\epsilon \sum_i s_i - \frac{1}{2}z\epsilon N, \tag{5.4}
$$

where z is the lattice coordination number introduced in Section 3.1 (equal to the number of nearest neighbours of each site). Making the same change of variables in Equation (5.1) we get

$$
\sum_i s_i = N(2\rho - 1), \tag{5.5}
$$

and substituting this into Equation (5.4) we then find that

$$
H = -\frac{1}{4}\epsilon \sum_{\langle ij \rangle} s_i s_j - z\epsilon N\rho. \tag{5.6}
$$

The second term is constant, since z, N and ρ all are. If we now define $J = \epsilon/4$ our Hamiltonian becomes

$$
H = -J \sum_{\langle ij \rangle} s_i s_j + \text{constant.} \tag{5.7}
$$

The additive constant makes no difference to the expectation value $\langle Q \rangle$ of any observable quantity Q, since it cancels out of Equation (1.7). And apart from the constant, this Hamiltonian for our lattice gas is identical to that of the Ising model, Equation (1.30), in zero magnetic field, just as we claimed above. The sum in Equation (5.5) is just the total (instantaneous) magnetization M of this model:

$$
M = N(2\rho - 1). \tag{5.8}
$$

(See Equation (1.33).) Since both ρ and N are constant quantities, so also is M, and for this reason we call this model the **conserved-order-parameter Ising model**, sometimes abbreviated to **COP Ising model**. (Recall that the magnetization M is also called the order parameter of the Ising model— see Section 3.7.1.) Throughout much of this chapter we will treat our lattice gas using this Ising representation. This will allow us to make use of many of the ideas we have developed in previous chapters to create algorithms for simulating the model.

Unlike the Ising model studied in previous chapters, not all states of
the spins in the COP Ising model are allowed; only those for which the
constraint (5.5) is satisfied are allowed. This means that the sum over states
involved in calculating an expectation value (Equation (1.7)) is a sum over
only these allowed states. When we come to design a Monte Carlo algorithm
for this model we will need to take this constraint into account. A related
point is that the COP Ising model has two free parameters we can specify.
In the normal Ising model in zero magnetic field the only parameter we can
vary is the temperature. In the COP Ising model on the other hand, we
have to specify both the temperature and the average density of particles ρ,
which in the language of the Ising model is the density of up-pointing spins
or equivalently the magnetization.

From our investigations in Chapter 3 we know that in the ferromagnetic
case ($J > 0$), the normal Ising model has an average magnetization which is
zero above the critical temperature T_c but non-zero below (see Figure 3.9, for
instance). Rearranging Equation (5.8) for ρ, this magnetization corresponds
to a preferred density of

$$\rho = \frac{1}{2}\left[1 + \frac{M}{N}\right] = \tfrac{1}{2}(1 + m), \tag{5.9}$$

where m is the magnetization per spin. In fact, m has two equilibrium values
below T_c, one positive and one negative, so there are really two preferred
densities for the model in this regime:

$$\rho_+ = \tfrac{1}{2}(1 + |m|), \qquad \rho_- = \tfrac{1}{2}(1 - |m|). \tag{5.10}$$

Suppose now that we fix ρ for the COP model somewhere in the range
between these two preferred densities: $\rho_- \leq \rho \leq \rho_+$. The system can then
arrange for the local density at every point on the lattice to take one or
other of the preferred values by **phase separating** into domains of these
two densities, and as we will see in Section 5.1 the COP model does exactly
this. When ρ is outside the range between ρ_- and ρ_+, it is still possible for
it to reach one of the preferred densities in some regions. For example, if
$\rho < \rho_-$ then the system has fewer up-pointing spins than it needs to reach
the preferred density, but it can still reach it in some portion of the lattice
by concentrating all its up-spins in local domains. This however would leave
the rest of the system starved of up-spins and even further from the preferred
density than it was before. It turns out that the energetic cost of doing this
is greater than the gain derived from forming the domains, so the system
prefers to be homogeneous.

Thus, when $J > 0$, our COP Ising model has two phases below the
critical temperature, one in which $\rho_- \leq \rho \leq \rho_+$ and there is separation
into domains of high and low density, and one in which ρ lies outside this
range and the system is homogeneous. The absolute magnetization $|m|$ of

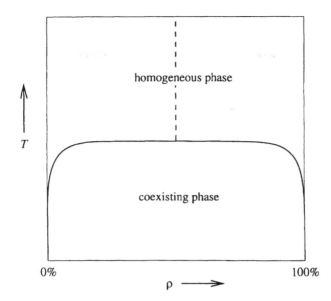

FIGURE 5.1 Phase diagram of the COP Ising model on a square lattice in two dimensions. Depending on the values of the temperature T and density ρ, the system will either be homogeneous or will separate into coexisting domains of the two preferred densities ρ_+ and ρ_-.

the normal Ising model gets smaller with increasing temperature, so that the range of ρ within which there is phase separation in the COP model, given by Equation (5.10), also gets smaller. Above T_c the range shrinks to zero and there is no phase separation. This behaviour is depicted in Figure 5.1, which shows the phase diagram for the two-dimensional COP Ising model.[1] Real fluid systems show a very similar phenomenon when they phase separate on cooling. H_2O at atmospheric pressure for example is a gas at high temperatures, but passes on cooling into a coexistence region in which it separates into distinct fractions of water vapour and liquid water. This lends support to our earlier contention that the COP Ising model shows behaviour qualitatively similar to that of real fluids.[2]

In this chapter we will show how the equilibrium properties of the COP Ising model can be efficiently simulated on a computer and discuss two ap-

---

[1] The line separating the two regimes, which is called the **binodal**, can be calculated exactly from the known magnetization $m = [1 - \mathrm{cosech}^2 2\beta J]^{1/8}$ of the normal Ising model in two dimensions, which was postulated by Onsager in 1949, though not proven until Yang tackled the problem three years later (Yang 1952).

[2] Note that the high-density phase of the COP Ising model is more like a solid than a liquid however, since the "molecules" in the condensed phase of the model are arranged on the regular grid imposed by the lattice.

plications of the model to the study of physical systems. In Section 5.1.1 we will use it to examine the behaviour of phase interfaces as temperature is varied and in Section 5.3 we look at the problem of calculating the equilibrium shapes of crystals. First however, we describe the standard Monte Carlo algorithm used for simulating the model, the Kawasaki algorithm.

5.1   The Kawasaki algorithm

Neither the single-spin-flip algorithms of Chapter 3 nor the cluster algorithms of Chapter 4 can be used for the simulation of the conserved-order-parameter Ising model, because flipping either a single spin or a cluster of similarly oriented spins would change the total magnetization, which is not allowed. Instead, our Monte Carlo moves must change the values of a number of spins, chosen in such a way that the total magnetization remains the same. The simplest such algorithm is the **Kawasaki algorithm** (Kawasaki 1965), which goes as follows. We choose at random a pair of adjacent spins on the lattice. Rather than flipping these spins, our basic Monte Carlo move is to *exchange* their values. In this way the total magnetization of the lattice will always be preserved. To decide whether to exchange a given pair of spins, we calculate the change in the energy $\Delta E = E_\nu - E_\mu$ between the states μ and ν of the system before and after, and accept or reject the move with the Metropolis acceptance probability[3]

$$A(\mu \to \nu) = \begin{cases} e^{-\beta \Delta E} & \text{if } \Delta E > 0 \\ 1 & \text{otherwise.} \end{cases} \tag{5.11}$$

(See Equation (3.7).)

Clearly this algorithm conserves the value of the order parameter M, but we still need to prove that it satisfies the dual criteria of ergodicity and detailed balance, so that we know that it will sample states with their correct Boltzmann probabilities. That the algorithm satisfies ergodicity can by seen most easily by considering the lattice gas representation of the model. In this representation our Monte Carlo move corresponds to the movement of a single particle from one site to another adjacent one. However, any configuration of the particles on a finite lattice can be reached in a finite number of steps from any other by moving each particle individually to its new position, and hence the algorithm is ergodic.

The demonstration of detailed balance is also very straightforward. On a lattice with N sites and lattice coordination number z, there are $\frac{1}{2}zN$ pairs

---

[3]In fact, Kawasaki originally proposed using the heat-bath acceptance probability, Equation (4.44), to perform the simulation. This yields a correct algorithm but, as discussed in Section 4.5.1, the Metropolis acceptance ratio is always larger than the heat-bath one and therefore gives a more efficient simulation, so most recent studies have used the Metropolis choice.

of nearest-neighbour sites. The probability of selecting any particular pair of spins at a given step of the algorithm is therefore $g(\mu \to \nu) = 2/(zN)$. The probability of selecting the same pair for the reverse move is exactly the same. The ratio of the rates for the forward and reverse moves is then required to satisfy Equation (3.3):

$$\frac{P(\mu \to \nu)}{P(\nu \to \mu)} = \frac{g(\mu \to \nu)A(\mu \to \nu)}{g(\nu \to \mu)A(\nu \to \mu)} = \frac{A(\mu \to \nu)}{A(\nu \to \mu)} = e^{-\beta \Delta E} \qquad (5.12)$$

As we argued in Section 3.1, the most efficient way to satisfy this condition is the choice given in Equation (5.11).

Implementing the Kawasaki algorithm is not complicated, but there are one or two tricks worth noting. First, if we pick two spins which are pointing in the same direction then clearly there is no point in exchanging their values, since this would make no change to the state of the lattice. So it is worth checking, before we calculate ΔE, whether the spins are indeed aligned with one another. If they are, we can save ourselves some work by forgetting them and going straight on to the next Monte Carlo step.

It is tempting, in fact, to think that we might improve the speed of the algorithm by ignoring pairs of spins which are aligned and choosing only amongst those pairs which are anti-aligned. This however would be a mistake, because the number of such pairs can change when we make a move, which means that the selection probabilities $g(\mu \to \nu)$ and $g(\nu \to \mu)$ for moves in opposite directions would not be equal any more, and so would not cancel out of Equation (5.12). Choosing our spins in this fashion would result in an algorithm which does not generate states with the correct Boltzmann probabilities.[4]

In theory calculating the energy difference ΔE means calculating the energies E_μ and E_ν of the system before and after an exchange and taking their difference. As with the Metropolis algorithm however, most of the spins on the lattice are the same in states μ and ν, and we can use this fact to derive a formula for ΔE which does not involve summing over the whole lattice. The derivation follows exactly the same lines as that for the Metropolis algorithm in Section 3.1.1. We won't go through it in detail,[5] but the end result is that

$$E_\nu - E_\mu = 2J\left[s_k^\mu \sum_{\substack{i \neq k' \text{ and} \\ \text{n.n. to } k}} s_i^\mu + s_{k'}^\mu \sum_{\substack{j \neq k \text{ and} \\ \text{n.n. to } k'}} s_j^\mu \right]. \qquad (5.13)$$

---

[4] Actually, it *is* possible to create a correct algorithm in this way, but one has to be quite crafty about it. And for equilibrium simulations there are, as we will shortly see, better ways than this to improve the efficiency of our algorithm. For out-of-equilibrium simulations, on the other hand, an approach which favours anti-parallel spin pairs can prove useful. This point is discussed in Section 10.3.1. (See also Problem 5.1.)

[5] The derivation appears at the end of the chapter as Problem 5.2.

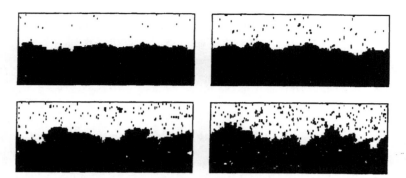

FIGURE 5.2   Typical interface configurations at several temperatures. From top left to bottom right the temperatures are T/T_c=0.6, 0.7, 0.8, and 0.9. The interface becomes fuzzier with increasing temperature and at the same time the particle densities ρ_- and ρ_+ approach each other.

Here k and k' are the two spins which we propose to exchange and the two sums are carried out over all the nearest neighbours of each spin except the two spins k and k' themselves. Notice that, like Equation (3.10) for the Metropolis algorithm, this expression only involves the values of spins in the initial state μ and not in the final state ν, which allows us to calculate the change in energy without actually having to carry out the spin exchange.

5.1.1   Simulation of interfaces

As an example of the use of the Kawasaki algorithm, we show in Figure 5.2 a set of simulations of a two-dimensional system in the phase coexistence regime in which the model separates into regions rich in particles and regions rich in vacancies. In this case, we were particularly interested in the behaviour of the interface between the two phases. In order to produce a straight interface and to prevent it from wandering we have given the system periodic boundary conditions only in the horizontal direction; at the top and bottom boundaries of the system we have fixed the spins to point down and up respectively so that the system divides into two domains with a clean interface between them.

The simulations were performed at 50% particle density (or $M = 0$ in Ising model terms) for a variety of different temperatures. As the figure shows, the interface is quite smooth at low temperatures, but becomes increasingly rough as the temperature of the simulation approaches T_c. A typical quantity of interest in such simulations is the thickness of the interface as a function of the temperature and system size. To quantify this thickness we measure the average particle density $\rho(y)$ as a function of the

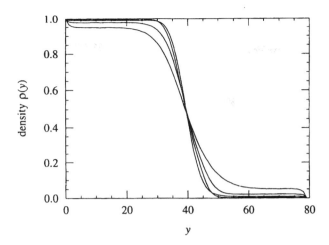

FIGURE 5.3   The average particle density $\rho(y)$ as a function of y for the four simulations shown in Figure 5.2.

coordinate y in the direction perpendicular to the interface. In Figure 5.3, for example, we show the measured values of $\rho(y)$ for the four temperatures used in Figure 5.2. A common definition of the width of the interface is the distance over which the particle density increases from $\rho_- + \frac{1}{10}\Delta\rho$ to $\rho_- + \frac{9}{10}\Delta\rho$, where $\Delta\rho = \rho_+ - \rho_-$.

In fact, the Kawasaki algorithm is not a particularly efficient way of simulating the equilibrium properties of the COP Ising model because the particles in the model move diffusively around the lattice, which is a slow way of reaching equilibrium. In the next section we develop a more efficient algorithm for the model in which particles make non-local moves.

5.2   More efficient algorithms for the COP Ising model

The speed of the Kawasaki algorithm is limited by its particular choice of Monte Carlo move. Kawasaki spin-exchange moves can only move a particle one lattice spacing at a time so that equilibration is by the diffusive motion of particles, which can be a slow process if the particles need to travel a long way. However, since we are not interested in the way in which we reach equilibrium but only in the final result, there is no need to confine ourselves to local Monte Carlo moves like this, and it turns out that we can do much better by using non-local moves. The simplest non-local move would be one in which we choose two lattice sites at random and exchange their spins, again using a Metropolis-type acceptance ratio, Equation (5.11). Although this algorithm is actually quite a good way of simulating the COP Ising model, we can make it better still by a small change.

If we choose two spins at random there is a reasonable chance that they will be pointing in the same direction, so that exchanging their values will have no effect on the state of the system. For randomly chosen pairs the probability of their being aligned is

$$p_\parallel = 1 - 2\rho + 2\rho^2, \qquad (5.14)$$

which has a minimum value of $\frac{1}{2}$ when $\rho = \frac{1}{2}$. Thus we are always wasting at least 50% of our moves doing nothing, and perhaps more, depending on the particle density ρ. If instead of picking pairs at random we pick one spin from the set of up-pointing spins and one from the set of down-pointing ones and exchange their values (or alternatively, flip them both over, which comes to the same thing) with a probability given by Equation (5.11), then no time is wasted on aligned pairs. This gives us an immediate improvement in efficiency by a factor of $1/(1 - p_\parallel)$, which is at least 2 and can become arbitrarily large for ρ close to zero or one. The proof of ergodicity for this algorithm follows exactly the same lines as for the Kawasaki algorithm, as does the proof of detailed balance once we note that the selection probability for every spin pair is the same, so that the factors $g(\mu \to \nu)$ and $g(\nu \to \mu)$ cancel out of Equation (5.12) again.[6]

Implementation is a little more complex for this algorithm than it was for the Kawasaki algorithm, since in order to pick an up-spin and a down-spin we need to know where all the up- and down-spins are. Simply picking spins at random and throwing them away if they don't have the desired direction would not work—the resulting algorithm would be no more efficient than the original one which we proposed. In order to achieve the full performance gain of our new algorithm we must keep a list of all the up- and down-spins and choose our pairs of spins at random from these lists. Luckily this is quite simple to do. Since the number of up- and down-pointing spins does not change during the simulation, two arrays of fixed length can store the two lists, and when the values of two spins are exchanged, we also exchange their entries in these lists. The computational overhead incurred in doing this is relatively small,[7] and the gain in efficiency of the algorithm certainly justifies the extra work involved.

This non-local spin-exchange algorithm is already a huge improvement on the Kawasaki algorithm. For example, we find that the equilibration of

---

[6]Note that this algorithm obeys the condition of detailed balance perfectly well, even though, as we pointed out in Section 5.1, the equivalent local algorithm, in which we choose at random amongst only those pairs of nearest-neighbour spins which are anti-aligned, does not.

[7]See the sample program given in Appendix B.

an interface simulation like the ones depicted in Figure 5.2 requires about $\frac{1}{40}$ as many Monte Carlo steps with the new algorithm as it does with the Kawasaki one. One reason for this is the non-diffusive nature of the particle motion. Another is that the non-local update move always picks pairs of spins which are anti-aligned. The Kawasaki algorithm, as we have pointed out, sometimes picks pairs which are aligned and thereby wastes time. As it turns out, in fact, the Kawasaki algorithm picks such aligned pairs almost all of the time when we are in the phase coexistence regime, which makes it quite inefficient. A glance once more at Figure 5.2 reveals the reason for this: the Kawasaki algorithm only selects adjacent pairs of spins, but most adjacent pairs of spins are pointing in the same direction when we are in the coexistence regime. To be fair however, the non-local algorithm is more complicated to program and in our implementation we find that a single step requires about twice as much CPU time as in the Kawasaki case. Even in terms of CPU time however, the new algorithm is still a factor of 20 faster than the Kawasaki algorithm.

Our non-local algorithm is still far from perfect however. The main problem with it is the perennial one of the acceptance ratio. If the acceptance ratio is low, we are wasting a large portion of our CPU time selecting pairs and then failing to exchange their values. In the simulations of interfaces described above, for example, we find that at $T = 0.8T_c$ the average acceptance ratio for the exchange of a pair of anti-aligned spins is only about 3%.

It would be satisfying if we could find an algorithm which only considered the exchange of anti-aligned spin pairs but had a higher acceptance ratio than this, perhaps even an acceptance ratio of one. It turns out that we can achieve this goal by using the continuous time Monte Carlo method described in Section 2.4.

5.2.1 A continuous time algorithm

The basic prescription for turning our non-local algorithm into a continuous time one is to ignore all moves which do not change the state of the system (which in this case means those which involve the exchange of two spins which are pointing in the same direction) and choose between the remaining moves in proportion to their mean rates $P(\mu \rightarrow \nu)$. These rates are dictated by the condition of detailed balance:

$$\frac{P(\mu \rightarrow \nu)}{P(\nu \rightarrow \mu)} = e^{-\beta(E_\nu - E_\mu)}. \tag{5.15}$$

In order to derive a general expression for the energy change $E_\nu - E_\mu$ it is helpful to define the **spin coordination number** for site i

$$n_i = \sum_{i \text{ n.n. to } j} \delta_{s_i s_j}, \qquad (5.16)$$

where δ_{ij} is the Kronecker δ-symbol, which is 1 when $i = j$ and zero otherwise. The spin coordination number is the number of nearest-neighbours of site i which have spins pointing in the same direction as s_i, i.e., the number of satisfied bonds between the spin on site i and its neighbours. The change in the total energy when we update a spin at site i is then just $-2J$ times the change in the number of satisfied bonds, or $-2J(n_i^\nu - n_i^\mu)$, where n_i^μ and n_i^ν are the values of n_i in the states μ and ν. The energy change on exchanging two spins i and j is

$$E_\nu - E_\mu = -2J(n_i^\nu - n_i^\mu) - 2J(n_j^\nu - n_j^\mu). \qquad (5.17)$$

Although it is not immediately obvious, this in fact is the appropriate generalization of Equation (5.13) to the case of non-local update moves. Plugging this into Equation (5.15) and rearranging a little, we get

$$\frac{P(\mu \to \nu)}{P(\nu \to \mu)} = e^{2\beta J[(n_i^\nu + n_j^\nu) - (n_i^\mu + n_j^\mu)]}. \qquad (5.18)$$

As we stressed in Section 2.3, detailed balance conditions such as this one only tie down the ratio of the rates for forward and backward moves, and not the rates themselves. But this equation does suggest a very simple choice for the rates:

$$P(\mu \to \nu) \propto e^{-2\beta J n_i^\mu} e^{-2\beta J n_j^\mu}. \qquad (5.19)$$

This choice is particularly elegant because it depends only on the initial state of the lattice μ and not on the final one, and also because it factorizes into a product of two terms, one for each of the two spins involved. As we will see in a moment, this simplifies the implementation of our algorithm.

Using this choice, our continuous time Monte Carlo algorithm is now as follows. We choose an up-pointing spin i at random from all the possible choices with probability proportional to $e^{-2\beta J n_i^\mu}$ and a down-pointing spin j with probability proportional to $e^{-2\beta J n_j^\mu}$. The joint probability of choosing the two is then given by Equation (5.19). Then we simply exchange the values of the two spins. In order to ensure that detailed balance is preserved we must also add an amount

$$\Delta t = \frac{1}{\sum_\nu P(\mu \to \nu)} \qquad (5.20)$$

to our time variable for each move we perform (see Equation (2.19)). Averages over the measured values of observable quantities then become time averages, with states weighted in proportion to the time Δt spent in them.

In practice, implementation of the algorithm involves creating separate lists of spins, divided according to whether they are pointing up or down, and by their spin coordination number. Suppose there are m_{up}^n members of the list of n-coordinated up-pointing spins, and m_{down}^n members of the list of n-coordinated down-pointing spins. Making a move involves choosing an up-pointing spin by first selecting at random one list of up-spins with probability proportional to $m_{up}^n e^{-2\beta Jn}$ and then choosing one of the spins in that list uniformly at random. We also choose a down-pointing spin in the same way from the lists of down-spins and then we exchange the two. To calculate the value of Δt, we substitute Equation (5.19) into Equation (5.20) to get

$$[\Delta t]^{-1} = \sum_{i \text{ up}} e^{-2\beta Jn_i^\mu} \sum_{j \text{ down}} e^{-2\beta Jn_j^\mu} = \sum_q m_{up}^q e^{-2\beta Jq} \sum_r m_{down}^r e^{-2\beta Jr},$$

(5.21)

(to within a multiplicative constant).

To compare the efficiency of this continuous time Monte Carlo algorithm with the previous two algorithms, we again turn to the simulation of an interface in two dimensions. Simulating the same systems as before, we find that the continuous time Monte Carlo algorithm requires about $\frac{1}{40}$ as many moves as our earlier non-local algorithm to bring the system to equilibrium, and about $\frac{1}{1500}$ as many moves as the Kawasaki algorithm. However, the continuous time algorithm is quite complicated to implement, and involves a lot of extra operations to keep track of the values of the sums appearing in Equation (5.21).[8] In our implementation each move of the continuous time algorithm takes about 20 times as much CPU as the simpler non-local algorithm and 40 times as much as the Kawasaki algorithm. Thus our overall gain in computational efficiency is about a factor of two compared with the non-local algorithm, and a factor of 40 compared with the Kawasaki algorithm.

5.3 Equilibrium crystal shapes

In the investigations of phase interfaces in Section 5.1.1, we kept our interfaces more or less straight by using a 50 : 50 ratio of particles to vacancies and by our choice of boundary conditions, with rows of fixed spins along the top and bottom of the lattice. These choices were appropriate for studying that particular problem, but the COP model behaves very differently if we use other densities of particles and other types of boundary conditions. For instance, on a two-dimensional lattice with periodic boundary conditions and a relatively small density of one of the two spin states, we find that

---

[8] Again, see the sample program given in Appendix B.

the preferred state of the model in the phase coexistence regime is one in which most of the particles in the system coalesce into a single domain or **droplet** which is usually approximately circular in shape. The reason is that the interface between domains of up- and down-pointing spins is energetically costly—the total energy of the system increases by $2J$ for every pair of anti-aligned nearest-neighbour spins along the interface. In effect, there is a **surface energy** or **surface tension** at the boundary of our droplet.[9] The system can therefore reduce its energy by reducing the length of the boundary. The lowest surface to area ratio in two dimensions is achieved when the droplet is a circle,[10] although, as we will see, it never actually takes this shape in practice because of the effects of the lattice.

Crystals display similar behaviour as they crystallize, choosing a shape which reduces their surface energy. In the limit of a large crystal this preferred shape is called the **equilibrium crystal shape** or ECS. In 1901 Wulff invented a geometrical construction for calculating equilibrium crystal shapes from a knowledge of the way in which the surface tension varies with the orientation of a surface. His construction shows that the ECS is indeed a circle (or a sphere in three dimensions) if the surface tension is isotropic. Because of the underlying crystal lattice, however, the surface tension is not isotropic and for real crystals the ECS is at best only approximately circular. Unfortunately, the orientation dependence of the surface tension is only known exactly in a very few cases, so that it is usually not possible to apply the Wulff construction to calculate the ECS. Instead we turn to Monte Carlo methods.

In order to perform a truly accurate calculation of an equilibrium crystal shape we would need to simulate in detail the interactions between the molecules in a solidifying solid. This however is not an easy thing to do. So instead many people have turned to the conserved-order-parameter Ising model as a simple model of the processes taking place. Like the real physical system, the COP Ising model shows "crystallization" in which particles coalesce out of the "melt" onto a regular crystalline lattice, and it possesses a surface tension which is orientation-dependent, giving rise to non-trivial equilibrium crystal shapes. It is true that the details of the interactions between the particles in our Ising model are very different from those in the true crystal, so we do not expect the model to give an accurate quantitative prediction of a real ECS, but, as with our earlier simulations of interfaces, the study of the COP Ising model can nonetheless shed light on the sort of processes taking place when a crystal solidifies.

---

[9]For simplicity we will use the three-dimensional term "surface tension" to refer to this phenomenon in both two and three dimensions, although in two dimensions it is strictly more correct to call it "line tension".

[10]Strictly, this is only true if the density ρ of particles is sufficiently low or sufficiently high. See Problem 5.3 for an exploration of this point.

FIGURE 5.4   Typical droplet configurations at $T/T_c = 0.25$ (left) and $T/T_c = 0.5$ (right). At low temperatures the boundary of the droplet is quite sharp and the droplet's shape resembles a square. With increasing temperature the boundary becomes fuzzier and the droplet's shape becomes round.

In order to carry out a Monte Carlo simulation of an equilibrium crystal shape we could start off by placing a certain fraction ρ of particles (i.e., up-pointing spins) at random on a lattice and allowing them to move about under the action of any one of the algorithms described in the earlier sections of this chapter at a temperature T below T_c, until they coalesce into a single domain. It turns out however, that it is a lot quicker to start them off in a single domain (say a square on a square lattice) and "warm them up" to temperature T. In Figure 5.4 we show two different droplet configurations from simulations of this kind for a two-dimensional COP Ising model[11] at two different temperatures below T_c. As we can see, the droplet remains rather square at low temperatures, but becomes rounder as T increases.

To get a more quantitative estimate of the ECS, we can run our simulation for a longer time, record the shape of the droplet at intervals and then take an average over the shapes found. Unfortunately, even if we start the droplet at the centre of the lattice, it has a tendency to drift away during the simulation, so in order to average the shapes correctly we translate them first so that their centres of mass coincide.[12] The average itself simply in-

---

[11] We illustrate the method using a two-dimensional system for clarity. At the end of this chapter we also give some results for three-dimensional systems. However, the two-dimensional results are not purely academic. In fact, the two-dimensional COP Ising model is quite a good approximation to the behaviour of adsorbed atoms on the surfaces of metals, and in certain temperature regimes one does indeed find droplets or **islands** of adatoms which look very similar to the ones depicted in Figure 5.4. This point is discussed further in Chapter 11.

[12] In fact, our system has four-fold rotational symmetry and inversion symmetry on the

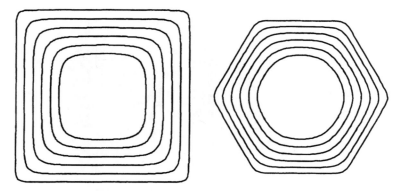

FIGURE 5.5   Equilibrium crystal shapes as a function of temperature
on a square lattice (left) and a hexagonal one (right). The outermost
droplet shapes correspond to inverse temperatures of $\beta J = 9.3$ in each
case. Going inwards, the values of β for the other shapes are factors of
2, 3, 4, 5 and 6 lower. To make the figure clearer we have rescaled the
overall size of the droplets so that the different curves do not overlap.

volves taking the mean of the particle density over all our measurements as
a function of position. At the centre of mass itself this mean takes a value
close to ρ_+ (see Equation (5.10)) and it remains roughly constant until we
approach the droplet interface, at which point it falls quickly towards ρ_-.
We define the edge of the droplet to be the point at which the density is
equal to $\frac{1}{2}$.

As an example of this kind of calculation we show in Figure 5.5 measurements of the average ECS of droplets on square and triangular lattices in
two dimensions. In these calculations we used the continuous time Monte
Carlo algorithm of Section 5.2.1, which again proves significantly faster than
either the Kawasaki algorithm or the non-local algorithm we introduced in
Section 5.2.

We can also use the Monte Carlo method to study the ECS of three-
dimensional systems. (Real crystals are after all three-dimensional.) The
behaviour of the ECS is richer in three dimensions than in two. In two dimensions the sides of the droplet are flat at zero temperature, but become
rounded for all finite temperatures. In three dimensions the story is more
complicated. On a cubic lattice for instance, the ECS is a cube at $T = 0$,
just as we would expect. As we raise the temperature above zero, the corners
and edges of the cube become rounded, but the droplet still has flat facets.

square lattice (the symmetry group is C_4), so we can reduce the errors on our average
by using each droplet configuration eight times over in each of its symmetry equivalent
orientations.

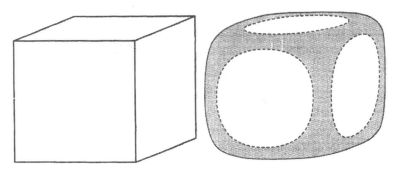

FIGURE 5.6   Schematic depiction of the equilibrium crystal shape for the COP Ising model on a cubic lattice at zero temperature (left) and at a non-zero temperature below the roughening temperature. Above the roughening temperature, the flat facets disappear.

Only when we reach a certain higher temperature, the so-called **roughening temperature** T_R, which is approximately half the critical temperature, do these facets disappear. This disappearance is called the **roughening transition** of the COP Ising model. In Figure 5.6 we give a schematic drawing of the ECS of a cubic system at zero temperature and at non-zero temperature below the roughening temperature. Above the roughening temperature, the droplet looks like a deformed sphere but has no flat surfaces.

The behaviour of equilibrium crystal shapes becomes richer still on more complex lattices and with the introduction of next-nearest-neighbour interactions. For example, Jayaprakash and Saam (1984) have studied the ECS on a face-centred cubic (fcc) lattice using a COP Ising model with both nearest- and next-nearest-neighbour interactions. In Figure 5.7 we show the ECS of the fcc crystal at zero temperature with only nearest-neighbour ferromagnetic interactions, and the ECS when small ferromagnetic next-nearest-neighbour interactions are also included. In the absence of next-nearest-neighbour interactions, the sharp corners of the ECS disappear as soon as the temperature becomes non-zero, but the facets persist up to a roughening temperature T_R, just as they do in the cubic case. Again T_R is about half the critical temperature. In the presence of ferromagnetic next-nearest-neighbour interactions, extra (110) facets appear which have a roughening temperature different from that of the (111) facets. With anti-ferromagnetic next-nearest-neighbour interactions, the corners disappear for all $T > 0$, but the edges do not disappear until we reach the melting point T_c.

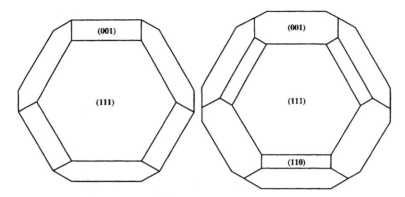

FIGURE 5.7 Equilibrium crystal shape of the COP Ising model on a face-centred cubic lattice at zero temperature without next-nearest-neighbour interactions (left) and after the introduction of small ferromagnetic next-nearest-neighbour interactions (right). After Jayaprakash and Saam (1984).

Problems

5.1 As we pointed out in Section 5.1, the variant of the Kawasaki algorithm in which we choose only among the anti-aligned nearest-neighbour spin pairs does not sample the Boltzmann distribution correctly. There is however a correct generalization of the Kawasaki algorithm which favours anti-aligned pairs and is more efficient than the standard algorithm when the density of particles ρ is far from $\frac{1}{2}$. In this variant, we choose at random a spin of the *minority* type (up or down, particle or vacancy) and then choose at random one of its neighbours, and exchange the values of the pair with the Metropolis acceptance ratio, Equation (5.11). Prove that this algorithm satisfies the conditions of ergodicity and detailed balance. Why is it more efficient than the standard Kawasaki algorithm?

5.2 Derive Equation (5.13).

5.3 Consider a COP Ising system on an $L \times L$ square lattice in two dimensions with a density of particles ρ. Calculate the total length of interface in the system (in terms of number of unsatisfied bonds) if (a) the particles coalesce into a single droplet and (b) they form a band with straight interfaces which stretches all the way across the system and wraps around the boundary conditions. Hence show that the droplet is not the preferred configuration of the system below T_c when $\frac{1}{4} < \rho < \frac{3}{4}$.

6

Disordered spin models

In the previous few chapters of this book we have studied a variety of different Monte Carlo algorithms for simulating the Ising model under various conditions. In Chapter 3 we looked at the Metropolis algorithm, as it applies to the Ising model, and in Chapter 4 we looked at a number of other algorithms, most of which are designed to simulate the model more efficiently in the critical region, the region near the phase transition between paramagnetism and ferromagnetism. In the last chapter, we looked at a slight variation on the Ising model theme, the conserved-order-parameter Ising model, which can be used as a model of a gas, or of atoms diffusing on a surface. As we have seen, for different regimes of temperature or different variations of the model, different algorithms are appropriate. The trick is to tailor the algorithm to the type of problem we wish to tackle. However, all the algorithms we have seen so far eventually boil down to one thing: they generate a series of states, a Markov chain, in which each state is derived from the previous one in the chain, and attempt to contrive that each should appear with a probability proportional to its Boltzmann weight. In this chapter, for the first time, we will come across systems for which an approach of this kind doesn't work. The problem, as we will see, is that for some systems, the so-called **glassy systems**, algorithms of this kind tend to get stuck in small regions of the state space from which they cannot escape. The reason for this is two-fold. First, the fact that each state is generated from the previous one in the chain means that there is a sense in which states are "close together". There are states which it is easy to get to from the present one in a small number of moves, such as states which differ by the flipping of just a few spins for a single-spin-flip algorithm, and there are other states which you can only get to by making a large number of moves. Even for algorithms such as the Swendsen–Wang algorithm of Section 4.4.1, which can in principle get from any state to any other in a single move, some states are much more likely to be generated from a given starting state than others, even though the states may have the same energy. The result is that, for almost all algorithms, there are pairs of states between which the only probable (or possible) routes are via a large number of intermediates. This in itself would not be a problem—all of the algorithms we have described for the Ising model have this property, and yet all of them work acceptably well, at least in some temperature regimes. For the glassy systems we will be interested in in this chapter however, it is a problem because the state spaces of these systems contain metastable "basins" in which all the states have relatively low energy, surrounded by states with much higher energy. In order to escape from such a basin, our Monte Carlo algorithm must pass through one of these states of higher energy. But if the algorithm samples states according to their Boltzmann weights, then the chances that it will ever make a move to one of these high-energy states is exponentially low, and so, at least at low temperatures, the algorithm gets stranded in the basin. We will have to wait a very long time indeed before it finds its way across the "energy barrier" into the world outside.

In fact, the existence of metastable energy basins is not unique to glassy models. For example, the normal Ising model of Chapters 3 and 4 also possesses such basins. Below the critical temperature, the Ising model has a spontaneous magnetization (see Section 3.7.1) so that a significant majority of its spins point in one direction or the other. In order to invert this spontaneous magnetization from (say) up to down, we must create a domain of down-spins amongst our up-pointing majority and grow it until it fills the entire system. It is not difficult to see that this implies that at some point we must have at least two domain walls between up- and down-spins which are of dimension at least the size L of the system. The cost of such a pair of walls is at least $4JL^{d-1}$, where d is the dimensionality of the system. This then is the energy barrier which must be crossed in order to invert the magnetization of the system. Since the probability of sampling a state with this energy is a factor of $\exp(-4\beta JL^{d-1})$ smaller than the chance of sampling a low-lying state of the system, the time to cross this barrier becomes exponentially long as the size of the system grows. We see an exactly similar exponential slowing down with system size in glassy models as well, and in fact this behaviour is often taken as definitive of a glassy system. What then is it that tells us that the Ising model is not a glassy system? The answer is that the two energy basins in the Ising model are symmetry equivalent—each state in one of them corresponds to a state in the other which has exactly the same spin configuration, except that each spin is inverted. This means that it does not matter which basin we are in below the critical temperature—we will get exactly the same answer for all measured properties of the system, except for a possible symmetry operation (such as the change of the sign of the average magnetization). By contrast, the basins in a glassy model are not related by any symmetry operations, so that making measurements in just one of them does not normally give us a good idea of the typical behaviour of the system. If we perform several different simulations of a glassy system, it may get stuck in different energy basins each time and give completely different answers for the measured properties of the model. This phenomenon is called "ergodicity breaking" and we discuss it further in Section 6.2 when we consider simulation techniques for glassy systems. As we argue there, if we wish to gain a good picture of the behaviour of a glassy model, we need to find an algorithm which will sample states from a number of different basins, and this means finding one which does not get stuck.

There are two common ways of tackling this problem. The first is to use an algorithm which samples high energy states with greater than the appropriate Boltzmann weight, making the system more likely to cross energy barriers. This technique is employed in the entropic sampling method, which we describe in Section 6.3. The other alternative is to generate states in a way which allows the system to jump to very different configurations in a single step, rather than making it pass through a whole set of intermediate states. In this way it is possible to skip over the high-energy barrier states altogether. This idea is the foundation for the technique of simulated tempering, which we discuss in Section 6.4. Before we come to the algorithms however, let us introduce some models which exhibit glassy behaviour.

6.1 Glassy systems

Glassy systems have presented statistical physics with some of its toughest challenges. Many of the physics world's brightest minds have studied these problems at various times, but there remain many secrets still to be uncovered. The most familiar glassy system for most of us is glass itself, although a physicist's definition of what constitutes a glass encompasses many other systems, including amorphous substances, metallic glasses and, as we shall see in this chapter, disordered spin models. The common factor connecting these systems is some sort of randomness in their composition. Ordinary window glass, for example, is not a solid like steel; it doesn't derive its rigidity from a crystalline lattice. Instead its atoms are arranged in a largely random fashion and they are, at least in principle, free to move past one another just like the atoms of a liquid. The reason that they don't is that, because of the nature of the interactions between them, there are large energy barriers which the atoms have to cross in order to move, and the probability of crossing such a barrier decreases exponentially with falling temperature. If we raise the temperature of the system, we can increase this probability, and indeed glass does flow as a liquid if you heat it up enough. But at normal temperatures the time taken for the atoms in the glass to move is so long that, for all practical purposes (such as building houses), we can consider it to be solid. The temperature at which the glass becomes solid is known as the **glass temperature**, usually denoted T_g. In fact, it is still a matter of debate whether this transition is a sharp one, like the phase transitions we saw in the last few chapters, or whether it occurs over a range of temperature even when the system is cooled infinitely slowly. For our purposes however, it is enough to note that there are temperatures sufficiently low that the dynamics becomes problematic, and others which are high enough that it doesn't.

In this chapter we will not be considering real glass. Real glass is a fantastically difficult system to study, and even after many decades of work its properties are not at all well understood. So instead, in order to get a handle on the basics of the problem, a number of simpler model systems have been developed which show many of the same features but which are mathematically easier to cope with. Here we will describe briefly two such, both of them variations on the Ising model of the previous chapters.

6.1.1   The random-field Ising model

The **random-field Ising model** is a model consisting of Ising spins $s_i = \pm 1$ on a lattice. Its Hamiltonian differs from that of the normal Ising model by the addition of a random field term. The full Hamiltonian looks like this:

$$H = -J \sum_{\langle ij \rangle} s_i s_j - \sum_i h_i s_i. \tag{6.1}$$

Here, just as before, the notation $\langle ij \rangle$ means that the sum runs over pairs of spins i, j which are nearest neighbours, and the variables h_i are **random fields**, local magnetic fields which each act on just one spin on the lattice, and whose values are chosen at random. There are many ways in which the random fields could be chosen. The simplest case, and the only one which has been studied in any detail, is the case in which their values have mean zero and in which the values on different sites are uncorrelated—they are independent random variables. Variously they are chosen to have a Gaussian distribution with some finite width σ, or to have values randomly $\pm h$ where h is a constant, or any of a variety of other possible alternatives. To a large extent the interesting properties of the model are believed to be independent of the exact choice of distribution (another consequence of the phenomenon of universality discussed in Section 4.1). However, to be definite about it, let us suppose that the fields h_i are chosen from a Gaussian distribution:

$$p(h_i) = \frac{1}{\sqrt{2\pi\sigma^2}} \exp\left[-\frac{h_i^2}{2\sigma^2}\right]. \tag{6.2}$$

The width σ of the Gaussian is often called the "randomness", since it measures how big the typical random fields are.

Strictly speaking the random-field Ising model is not a true glassy system. As discussed below, its slow equilibration arises as a direct result of the onset of ferromagnetism below T_c, so that it doesn't really have a glass temperature, only a critical temperature.[1] However, for the purposes of Monte Carlo simulation we may as well consider it in the same category as the true glasses, since it presents all the same problems that true glasses do. The model also has the nice property that it is particularly easy to see how the large energy barriers arise which make simulation hard at low temperatures.

Consider first the ordinary Ising model, familiar to us from Chapters 3 and 4 of this book. As we know, in two or more dimensions this model possesses a critical temperature T_c above which it is paramagnetic and below which it is ferromagnetic, possessing some spontaneous, non-zero magnetization, either up or down. Imagine doing a Metropolis Monte Carlo simulation with single-spin-flip dynamics in which we cool the Ising model down quickly from some high-temperature state (all spins random, for instance) to a temperature below T_c. What we see is depicted in the frames of Figure 6.1. Quite quickly there form domains of moderate size in which all the spins are either up or down and as time progresses the smallest of these domains shrink and vanish, closely followed by the next smallest, until eventually most of the spins on the lattice are pointing in the same direction. The reason for this behaviour is that the domains of spins possess a surface energy—there is an energy cost to having a domain wall which increases with the length of the wall, because the spins on either side of the wall are pointing in opposite directions. The system can therefore lower its energy by flipping the spins around the edge of a domain to make it smaller (and its wall shorter). Thus, the domains "evaporate", leaving us with most of the spins either up or down.[2]

However, the story is different when we come to the random-field Ising model. First of all, the random-field Ising model doesn't actually possess a phase transition in two dimensions, so Figure 6.1 is not really valid in this case. If we go to three dimensions it does have a transition and it is ferromagnetic below its critical temperature T_c, but the physics of the model is nevertheless very different from that of the normal Ising model. In the random-field model domains still form in the ferromagnetic regime, and there is still a surface energy associated with the domain walls, but it is no longer always possible to shrink the domains to reduce this energy. For each spin in a random-field Ising model, the random field acting on it means that the

---

[1] Also unlike true glassy systems, it usually has a unique ground state and doesn't possess a residual entropy in the limit $T \to 0$. However, this is all rather beside the point for our purposes.

[2] These and other processes involved in the equilibration of the normal Ising model are discussed in greater detail in Chapter 10.

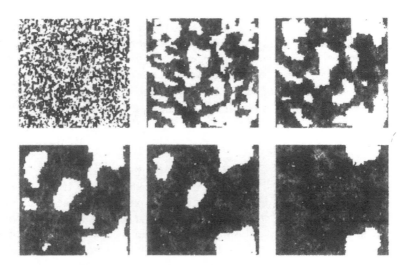

FIGURE 6.1   When the normal Ising model is cooled from an initial high temperature state to a low temperature one ($T = 1.2J$ in this case), we see domains of aligned spins forming which one by one evaporate, leaving the lattice with most of its spins pointing in the same direction.

spin has a preferred direction. Furthermore, at occasional sites there will be a very large local field h_i pointing in one direction, up say, which means that the corresponding spin will *really* want to point up and it will cost the system a great deal of energy if it is pointing down. Suppose now that a domain shrinks by moving its domain wall until one part of the wall encounters such a spin. If we want to move our wall beyond this point, we are going to have to flip this spin. Although in the long run this may be an advantageous thing to do, in the short run it is a very unlikely event: the acceptance ratio for flipping this spin contains a factor $\exp(-2\beta|h_i|)$, which is a very small number if $|h_i|$ is large. We say that the domain wall is **pinned** by the local field; it is prevented from moving by an energy barrier produced by the large random field. Such energy barriers make the Metropolis simulation of the random-field Ising model very time-consuming and severely constrain the sizes of the systems we can study. (Notice however, that this only applies below T_c. As we pointed out above, the critical temperature plays the role of the glass temperature for the random-field Ising model.)

You might well object to the argument we have given here, saying that the problem only arises because we have chosen to do single-spin-flip dynamics. What if instead we flipped groups of spins at a time, as we did in the cluster-flipping algorithms of Chapter 4? This is a good point. However, as it turns out it doesn't make much difference. One might think that if we were to flip, say, clusters of eight spins (in the three-dimensional case), we could get around the problem because if one of the local fields in the cluster had a large value in one direction, there is good chance that others would have opposite values and cancel it out, so that the overall cost in energy of flipping the spins in the cluster would be small. Unfortunately, this is not the case. The overall energy change when we flip the spins in a cluster depends on the sum $\sum h_i$ of the local fields in the cluster, which is also a random variable.[3] We can repeat the argument we made above for this new variable to show that if we flip groups of eight spins, the domain walls can still become pinned at places where the sum of eight local fields is unusually large, and so the acceptance ratio will still become exponentially small.

One possible way around this problem is to consider flipping clusters on all different length-scales. If a certain region of spins is pinned because the sum of the local fields is of unusually large magnitude, it may be that when we consider that region as part of a larger one, the local fields cancel out and the acceptance ratio becomes reasonable again. Or conversely, perhaps it might help to break the pinned region up into smaller ones, some of which could be quite easy to flip, even though the whole region is not. Newman and Barkema (1996) came up with a cluster-flipping algorithm of this kind for the random-field Ising model, based loosely on the Niedermayer algorithm of Section 4.4.2. However, their algorithm is tailored very much to the particular problems of the random-field model and does not work well for other glassy spin models. In this chapter we want to concentrate instead on universal algorithms which are good for a wide variety of systems.

6.1.2 Spin glasses

Another variation on the Ising model theme which shows glassy behaviour is the **Ising spin glass**. An Ising spin glass is an Ising model in which the value of the spin–spin interaction J varies randomly from bond to bond. In general, to get glassy behaviour there must be both positive and negative interactions in the system. If only positive ones are present, for example, the model is not glassy. The simplest and best studied Ising spin glass is the one described by the Hamiltonian

$$H = -\sum_{\langle ij \rangle} J_{ij} s_i s_j, \tag{6.3}$$

where the variables J_{ij} are the random bond strengths. This model is one particular instance of the general class of models known as **Edwards–**

---

[3] In fact, the central limit theorem tells us that it will be a Gaussianly distributed random variable, regardless of the distribution of the underlying random fields. Coupled with renormalization group arguments it is this fact which tells us that the properties of the model should not depend on the distribution we choose for the random fields.

Anderson spin glasses (Edwards and Anderson 1975). Various choices for J have been investigated. A common one is to choose $J_{ij} = \pm 1$ randomly for each bond on the lattice. Another is for the bond strengths to be uncorrelated Gaussian random variables with mean zero, just as our random fields were in the last section.

It is less clear in the case of this spin glass than it was in the case of the random-field model why there are energy barriers in the system—why getting from one state to another may require you to go through intermediate states with much higher energies. However, we can see that there can be states with very different spin configurations which have similar low energies, since we know that we can get from any state of the lattice to any other by flipping one or more clusters of nearest-neighbour spins. Except under very unusual circumstances, it should be possible to find a set of such clusters for which the sum $\sum J_{ij}$ of all the interactions around their surfaces adds up to a number less than or close to zero. In this case, we can get from our low-energy state to another with comparable or lower energy just by flipping this set of clusters. However, if we now try to get from one of these low-energy states to the other using, for example, the single-spin-flip Metropolis algorithm, we get exactly the same problem as we did in the random-field case. Getting from the first state to the second involves flipping in turn each spin by which they differ and it is very likely that one or more of the spins will be unwilling to flip, because the bonds joining it to its neighbours are unusually strong. In effect, there is an energy barrier between the two states and the time-scale for our Monte Carlo simulation to get from one side of the barrier to the other will be exponentially long, which makes the simulation of large, or even moderate-sized spin glass systems prohibitively slow. As we pointed out before, we can always overcome the barriers in the system if we make the temperature high enough. But when kT falls below the height of the largest barriers in the system we lose ergodicity and things become difficult. This is the glass transition which we spoke of in Section 6.1, which occurs around the glass temperature T_g, and we will assume in this chapter that it is in this low-temperature regime that we are interested in simulating the model.

There is one more point which is worth making before we go on to talk about algorithms: both the random-field Ising model and the Edwards–Anderson spin glass exist in real life. They are not just theoretical inventions. There are real magnetic substances which show the critical properties of each of these systems, which makes studying the models worthwhile in its own right, in addition to shedding light on other glassy systems. To readers interested in learning more about glassy spin systems we recommend the book *Spin Glasses* by Fischer and Hertz (1991), which gives a clear account of both theoretical results and experimental data.

6.2   Simulation of glassy systems

Before we get into the details of algorithms for the Monte Carlo simulation
of glassy systems, we need to consider exactly what it is we want to do
with our simulation. Real spin glasses are presumably passing continually
through large numbers of states, each of which appears with its appropriate
Boltzmann weight. If however, as we have argued above, these systems
require exponentially long times to get over the energy barriers posed by the
randomness in their Hamiltonians, then presumably this phenomenon will be
visible in their measured properties. The effect we are talking about is called
ergodicity breaking. As discussed in the introduction to the chapter, the
system fails to sample all states possessing the same energy with the same
probability, because it simply can't get to some of them during the course
of an experiment. There are energy barriers in the system so high that the
system is not likely to cross them in a minute, or a year, or a century. We can
observe the system in our laboratory for as long as we like and still find it in
the same basin of low energy states, never getting beyond the surrounding
energy barriers to sample other portions of its state space.

Thus we should not necessarily count it a failing of a Monte Carlo algorithm if in simulating a glassy system it too gets stuck in an energy basin
and does not sample the whole of the state space. Perhaps this is exactly
what we want it to do in order to mimic the real system. However, there
are a number of problems with this interpretation. The first is that both
the nature of the energy barriers and the regions that a Monte Carlo algorithm will sample depend on the particular dynamics of the algorithm. As
we pointed out in Section 6.1, cluster algorithms for the random-field Ising
model may be able to get over energy barriers which the Metropolis algorithm cannot, for example. Thus the subset of states which get sampled, and
therefore the observed value of any particular quantity (like the magnetization or the specific heat of the model), depend on the particular algorithm
we use. This is clearly unsatisfactory: there is no way the different answers
given by different algorithms can all agree with the results of our laboratory
experiments, and how are we to know which algorithm is right and which
is wrong? It is much better to find some observable property of our model
which is independent of the algorithm we use, so that we are free to choose
any algorithm we think will be efficient, without worrying about whether it
is the "right one".

Another problem concerns the particular values of the random interactions in the Hamiltonian—the **realization** of the randomness, as it is commonly called. The measured properties of any finite system are clearly going
to depend on how we choose these values. For some models this variation
averages out as the size of the system becomes larger, although for some it
does not. (The system is then said to be **non-self-averaging**.) However,

even in the cases where it does, it proves so hard to simulate glassy systems that we rarely get around to systems large enough for this to be a factor.

As well as these problems, there is an additional one that a glassy system is **history dependent**, which means that even for a given algorithm and a given realization of the randomness the simulation can give different results just because we start it in a different spin configuration, or with a different random number seed (see Section 16.1.2). Depending on how we start the simulation, the system can find its way into different energy basins, and so give entirely different measurements of the magnetization or any other parameter.

In order to avoid these various problems, calculations and simulations of glassy systems tend to concentrate on average properties of the systems. In particular we are usually interested in the values of quantities averaged over many different realizations of the randomness, and in thermal averages over the whole of phase space (to get around the problems of ergodicity breaking and history dependence). In some sense this is rather unrealistic. Real glassy systems are single realizations of the randomness which certainly display both ergodicity breaking and history dependence. However, we can cope with this by performing experiments on many different samples of a spin glass (or one large sample, for systems which are self-averaging), or by heating and cooling our samples to coax them into different basins of low energy, and the advantages of having well-defined physical properties which are independent of dynamics and starting conditions (and which therefore allow us to compare simulation and experiment) are great enough to make it worthwhile.[4] As far as the simulation of glassy spin systems is concerned then, what we want to find is an algorithm which can sample over the whole of the state space without getting stuck in the metastable regions formed by local basins of low energy. We will also want to perform our simulation many times with different realizations of the randomness and average over the results. This second step is trivial (although it can be time-consuming), but the first—creation of the algorithm—is a problem which has puzzled physicists for decades.

---

[4]There are some glassy systems for which this is not the case. A good example is the folding of proteins. Although at first it may not be obvious, the dynamics of this problem is glassy in nature. But in proteins the realization of the randomness (which depends on the sequence of amino acids), the starting conditions (which depend on the protein translation mechanism, as well as the possible presence of "helper enzymes" which aid the initial folding of the protein), and the kinetics of the folding process—the "algorithm" which the protein uses to fold—are all very important in determining the conformation of the protein. It would be quite wrong in this case to concentrate on properties which are independent of these factors, since the measurement of interest is the final (folded) state of the protein, rather than any thermally averaged quantity like an internal energy or a specific heat.

6.3 The entropic sampling method

The first algorithm we will study is the **entropic sampling algorithm** invented by Lee (1993). The method is actually charmingly simple, although getting it to work for any particular model can take some effort. The basic idea, as we hinted at the beginning of this chapter, is to sample from a distribution other than the Boltzmann distribution. Equation (2.2), which we repeat here for convenience

$$Q_M = \frac{\sum_{i=1}^{M} Q_{\mu_i} p_{\mu_i}^{-1} e^{-\beta E_{\mu_i}}}{\sum_{j=1}^{M} p_{\mu_j}^{-1} e^{-\beta E_{\mu_j}}} \tag{6.4}$$

tells us that we can in fact sample states from any probability distribution p_μ and still get an estimate Q_M of the observable Q which we are interested in, provided we divide out the sampling distribution and then multiply by the Boltzmann weights $e^{-\beta E_\mu}$. In the algorithms we have seen so far in this book, we chose the probabilities p_μ to be the Boltzmann weights, which meant that everything cancelled out nicely in Equation (6.4) and we were just left taking the average over our measurements Q_{μ_i} to get an estimate of Q. This is a very convenient route to take in most cases but, as we have argued above, it is doomed to failure in the case of glassy systems. So, as an alternative, Lee proposed that we sample from a new distribution defined as follows. Let $\rho(E)\,dE$ be the number of states of our system in the interval between E and $E + dE$. The quantity $\rho(E)$ is the density of states for the system, which we encountered previously in Section 4.5.3. In the entropic sampling method, instead of sampling states with probability proportional to $e^{-\beta E}$, we sample with probability proportional to $[\rho(E)]^{-1}$, the reciprocal of the density of states. In other words, states in ranges of E where there are many states, so that the density of states is high, are sampled with lower probability than those in ranges where there few states. (As discussed below, the big problem with the method is knowing what the density of states actually is for a given energy. However, for the moment let us assume that, by some magical means, we have this information.)

Why do we do it this way? Well, consider now the probability of sampling a state in a given small region of energy in any particular Monte Carlo step. The number of states in our region is just $\rho(E)\,dE$, and we sample each of them with a probability

$$p(E) = \frac{1}{K\rho(E)}, \tag{6.5}$$

where K is an E-independent normalizing constant given by

$$K = \int \frac{dE}{\rho(E)} \tag{6.6}$$

which plays a similar role to the partition function in ordinary importance sampling. (As with the partition function, we will see that we never actually need to calculate K if we are only interested in measuring the magnetization or the internal energy, or something like that.) Then the probability of choosing a sample in this region on this step is just the product

$$\rho(E)\,\mathrm{d}E\,p(E) = \rho(E)\,\mathrm{d}E\frac{1}{K\rho(E)} = \frac{\mathrm{d}E}{K}. \tag{6.7}$$

In other words, we sample energy values with a completely flat distribution. In any given step during a simulation the chances of picking a state with one energy are the same as for any other energy. (Note that this is not the same thing as picking every *state* with equal probability, which would produce more samples in ranges of energy for which there are many states than it would in those for which there are few.)

It is easy to see why this might give good results with glassy systems. The problem with standard Boltzmann importance sampling in glassy systems is that the states at the top of the energy barriers are very unlikely to be sampled and so our chances of passing through those states to the other side of the barrier are slim. With the entropic sampling technique on the other hand, any energy is as likely as any other to be selected, so we cross barriers easily. There is no Boltzmann weight making it hard to get over the barriers.

Implementation of the entropic sampling method is straightforward. We simply employ the condition of detailed balance, Equation (2.14), except that now, instead of putting in the Boltzmann weight for p_μ, we put in our new probability, Equation (6.5), thus:

$$\frac{P(\mu \to \nu)}{P(\nu \to \mu)} = \frac{p_\nu}{p_\mu} = \frac{\rho(E_\mu)}{\rho(E_\nu)}. \tag{6.8}$$

All the proofs of Chapter 2 concerning equilibrium distributions and the condition of detailed balance apply just as well to this choice as to the more conventional one. Thus, all we have to do to ensure states are chosen from the distribution in Equation (6.5) is choose transition probabilities for states to satisfy (6.8). Shortly we will give an explicit example of how this can be done for the random-field Ising model.

6.3.1   Making measurements

Making measurements using the entropic sampling method is a straightforward application of Equation (6.4). Plugging Equation (6.5) into it gives us

$$Q_M = \frac{\sum_{i=1}^{M} Q_{\mu_i}\rho(E_{\mu_i})\,\mathrm{e}^{-\beta E_{\mu_i}}}{\sum_{i=1}^{M} \rho(E_{\mu_i})\,\mathrm{e}^{-\beta E_{\mu_i}}}. \tag{6.9}$$

There's bad news and good news about this equation. The bad news is that for a given number of samples it doesn't give as accurate an answer as normal importance sampling using Boltzmann weights. The reason is that, since the entropic sampling method samples states of all energies, it will certainly sample some states whose energies are high enough that the factor $e^{-\beta E_{\mu_i}}$ in this equation is tiny. These states will make a negligible contribution to Q_M and therefore we are, in a sense, wasting CPU time putting them in. On the other hand, these states were the very reason we resorted to entropic sampling in the first place; they, or some of them at least, are precisely the high-energy states at the top of the energy barriers which we need to be able to reach in order to cross from one basin to another in the state space. Thus there is really no escaping them—we have them there for a good reason, even though they don't contribute any accuracy to our final estimate of the observable Q. There can also be states for which $\rho(E)$ is very small, and these also make a small contribution to Q_M. Whether this is a problem or not depends on the form of the density of states.

And how about the good news? Well, to understand this take another look at Equations (6.5) and (6.9). Notice that the sampling probability p_μ is independent of temperature. This means that the sample of states which the algorithm chooses does not depend on what temperature we are interested in. The temperature only enters the calculation in Equation (6.9), where we reweight our samples to get an answer. However, we can apply this equation after our simulation is finished and all the samples have been taken. In other words, simply by entering different values of β into Equation (6.9), we can calculate the value of Q for any temperature we please from the same set of samples. We only have to run the simulation once, and we get data for the entire temperature range from zero to infinity. This makes up for the decreased accuracy of each estimate, because we are usually willing to perform a longer simulation to recover the lost accuracy if we know that we will only have to perform one such simulation, rather than one for every temperature we are interested in.

6.3.2   Internal energy and specific heat

The internal energy and specific heat are special cases in entropic sampling calculations, because they are functions only of the energy E. This means that, given the density of states $\rho(E)$, we can calculate the two of them from

$$U = \langle E \rangle = \frac{\int E\rho(E)\,e^{-\beta E}\,dE}{\int \rho(E)\,e^{-\beta E}\,dE}, \tag{6.10}$$

$$\frac{C}{\beta^2} = \langle E^2 \rangle - \langle E \rangle^2 = \frac{\int E^2\rho(E)\,e^{-\beta E}\,dE}{\int \rho(E)\,e^{-\beta E}\,dE} - U^2, \tag{6.11}$$

where we have used Equation (1.19) (with $k = 1$) to calculate the specific heat. Notice that these equations don't involve the samples from our simulations—they are exact expressions for these quantities. In other words, we don't even need to perform the simulation to measure U and C! This should raise a few eyebrows around the table. Clearly something is suspicious here. You never get something for nothing. The point is that $\rho(E)$ contains enough information to calculate the internal energy and the specific heat, and so obviously it is a complicated thing to calculate. Except for the most primitive systems, it's not given by some simple expression which we can write down easily. We have to perform a whole Monte Carlo simulation, or more often several, just in order to calculate $\rho(E)$. And this is the real problem with the entropic sampling method. In order to use the method we need to know what the density of states is. But if we knew that, we wouldn't even have to perform the simulation, as the equations above show. So, how do we proceed?

6.3.3 Implementing the entropic sampling method

The basic idea in implementing the entropic sampling method is to calculate the density of states at the same time as we perform the simulation itself, using an iterative method. We start off with some crude guess at the density of states—uniform in E, say. We perform a simulation using this guess and make a histogram of the energies of the sampled states. If we had the density of states correct, this histogram would be entirely flat (see Equation (6.7)), but in practice of course it won't be, because our density of states was a poor approximation to the true answer. Based on the histogram however, we can make a better guess at the density of states, and then we repeat the whole simulation. We keep on doing this until we have a nice flat histogram. When we are all done we are left with both a good estimate of the density of states of the system and a set of samples—the ones from the last iteration—which were sampled using that estimate and which we can use in Equation (6.9) to estimate any quantities of interest. In detail, here is how the calculation goes.

Let us call our initial guess for the density of states $\rho_1(E)$. This can be a fairly poor estimate of the true $\rho(E)$. It doesn't matter. We will improve it as we run the algorithm. Now we use this guess to perform an entropic sampling calculation by feeding it into the detailed balance condition, Equation (6.8). This gives us a set of states which are sampled with probability $p_\mu = [K\rho_1(E_\mu)]^{-1}$, where K is an unknown constant. We make a histogram $h_1(E)$ of the energies of these states. The number of states in a certain bin of that histogram is proportional to the number of states of the system which fall in the region of energy it covers multiplied by the probability p_μ with which we sampled each of them. In other words

$$h_1(E) = \frac{\rho(E)}{\rho_1(E)}, \tag{6.12}$$

to within a multiplicative constant. Rearranging for $\rho(E)$, we get

$$\rho(E) = \rho_1(E)h_1(E). \tag{6.13}$$

There are a number of reasons why in practice this expression gives a poor result for the density of states (see below), but in most cases it is at least a better estimate than the one we started off with. So we can use this new estimate—let's call it $\rho_2(E)$—as the starting point for another simulation, and repeat the process over and over until we are satisfied with the results. In general, the density of states at the $(n+1)^{\text{th}}$ step is related to that at the nth by

$$\rho_{n+1}(E) = \rho_n(E)h_n(E). \tag{6.14}$$

Simple though it appears in theory, there are several practical problems with this procedure. First there is the question of the dynamic range of $\rho(E)$. For most systems one finds that the density of states has a very large dynamic range; its value can vary over hundreds of orders of magnitude. Computers are poor at dealing with numbers like this because of numerical inaccuracies, so almost always one deals instead with the logarithm of the density of states. Defining the quantity $S(E)$ to be

$$S(E) = \log \rho(E), \tag{6.15}$$

Equation (6.14) becomes

$$S_{n+1}(E) = S_n(E) + \log h_n(E). \tag{6.16}$$

$S(E)$ is known as the **microcanonical entropy**. It is the log of the number of states which are accessible to the system if it is restricted to states with energy in some small interval close to E. In fact, when the entropic sampling algorithm was originally presented by Lee, it was entirely expressed in terms of this entropy, hence the name "entropic sampling". In terms of $S(E)$, the probability p_μ of sampling a particular state μ is proportional to $e^{-S(E_\mu)}$. Although this obscures the connection to the density of states somewhat, it does make clear the analogy between this method and ordinary importance sampling—here the entropy is playing the role which energy plays in the previous algorithms we have considered.

Although it solves our problems with manipulating numbers with a large dynamic range, Equation (6.16) also introduces a small problem. Given that we take a finite number of samples in our simulation, it is possible for one or more of the bins in the histogram $h(E)$ to contain no samples, which makes the term $\log h(E)$ infinite. A simple way around this is simply to leave the value of $S(E)$ untouched in this case. In other words, the actual algorithm is

$$S_{n+1}(E) = \begin{cases} S_n(E) + \log h_n(E) & \text{if } h_n(E) > 0 \\ S_n(E) & \text{otherwise.} \end{cases} \qquad (6.17)$$

A more serious problem with the entropic sampling method is its rate of convergence, which in many cases is extremely poor indeed. It is clear that if, as we have claimed, the density of states does have a very large dynamic range, and if we start off by approximating it with a flat distribution, then there are inevitably going to be parts of the function which are overestimated by many orders of magnitude, relative to other parts. Bearing in mind that the magnitude of the function $h(E)$ varies between its most and least populated bins by at most a factor of the number of samples taken in one iteration of the calculation, we can see that it could take many such iterations to arrive at the true functional form for $\rho(E)$. There are sometimes a few tricks one can play to get around this problem, but by and large the method simply is slow to converge, and this is its primary shortcoming.

6.3.4   An example: the random-field Ising model

So far we have discussed the entropic sampling method in general terms. In this section we will apply it to a particular problem, the random-field Ising model of Section 6.1.1. We will see that the algorithm does indeed overcome the problems of lack of ergodicity in the model, crossing energy barriers easily, but that it does so at the expense of slow convergence, which limits its applicability to rather small lattices.

Each step of our entropic sampling algorithm for the random-field Ising model involves first selecting a new state of the lattice ν given the state μ we are in at the moment, with some selection probability $g(\mu \to \nu)$. We then accept or reject that state with an acceptance ratio $A(\mu \to \nu)$ such that the condition of detailed balance is obeyed thus:

$$\frac{P(\mu \to \nu)}{P(\nu \to \mu)} = \frac{g(\mu \to \nu)A(\mu \to \nu)}{g(\nu \to \mu)A(\nu \to \mu)} = \frac{\rho(E_\mu)}{\rho(E_\nu)}. \qquad (6.18)$$

We note that states μ and ν with similar energies are likely to have similar values of $\rho(E)$, which makes it possible for both of the acceptance ratios in this equation to be close to one. In other words, it is a good idea to choose moves which take us to states with energies close to the energy of the current state. As with the algorithms of Chapter 3, a simple way to achieve this is to use single-spin-flip dynamics. (Other choices are possible, but this one will do fine for now.) In other words, our new state ν is generated by taking the present state μ and flipping just one spin, chosen at random. Each of the N possible such states are generated with equal probability $1/N$, and all

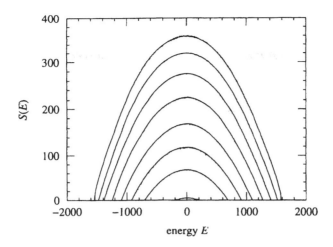

FIGURE 6.2   The logarithm $S(E)$ of the density of states of the three-dimensional random-field Ising model on an $8 \times 8 \times 8$ cubic lattice with random fields $\pm h$ with $h = 2.0$. The curves show the best estimate of $S(E)$ at intervals of about 20 iterations, with the top curve being the final converged result. The entire calculation took about 150 iterations.

the other states of the lattice are generated with probability zero. Then the selection probabilities $g(\mu \to \nu)$ and $g(\nu \to \mu)$ in Equation (6.18) drop out and we are left with

$$\frac{A(\mu \to \nu)}{A(\nu \to \mu)} = \frac{\rho(E_\mu)}{\rho(E_\nu)}, \tag{6.19}$$

which can be most efficiently satisfied by choosing

$$A(\mu \to \nu) = \begin{cases} \rho(E_\mu)/\rho(E_\nu) & \text{if } \rho(E_\mu) < \rho(E_\nu) \\ 1 & \text{otherwise.} \end{cases} \tag{6.20}$$

This then defines the algorithm. We take our first estimate of the density of states, start flipping spins and accepting the flips according to this acceptance ratio. When we have accumulated enough samples to make a reasonable histogram we stop, recalculate $\rho(E)$ from Equation (6.14) (or alternatively Equation (6.17)) and then repeat the whole process. In practice a good criterion for telling when to finish a particular iteration is to count the average number of samples in each non-empty bin. (It is normal for many of them to be empty, especially in the early stages of the calculation.) When this number reaches a certain predetermined level, the iteration ends and you recalculate $\rho(E)$. In practice, quite small numbers of samples give perfectly adequate results—ten or so per bin is a reasonable figure. The reason is that it is more important to get through a large number of iter-

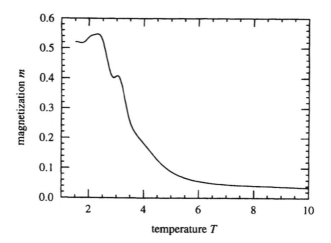

FIGURE 6.3   The magnetization m of our $8 \times 8 \times 8$ random-field Ising model with random fields $\pm h$ with $h = 2.0$ calculated using the entropic sampling method.

ations, so as to find a decent approximation to the density of states, than it is to get great accuracy on any particular iteration. Usually, when you have converged to a stable estimate of $\rho(E)$, you should do one last, longer simulation using that estimate of $\rho(E)$ to get a large set of samples which can be used for calculating the values of observable quantities.

There are a variety of tricks one can use to speed up the convergence of the algorithm, such as using bins of different sizes in different portions of the energy scale, or starting off with larger bins and making them smaller as the calculation progresses. Here, however, we apply the algorithm in its simplest form, just as we described it above. In Figure 6.2 we show a series of curves of the logarithm $S(E)$ of the density of states of a random-field Ising model on an $8 \times 8 \times 8$ cubic lattice taken at intervals of about 20 iterations. We can see that on later iterations the algorithm samples over a wider range of energies, as our approximation to the density of states becomes better, until, after about 150 iterations, the curve stops changing. At this point we stop iterating and sample one more, larger set of states for use in making estimates of other properties of the model. In Figure 6.3 we show our estimate of the magnetization of the system using Equation (6.9). Although we have not done it here, we could now use the bootstrap resampling technique of Section 3.4.3 to calculate the errors on the calculation.

This however is not quite the end of the calculation. As we pointed out in Section 6.2, we still have to average our calculation over the randomness, which means performing the whole simulation many times over with different random fields. Note that this averaging will also contribute an error to our value for the magnetization (or any other quantity). Ideally we should divide the available CPU time between our many simulations in such a way that the statistical errors introduced by each individual run are of the same order of magnitude as those introduced by averaging over the randomness. It would be a waste of CPU time to perform a simulation long enough to reduce the sampling error to a level where it was swamped by variations due to the randomness. And conversely it would be a waste to perform so many runs that the average over the realizations of the randomness was known much more accurately than the statistical error in the individual measurements.

6.4 Simulated tempering

A different approach to the simulation of glassy systems is the **simulated tempering method** invented by Marinari and Parisi (1992). We will look in particular at the variation they refer to as **parallel tempering**, which is the simplest and most general form of simulated tempering.[5] This method is, like the entropic sampling scheme of Section 6.3, quite simple in concept, and unlike entropic sampling it does not require us to evaluate the density of states or any other quantity iteratively. On the other hand, it does require us to perform more than one simulation to get an answer—the simulations are performed concurrently, rather than one after another as with entropic sampling—and it only gives us the properties of the system at a finite number of temperatures, rather than at every temperature. The method does have the advantage that it can be implemented in a very straightforward fashion on a parallel computer. On balance, like so many things, it's horses for courses. You have to decide which is the best algorithm for the problem in hand. However, simulated tempering has had a lot of success in the simulation of spin glasses. At the end of this section we give an example of its application to an Ising spin glass.

6.4.1 The method

The basic idea behind the simulated tempering method is to perform several different simulations simultaneously on the same system (with the same realization of the randomness), but at different temperatures. Every so often, one swaps the states of the system in two of the simulations with a certain probability which is chosen so that the states of each system still follow the Boltzmann distribution at the appropriate temperature. By swapping the states, we mean that one sets the values of the spins (or other variables)

---

[5]This in turn is a variation on the general method known as the **multicanonical ensemble**, which is itself a subset of **umbrella sampling**. See, for example, Berg (1993).

in each of the two simulations to those in the other. The result is that the higher-temperature simulation helps the lower-temperature one across the energy barriers in the system. The higher-temperature one crosses barriers with relative ease, and when it gets to the other side, we swap the states of the two models, thereby "carrying" the lower-temperature model across a barrier which it otherwise would not have been able to cross. We now describe the algorithm in detail, for the simplest case in which we do just two simulations of the system at the same time.

Consider a glassy system, the random-field Ising model of the last section, for example, or the Ising spin glass described by the Hamiltonian (6.3). The system has a temperature—T_c in the case of the random-field model, or the glass temperature T_g in the spin glass case—below which glassy behaviour sets in and simulation becomes difficult. If we are interested in measuring properties of the system at some temperature T_{low} below this transition, we can do so by performing, simultaneously, two simulations, one at T_{low} and another at a higher temperature T_{high}, which is above the transition. By definition, the higher-temperature simulation will not show ergodicity breaking, and we assume therefore that it can sample uniformly over the state space of the model with reasonable ease. The lower-temperature one on the other hand will probably get stuck in some basin of low energies and be unable to get out. Our algorithm then is as follows. At each time step we do one of two things. On the majority of time steps, we simply do one step in the simulation of each of the two systems. For example, with our spin models we could do one single-spin-flip Metropolis Monte Carlo step, exactly as in Chapter 3. However, every so often (exactly how often is discussed below) instead of doing this we calculate the difference in energies of the current states of the two simulations, $\Delta E = E_{\text{high}} - E_{\text{low}}$, and we swap the values of every spin in the two with an acceptance probability

$$A = \begin{cases} e^{-(\beta_{\text{low}} - \beta_{\text{high}})\Delta E} & \text{if } \Delta E > 0 \\ 1 & \text{otherwise.} \end{cases} \tag{6.21}$$

The first step in understanding why this algorithm works is to prove that it satisfies ergodicity and detailed balance.

Proving ergodicity is straightforward. We already know that if we simulate a single system using the Metropolis algorithm we achieve ergodicity. If we therefore follow the state of one of our two systems until a swap occurs, and then follow the other and so on, it is clear that we can reach any state in a finite number of moves. However, if one system can reach any state in a finite number of moves, then so by symmetry can the other, and we are guaranteed ergodicity. The same argument also applies if we use an algorithm other than the Metropolis algorithm for the individual steps of the two simulations.

Detailed balance is a little more tricky. To prove it, let us consider the joint probability $p_{\mu\nu}$ that the low-temperature system is in state μ and the high-temperature one in state ν. We want the equilibrium value of this probability to reflect the desired Boltzmann distribution of states in both systems. In other words, we would like to have *independent*

$$p_{\mu\nu} = \frac{1}{Z_{\text{low}} Z_{\text{high}}} e^{-\beta_{\text{low}} E_\mu} e^{-\beta_{\text{high}} E_\nu}, \tag{6.22}$$

where Z_{low} and Z_{high} are the partition functions of the two systems. The condition of detailed balance, Equation (2.12), tells us that we can achieve this if we ensure that the transition probabilities $P(\mu\nu \to \mu'\nu')$ and $P(\mu'\nu' \to \mu\nu)$ satisfy

$$\frac{P(\mu\nu \to \mu'\nu')}{P(\mu'\nu' \to \mu\nu)} = \frac{p_{\mu'\nu'}}{p_{\mu\nu}} = \frac{e^{-\beta_{\text{low}} E_{\mu'}} e^{-\beta_{\text{high}} E_{\nu'}}}{e^{-\beta_{\text{low}} E_\mu} e^{-\beta_{\text{high}} E_\nu}}. \tag{6.23}$$

We have three kinds of moves in our algorithm—normal Monte Carlo moves on either of the two systems, and the swap moves—and we need to prove that Equation (6.23) is satisfied for each of them. The normal Monte Carlo moves are easy. Consider for example a move in which the low-temperature system goes from state μ to μ', while the high-temperature one stays in state ν. Setting $\nu' = \nu$ in Equation (6.23), we get

$$\frac{P(\mu\nu \to \mu'\nu)}{P(\mu'\nu \to \mu\nu)} = \frac{e^{-\beta_{\text{low}} E_{\mu'}}}{e^{-\beta_{\text{low}} E_\mu}}, \tag{6.24}$$

which is just the normal detailed balance condition for a single simulation. Thus any Monte Carlo move which satisfies this condition, such as the single-spin-flip Metropolis one, will correctly preserve detailed balance. The proof that ordinary Monte Carlo steps on the high-temperature system also satisfy Equation (6.23) is identical.

The swap moves are only a little more complicated. In these moves we are swapping the states of the two systems, so $\mu' = \nu$ and $\nu' = \mu$. Thus our equation becomes

$$\frac{P(\mu\nu \to \nu\mu)}{P(\nu\mu \to \mu\nu)} = \frac{e^{-\beta_{\text{low}} E_\nu} e^{-\beta_{\text{high}} E_\mu}}{e^{-\beta_{\text{low}} E_\mu} e^{-\beta_{\text{high}} E_\nu}} = \frac{e^{-\beta_{\text{low}} \Delta E}}{e^{-\beta_{\text{high}} \Delta E}}, \tag{6.25}$$

where $\Delta E = E_\nu - E_\mu$. The transition rate $P(\mu\nu \to \nu\mu)$ is, as discussed in Section 2.3, just the product of the selection probability $g(\mu\nu \to \nu\mu)$ for the swap move and the acceptance ratio $A(\mu\nu \to \nu\mu)$ for that move. If we make the selection probability a constant, for example by attempting a swap at regular intervals throughout the simulation, then the selection probabilities cancel out of the equation and we are left with

$$\frac{A(\mu\nu \to \nu\mu)}{A(\nu\mu \to \mu\nu)} = \frac{e^{-\beta_{\text{low}} \Delta E}}{e^{-\beta_{\text{high}} \Delta E}}. \tag{6.26}$$

It takes only a moment to verify that the acceptance ratio given in Equation (6.21) satisfies this condition.

Thus each of the three types of move in the algorithm satisfies detailed balance, and so, therefore, does the entire algorithm. All the proofs of Chapter 2 then apply to the equilibration of the algorithm and at equilibrium we know that any joint state $\mu\nu$ of the combined systems will appear with the probability given in Equation (6.22). Thus, each state in each of the simulations appears with exactly its correct Boltzmann weight, so we can make measurements in tempering simulations in the normal fashion—we sample the observable of interest at intervals throughout the simulation (after waiting a suitable time for equilibration) and take the mean of those samples.

The reason why the method gets us over energy barriers is intuitively clear. Since the high-temperature simulation is performed at a temperature above the glass transition of the system, it shows no ergodicity breaking, and can find its way about the state space of the system on a time-scale similar to that of a simulation of a normal, non-glassy system. If we start off both our simulations in a deep energy basin then the low-temperature one will remain stuck there, while the high-temperature one moves over the surrounding energy barriers and into other regions of state space. Now suppose we attempt to swap the states of the two simulations. Equation (6.21) says that if the swap would increase the energy of the low-temperature simulation by a great deal then it is unlikely to be accepted. On the other hand, if, as it will from time to time, the high-temperature system finds its way into a region of low energy—another basin—then the move will quite likely be accepted, and we will perform the swap. Thus the low-temperature system is transported in one move to another energy basin, and the high-temperature one finds itself back in the basin that it started in. Now the process repeats, and over a long simulation the low-temperature simulation is moved repeatedly to new energy basins. Thus the simulated tempering algorithm effectively overcomes the problems of barrier crossing which make simulation of glassy systems so hard and allows us to sample a significant fraction of our state space, while still sampling with the correct Boltzmann weights for a temperature below the glass transition.

One question which remains to be settled is how often we should perform the swap moves. It is clearly not a good idea to perform them too often: if we swap the states of our two simulations, and then immediately swap them back again, it is very likely that the low-temperature one will be dumped back in the basin that it only just escaped from, which defeats the point of the exercise. On the other hand, we want to perform swaps as often as is practical, since otherwise we waste time doing the high-temperature simulation and not making use of the results. (The only reason we perform the high-temperature simulation is to help us in performing the low-temperature one; if we were actually interested in simulating the model above the glass temperature then we could just use a normal Monte Carlo algorithm—there are no particular problems with simulations above T_g.) Basically, we want the high-temperature simulation to move away from the basin it is in immediately following a swap, cross a barrier and find another basin. One criterion for choosing how long we should wait before the next swap is therefore to follow the energy of the high-temperature simulation until it has passed through values significantly different from the initial value. But we already have a way of determining the time-scale for this to happen; we calculate the **energy autocorrelation function**:

$$
\chi_E(t) = \frac{1}{t_{\max} - t} \sum_{t'=0}^{t_{\max}-t} E(t')E(t' + t)
$$

$$
- \frac{1}{t_{\max} - t} \sum_{t'=0}^{t_{\max}-t} E(t') \times \frac{1}{t_{\max} - t} \sum_{t'=0}^{t_{\max}-t} E(t' + t). \quad (6.27)
$$

(See Equation (3.21).) As we showed in Section 3.3, such an autocorrelation function measures the extent to which the values of the energy are correlated at times t apart. And as we showed in Section 3.3.2, all such autocorrelations have the same long-time behaviour, decaying as $\exp(-t/\tau)$, where τ is the correlation time.[6] We can extract τ from a semi-logarithmic plot of χ_E against t, just as we did in Figure 3.6, or calculate the integrated correlation time using Equation (3.20). Our tempering algorithm should then attempt one swap move approximately every τ sweeps of the lattice, and the rest of the time perform normal Monte Carlo moves on each of the two simulations separately.[7]

One further point, which is important when it comes to implementing the simulated tempering algorithm, is that it is usually rather inefficient to actually interchange the values of all the spins in the two systems when a swap move is accepted. In a system of a million spins, such an operation could use a significant amount of CPU time. So instead what one normally does is to interchange the temperatures at which the simulations are carried out, leaving the spin states the same. This of course is entirely equivalent to

---

[6]At shorter times, autocorrelation functions of different quantities behave differently because of different values of the constants b_i in Equation (3.35). Thus if, for example, you calculate the correlation time using the integral formula (3.20), you may get a different result using the energy autocorrelation from what you would get using, say, the magnetization. For this reason, some people define different correlation times τ_E, τ_M, and so on for different quantities. Strictly however, it should not matter which you use if you do the calculation correctly.

[7]In fact, if the mean acceptance ratio $\bar{A}$ for the swap moves is low, we may wish to attempt them more often than this. Strictly, we should attempt swaps every $\bar{A}\tau$ sweeps, in order that the rate at which they are actually performed should be equal to the correlation time.

interchanging the spin states and leaving the temperatures unchanged, and is much faster.[8]

6.4.2 Variations

A number of variations are possible on the basic scheme described above. One common one is to perform the basic Metropolis or other Monte Carlo steps on the two systems at different rates. We have to wait about one correlation time of the high-temperature system between swap moves, but there is no reason why it should be optimal to perform the same number of moves on the low-temperature system during this time. In particular, the basins which the low-temperature system finds itself in may be very small, which means that it does not take very many moves to sample them thoroughly, and performing any more is a waste of computer resources. In this case it is desirable to perform a smaller number of moves in the low-temperature system than in the high. For example we might perform one Monte Carlo step on the low-temperature system for every two in the high-temperature one. An extreme example of this problem is when the low-temperature system is at $T = 0$, or close to it. In this case, the low-temperature system will simply sink to the lowest energy state in the local basin and then stay there, unmoving, until a swap move occurs and moves it to another basin. Clearly, once the local minimum of energy has been found there is no point in expending any more time on the low-temperature system, since we know that all moves are going to be rejected. So we can save time by simply forgetting the low-temperature system and simulating the high-temperature one until one correlation time has passed, and then attempting a swap. This is not a very likely scenario. We are not often interested in the behaviour of glassy systems at absolute zero. But it serves to illustrate the point.

Another common variation on the algorithm is to simulate at more than just two temperatures. There is in many cases a distinct advantage to adding extra simulations at temperatures intermediate between the two discussed so far. The reason is that, as pointed out in Section 1.2.1, the range of energies sampled by a system at a given temperature is usually quite small, and decreases with increasing system size as $1/\sqrt{N}$. However, as Equation (6.21) indicates, for the swap moves in the simulated tempering algorithm to work, the two states to be swapped must be of comparable energies—the state of the high-temperature system must be a viable state at the lower temperature in order to be accepted. In effect, this means that the ranges of energy sampled by the simulations at the two different temperatures must overlap. As system sizes get larger and the ranges get smaller, this becomes less and

---

[8] An alternative in programming languages such as Pascal and C which provide pointers is to interchange the values of pointers to the spin arrays, leaving the arrays themselves unchanged.

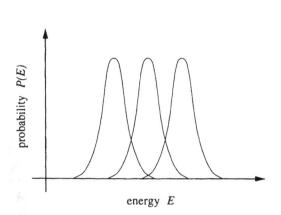

FIGURE 6.4 The probability of sampling states of particular energy for simulations at three different temperatures. The areas of overlap between the three distributions represent the states in which a swap is likely to be accepted. As the figure shows, including a simulation at a temperature intermediate between two others can greatly increase the probability of swaps taking place.

less likely, unless the temperatures of the two are very close together. We can get around this problem by adding one or more extra simulations at temperatures in between the first two. As Figure 6.4 shows, a new middle simulation can sample states over a range of energies which permits it to overlap with the ranges of both the other simulations and so to swap states with them easily. The middle simulation (or simulations) effectively acts as a "conveyor" of states between the simulations at the highest and lowest temperatures. Generalization of the algorithm to more than two simulations is very straightforward. Each simulation attempts to swap states with the one at the next highest temperature at regular intervals determined by the correlation time of the hotter of the two. The acceptance ratios are exactly as in Equation (6.21). Of course, the more simulations you do at once, the more computer time the method demands, which again places limits on the size of the system you can study. However, the method correctly samples the Boltzmann distribution at all of the temperatures simulated, and so gives results at as many temperatures as there are simulations, so the extra computer time is by no means wasted. It does not give results for the entire continuum of temperatures in the range simulated, as the entropic sampling method does. However, it is possible to interpolate between temperatures using the "multiple histogram method" described in Section 8.2, which gives results of accuracy comparable to entropic sampling.[9]

Another slant on the method is to implement it on a parallel computer. The algorithm parallelizes very nicely, since the individual simulations can

---

[9] In fact, as we will see in Section 8.2.1, the histogram interpolation method gives accurate interpolation only between simulations which sample overlapping sets of states. Since this is also precisely the requirement for the simulated tempering method, the histogram method almost always works well with results from simulated tempering calculations.

be done on separate processors, and every so often they make a decision about whether a swap is in order. If it is, then rather than communicating the values of all the spin variables on the two lattices to each other, the two processors simply swap the temperatures at which they are performing their simulations, so that the previously high-temperature lattice becomes the new low-temperature one, and *vice versa.* Even on a computer with slow inter-processor communications, this algorithm can show speedups close to linear in the number of processors used. We discuss parallel Monte Carlo simulation in more detail in Chapter 14.

Hukushima and Nemoto (1996) have applied the parallel tempering technique to the three-dimensional Edwards–Anderson Ising spin glass described in Section 6.1.2, with nearest-neighbour spin interactions chosen to have strengths $J_{ij} = \pm 1$ randomly. In their calculation they used a 2 : 1 ratio of single-spin update moves and exchange moves in which the temperatures of adjacent simulations are swapped (although their measured values for the correlation times of the simulation suggest that a higher ratio might have been more efficient). They also performed Monte Carlo steps at each temperature at the same rate. The entire calculation consisted of systems at 32 different temperatures which were simulated simultaneously using multispin coding techniques similar to those discussed in Chapter 15.[10] The run was repeated for ten different realizations of the randomness and for lattices of up to $16 \times 16 \times 16$ spins. The calculation appears to work very well. The acceptance ratio of the individual single-spin Monte Carlo moves is measured to be in the vicinity of 50% or higher, depending on the temperature and the system size. The autocorrelation functions for the simulations below the glass transition are found to fall off roughly speaking exponentially, which is unusual for glassy systems; usually the autocorrelations fall off slower than this. This exponential fall-off is a good sign for the rapid equilibration of the algorithm.

In Figure 6.5 we reproduce Hukushima and Nemoto's measurements of the correlation time as a function of temperature. The transition to the glassy phase is clearly visible in this figure as a sharp increase in the value of τ with falling temperature. However, and this is the crucial point, the correlation time does not continue to increase sharply below this transition. It goes up a little further with falling T, but not enough to make simulations difficult at lower temperatures. Although Hukushima and Nemoto did not extend their calculation to experimentally measurable quantities such as the specific heat of the spin glass, these results do indicate that the parallel

---

[10]This explains why Monte Carlo moves were carried out on all systems at the same rate, since the multispin coding method requires it. It is possible that an algorithm which updated different systems at different rates could make more efficient use of CPU time, but the speed advantage of using multispin coding is a significant one, so one should not assume this.

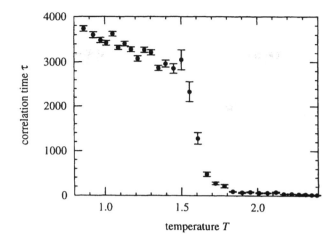

FIGURE 6.5 The correlation time of the parallel tempering algorithm applied to the Edwards–Anderson Ising spin glass on a $12 \times 12 \times 12$ cubic lattice. After Hukushima and Nemoto (1996).

tempering method is successful at overcoming the problems of Monte Carlo simulation in these systems.

Problems

6.1 Consider a simple system with only three states, such that one state forms an energy barrier between the other two thus:

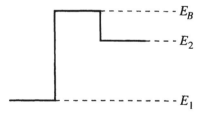

Imagine performing a Monte Carlo simulation of this system, starting in state 1 with energy E_1. If the system can only reach state 2 by passing through the intermediate barrier state, show that it takes at least $e^{-\beta B_1}$ Monte Carlo steps to make a transition to state 2, where $B_1 = E_B - E_1$ is the height of the energy barrier for the transition $1 \rightarrow 2$.

6.2 In Section 6.1.2 we described a version of the Edwards–Anderson spin glass in which adjacent pairs $\langle ij \rangle$ of Ising spins interact with strengths J_{ij} which are randomly ± 1. One particular choice of interaction constants which fits this description, proposed by Mattis (1976), is to let $J_{ij} = \xi_i \xi_j$, where $\xi_i = \pm 1$ are random constants defined on the sites of the lattice. Show that this choice of J_{ij} does not give a glassy model. Show further, however, that if we just choose $J_{ij} = \pm 1$ at random, the chances of coincidentally making the Mattis choice go down exponentially with increasing system size.

6.3 Calculate the density of states for a one-dimensional Ising model with N sites and periodic boundary conditions. (Hint: you may want to refer to the solution of the model in Problem 1.4.) Use this to calculate the appropriate acceptance ratio for a single-spin-flip entropic sampling algorithm for the model. (If you feel like it, you could write a program to implement this algorithm and check that it does indeed sample all energies with equal probability.)

6.4 In Section 6.4.2 we explained that in order for the parallel tempering algorithm to work there needs to be significant overlap of the energy ranges sampled by the different simulations. How does the number of simulations necessary to satisfy this requirement vary with the size of the system being simulated?

6.5 Write a program to apply the simulated tempering algorithm to the random-field Ising model in three dimensions. (Hint: the simplest approach is probably just to use a single-spin-flip Metropolis algorithm for the individual simulations. Try the calculation on a relatively small system at first—say $4 \times 4 \times 4$—so that you don't have to perform too many simulations at once.)

7

Ice models

Most of the models we have studied in the preceding chapters of this book have been variations of the Ising model. Although Ising models are the most important and best-studied class of models in statistical mechanics, it is useful for us to study other classes as well, since there are many concepts in the design of Monte Carlo algorithms which only arise when we look at other models. In this chapter we introduce the so-called ice models, which, as the name suggests, model some of the thermodynamic properties of ice, though they do so only in a rather primitive fashion. Ice models are similar to Ising models in that the microscopic variables making up the model are discrete quantities having just two possible values, positioned on a lattice. However, unlike the models of the previous chapters, these variables are now located on the *bonds* of the lattice, rather than the sites, and, as we shall see, this, in combination with some other features of ice models, makes for some interesting challenges in the design of Monte Carlo algorithms.

7.1 Real ice and ice models

Ice, the solid form of H_2O, is a very common substance in our surroundings and it is therefore not surprising that a large amount of research has been devoted to understanding its behaviour. Ice has occupied a particularly prominent position in the literature on computer simulation because its building blocks, water molecules, are small enough that, even in the early days of computers, they could already be modelled quite accurately using numerical methods. Especially at the end of the 1960s and the beginning of the 1970s, interest in ice research was strong, and many conferences and workshops have been devoted purely to the study of ice.

By the end of the nineteenth century, it was well known that snowflakes often show a hexagonal symmetry, a fact which hinted at the underlying

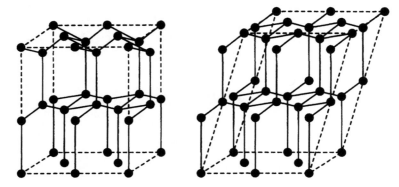

FIGURE 7.1   The structure of the oxygen atoms and the hydrogen
bonds between them in hexagonal ice I_h (left) and cubic ice I_c (right).

structure of ice, and in the early years of the twentieth century Bragg and
others started experimenting with X-ray diffraction to try to determine this
structure. The crucial breakthrough was made by Barnes (1929), who was
the first to map the structure now known as **hexagonal ice** or I_h, which
is the normal form of ice at atmospheric pressure. The arrangement of the
oxygen atoms in hexagonal ice is shown on the left-hand side of Figure 7.1.
Since X-rays are scattered primarily by the electron distribution in a crystal,
the experimental signal measured in scattering studies is mostly due to the
oxygen atoms, and it is difficult to detect the small contribution from the
electron cloud around the hydrogen atoms (protons). However, since the X-
ray results show no evidence of any superlattice lines, we can conclude that
the protons are not arranged in a unit cell which is larger than that of the
oxygens. Given that the unit cell of the oxygens contains only four oxygen
atoms, this does not leave room for a very large number of different proton
arrangements, and Barnes suggested the most probable arrangement to be
one in which each proton is located exactly half-way along the line joining
the centres of two adjacent oxygens.

Barnes' conjecture has not stood the test of time however. Later and
more sensitive scattering experiments indicated clearly that protons are not
shared equally between two oxygen atoms, but that each proton stays close to
one oxygen atom only. Furthermore, two and only two protons are associated
with each oxygen, so that the integrity of the water molecules as structural
units is preserved. Building on these experimental results, Bernal and Fowler
in 1933 proposed a number of alternative proton arrangements. They were
however unable to find any periodic arrangement which had a repeat size as
small as the measured unit cell of the oxygen structure; all their proposed
structures had a larger unit cell, in direct opposition to earlier experimental findings. This apparent contradiction ultimately led Bernal, Fowler and

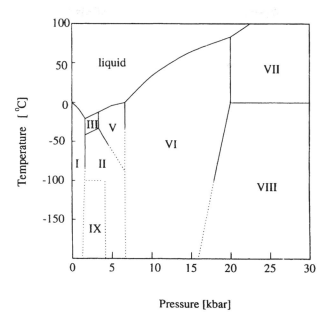

Pressure [kbar]

FIGURE 7.2   The phase diagram of ice. Dotted lines indicate first-order phase boundaries dividing pairs of phases each of which can exist metastably in the region of stability of the other.

others to conclude that the protons in ice cannot in fact be arranged periodically at all, but must instead be disordered. In their paper, Bernal and Fowler wrote: "It is quite conceivable and even likely that at temperatures just below the melting point the molecular arrangement is still partially or even largely irregular, though preserving at every point tetrahedral coordination and balanced dipoles. In that case ice would be crystalline only in the positions of its molecules but glass-like in their orientation."

Since then, it has been shown that ice can assume a large number of other stable structures at different temperatures and pressures. On the right-hand side of Figure 7.1, for example, we show the oxygen structure of cubic ice (I_c), which is a metastable structure which exists in approximately the same temperature and pressure regimes as hexagonal ice. The many different ice structures all have one important feature in common: they obtain their strength from the existence of hydrogen bonds between the oxygen atoms—four hydrogen bonds per oxygen. A phase diagram of ice is sketched in Figure 7.2. This figure may well be incomplete. It includes the most widely accepted phases known at the time of the writing of this book, but a number of others have been tentatively identified in the literature, and some of these may well have undergone more thorough experimental verification by the time you read this. For example, it has recently been suggested that at very low temperatures ice I_h becomes a metastable phase, and that given enough time it will transform into another phase, called ice XI. Also at low temperatures it has been suggested that ice VI, which is tetragonal, transforms into an orthorhombic phase called ice VI'. Most transitions that occur under a lowering of the temperature involve only an ordering in the proton arrangement, while the oxygen structure stays unaltered. This is true for the transition from ice VII to ice VIII, from ice II to ice IX, and from ice I and VI to ice XI and VI' respectively if the latter exist. These transitions are very difficult to observe experimentally, since the proton configuration becomes frozen below about 100 K and some trickery has to be employed if equilibrium is ever to be reached.[1]

7.1.1 Arrangement of the protons

The protons in ice are located between pairs of nearest-neighbour oxygen atoms, forming hydrogen bonds. As we now know, the potential energy of such a proton has two minima as a function of position along the line joining the two oxygens. In hexagonal ice the distance between neighbouring oxygen atoms is about 2.75 Å, and the two most energetically favourable positions for the intervening hydrogen atom are about 1 Å away from the centre of each of the oxygens. Each proton must thus choose between two positions about 0.75 Å apart. Although, as discussed in Section 7.1, the protons' choice of positions does not show long-range order, it is not completely random either, since each oxygen is normally associated with only two protons. Bernal and Fowler (1933) and Pauling (1935) proposed that the protons are arranged on the bonds of the lattice formed by the oxygen atoms according to two rules:

1. There is precisely one hydrogen atom on each hydrogen bond.

2. There are precisely two hydrogen atoms near each oxygen atom.

These two constraints are known as the **ice rules**. In real ice, the ice rules are sometimes violated. In violation of the first rule, it can sometimes happen that a hydrogen bond is populated either by zero or by two protons. These violations are known as Bjerrum L and D defects, respectively. They are energetically highly unfavourable, and therefore occur rarely. It has been estimated that already at $-10\,^\circ C$ there is only about one such defect per five million bonds, and the concentration decays rapidly with decreasing

---

[1]In fact there are other crystalline structures besides ice which have four hydrogen bonds per molecule, and some of these are easier to study experimentally than ice itself. KH_2PO_4 for instance, is a particularly popular alternative. It may turn out that experiments on these alternative ice-like compounds will yield the first observations of some of the more exotic phases proposed.

temperature. The second ice rule is violated if either one or three protons are located near an oxygen atom, a situation which we call an ionic defect. Ionic defects cost even more energy than Bjerrum defects and are therefore even more scarce.

In this chapter we study the class of models known as **ice models**, which are models which obey the ice rules exactly. The low frequency with which the ice rules are violated makes these models rather a good description of the behaviour of the protons in real ice. As we will see however, although models based on the rules are easily constructed, it is by no means straightforward to determine their properties. It took over thirty years from the first formulation of the ice rules before an exact solution to even the simplest ice model was found, and most ice models are not solved to this date. Instead, therefore, researchers have turned to computer simulation, and in this chapter we study the problem of designing an efficient Monte Carlo algorithm for an ice model, and give a selection of results to demonstrate the kind of insight we can gain from our simulations.

7.1.2 Residual entropy of ice

Many systems have a unique ground state—just one state in which the energy of the system is at a minimum—so that the entropy, which is loosely the logarithm of the number of states available to the system at a given temperature, is zero in the limit of zero temperature. Sometimes a system may have several degenerate ground states, such as the Ising model of Section 1.2.2 for instance, which has two. But even in this case, the zero-temperature entropy *per site* is still zero in the limit of large system size. (This is the reasoning behind the third law of thermodynamics, which we invoked in Equation (1.12) in order to fix the value of the integration constant.) Ice however is different, because of the many possible configurations of the protons discussed in the last section. To a good approximation, every such configuration has the same energy as every other, so there are as many different ground states of the system as there are configurations of the protons. This number increases exponentially with the size of the system, which means that even in the thermodynamic limit the entropy per site is finite as the temperature of the system tends to zero. This zero-temperature entropy is known as the **residual entropy** of the ice. Giauque and Stout (1936) measured the residual entropy of hexagonal ice I_h at 10 K to be $S = 0.82 \pm 0.05$ cal K^{-1} mole^{-1}.

The first serious attempt at estimating the residual entropy of ice theoretically was made by Linus Pauling (1935), who gave the following argument. If we ignore the first ice rule and allow a hydrogen bond to be either empty or doubly occupied, but we enforce the second ice rule and only allow exactly two protons to be near each oxygen atom, then there are six possible states per oxygen atom. Thus the number of states that obey the second ice rule

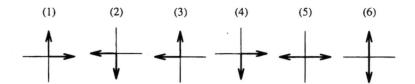

FIGURE 7.3   The six possible vertex configurations.

is 6^N, where N is the number of oxygen atoms. This is not a good estimate of the number of ground states of the system however; ignoring the first ice rule introduces a big error, since only a small fraction of these 6^N states do not violate the first ice rule. This is easy to see, since in any one of the states, the probability for any hydrogen bond to be occupied by exactly one proton is only $\frac{1}{2}$, and if any single bond is either doubly occupied, or not occupied at all, that state should be disallowed as a possible ground state of the system. However, this error should be easy to correct for. First note that there are $2N$ hydrogen bonds on any ice lattice, since each oxygen has four bonds and each bond is shared between two oxygens. Therefore, if the probability of any one bond being correctly occupied by a single proton is $\frac{1}{2}$, then the probability of all of them being so occupied is 2^{-2N}. Thus, we estimate the number of states obeying both ice rules to be $6^N/2^{2N}$ and so the residual entropy is

$$S \simeq \log\left(\frac{6^N}{2^{2N}}\right) = N \log \tfrac{3}{2}. \tag{7.1}$$

This value agrees well with experimental values, and for a long time it was believed to be an exact answer. As we now show however, this is not the case.

   To investigate Pauling's argument in more detail, we will look at the simplest model system obeying the ice rules, **square ice**.[2] In square ice, the oxygens are arranged on the vertices of a square grid—the simplest lattice with the required coordination number of four—and each oxygen has a hydrogen bond with its neighbours to the left, right, up and down, represented by the lines of the grid. Commonly, an arrow is drawn on each bond to represent the state of the corresponding proton: the arrow points

---

[2]The name "square ice" is used specifically to refer to the ice model on a square lattice in which all configurations are assigned the same energy. In Section 7.6 we will study a number of other ice models, such as the KDP and F models, in which different configurations are assigned different energies. Although these models are also ice models and are also defined on a square lattice, they are not also called square ice. This is somewhat confusing, but since the terminology is widely used we will follow convention and use it here too.

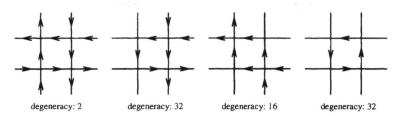

FIGURE 7.4 The 82 different ways in which the six types of vertices can be combined on a small 2 × 2 segment of a larger lattice. All of the 82 configurations can be transformed into one of the four shown here by rotation or reflection, or a combination of the two.

towards the end of the bond nearest the proton. The second ice rule then requires that each vertex have exactly two arrows pointing towards it, and two pointing away. This allows for exactly six types of vertices, as we pointed out above, and has led to the alternative name **six-vertex models** being applied to ice models, a name which you may well run across in the literature from time to time. The six vertices are illustrated in Figure 7.3.

Let us now zoom in on a small piece of the lattice, containing only four sites arranged in a square (see Figure 7.4). The arrows around any one of these four vertices can take any of the six allowed configurations, giving a total of 6^4 possible states overall. However, four of the bonds in this small piece of lattice are shared between two sites, so that choosing the directions of the arrows around one vertex affects the possible directions at the neighbouring vertices. In the language of Pauling's argument, we can correctly assign two protons to each of the four vertices, which is equivalent to drawing two arrows pointing in towards each one and two pointing out, but that does not guarantee that there will be only one arrow on each bond. In fact there is only a 50% chance that any given bond will have exactly one arrow on it, and thus only a fraction 2^{-4} of the possible arrangements of the arrows correctly obey the first ice rule. Thus we expect to find a total of $6^4/2^4 = 81$ ways to arrange the arrows. However, if we count the possible configurations explicitly we find that there are in fact 82 of them, as Figure 7.4 demonstrates. This is a strong indication that Pauling's reasoning is not correct. In fact there *is* a flaw in his argument, which is that it assumes—incorrectly—that the probabilities for different hydrogen bonds to be singly occupied are independent of each other.

Although this argument demonstrates that Pauling's formula for the entropy is only approximately correct, it does not tell us how to put the formula right, and in fact a general answer to this problem still escapes us. However, in 1967, Lieb gave the solution for square ice, employing a rather lengthy argument involving transfer matrices. For the details of Lieb's derivation the interested reader is referred to his paper (Lieb 1967). Here we just quote his result:

$$S = N \log \left[\left(\frac{4}{3} \right)^{\frac{3}{2}} \right] \simeq N \log(1.5396007\ldots). \tag{7.2}$$

Structures other than square ice are expected to have entropies differing slightly from this figure. Although we don't have an exact solution for any three-dimensional case, some very accurate estimates have been made using series expansions. For instance, Nagle (1966) obtained the figure

$$S = N \log (1.50685 \pm 0.00015) \tag{7.3}$$

for the entropy of hexagonal ice I_h. When converted into the units used in the experiments, this comes to $S = 0.8154 \pm 0.0002$ cal K^{-1} mole^{-1}, which is in excellent agreement with the experimental result of Giauque and Stout quoted earlier. Note also that Pauling's estimate is only a few tenths of a per cent different. At the time of the writing of this book the accuracy of even the most recent experiments is insufficient to prove him wrong.

7.1.3   Three-colour models

In the next section we will start to look at ways of simulating ice models using Monte Carlo methods. An important result about square ice which will help us in the design of an algorithm was given by Lenard (1961) who demonstrated that the configurations of the arrows on a square lattice can be mapped onto the configurations of a lattice of the same size in which the squares or **plaquets** between the bonds are coloured using three different colours, with the restriction that no two nearest-neighbour plaquets have the same colour. It is actually not very difficult to demonstrate this equivalence. Given a suitable colouring of the plaquets we can work out the corresponding configuration of arrows using the procedure illustrated in Figure 7.5, in which the three colours are represented by the numbers 1, 2 and 3. The rule is that we imagine ourselves standing on one of the squares of the lattice and looking towards one of the adjacent ones. If the number in the adjacent square is one higher (modulo three) than the number in the square we are standing on, we draw an arrow on the intervening bond pointing to the right. Otherwise we draw an arrow to the left. The procedure is then repeated for every other bond on the lattice.

Clearly the resulting configuration of arrows obeys the first ice rule; since neighbouring plaquets must have different colours the prescription above will place one and only one arrow on every bond. The second ice rule requires that each vertex has two ingoing and two outgoing arrows. If we walk from square to square in four steps around a vertex, then each time we cross a bond separating two squares the colour either increases or decreases by one,

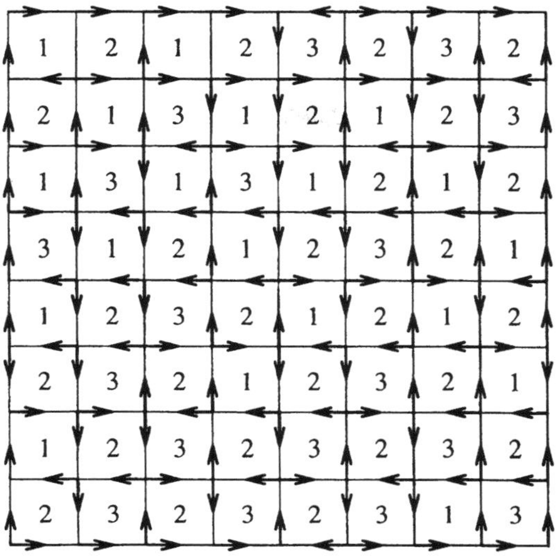

FIGURE 7.5   A
three-colouring of a square
lattice and the
corresponding configuration
of the arrows in square ice.

modulo three. The only way to get back to the colour we started with when
we have gone all the way around is if we increase twice and decrease twice.
This means that the vertex we walk around must have two ingoing and
two outgoing arrows, exactly as we desire. Thus each configuration of the
three-colour model corresponds to a unique correct configuration of square
ice.

We can also reverse the process, transforming an ice model configuration
into a three-colouring. We are free to choose the colour of one square on
the lattice as we wish, but once that one is fixed, the arrows on the bonds
separating that square from each of its neighbours uniquely determine the
colour of the neighbouring squares, and, by repeated application of the rule
given above, the colour of all the rest of the squares in the lattice. Thus,
the number of ways in which the squares of the lattice can be coloured is
exactly the number of configurations of the ice model on the same lattice,
except for a factor of three.

7.2   Monte Carlo algorithms for square ice

Having introduced the physics of ice models, we now look at how we can
simulate these models using a Monte Carlo method and what results we
can get from such a simulation. We will start with square ice, which is the
simplest case and also the easiest to visualize. Many of the properties of
square ice are known exactly from Lieb's solution of the model. However, it
is a good first step to reproduce these known results using a Monte Carlo
method, in the same way as it was a good first step in Chapter 3 to study

the two-dimensional Ising model, for which we also have an exact solution. In the later sections of this chapter we will develop algorithms for more complicated ice model variations.

As with the Ising model, the first step in designing a Monte Carlo algorithm is to choose a set of moves which will take us from one configuration of our ice model to another. In the case of the ordinary Ising model the obvious Monte Carlo move was just to flip a single spin. Unfortunately, there is no such obvious move for an ice model. We cannot simply reverse an arrow, or change the configuration of arrows around a vertex in an ice model, since that would affect the configuration of arrows at all the neighbouring vertices and we would surely end up in a state that violated the ice rules. So what elementary move can we devise which will take us from one configuration that obeys the ice rules to another? As we will see, there are a number of candidates, and of course it is part of our goal here to explore all the possibilities to see how we can make algorithms that are as efficient as possible. We begin, however, as we did with the Ising model, by describing the standard algorithm for this problem, due to Rahman and Stillinger (1972), which involves reversing the arrows around loops on the lattice.

7.2.1   The standard ice model algorithm

It is clear that one possible move which takes us from an allowed configuration of an ice model to another is the reversal of all the arrows in a loop chosen such that all arrows point in the same direction around the loop. Such a loop has one arrow pointing in and one pointing out of each participating vertex, so that the reversal of all of them preserves the number of arrows going in and out of each vertex. How do we find a loop of this kind? The most straightforward method is illustrated in Figure 7.6. Starting with a correct configuration of the model (a), we choose a single vertex at random from the lattice. We then choose at random one of the two outgoing arrows at this vertex and reverse it (b). (We could just as well choose an ingoing arrow—either is fine.) This creates violations of the second ice rule—ionic defects—at the initial vertex (which now has only one outgoing arrow instead of two) and at a new vertex at the other end of the reversed arrow (which now has three). We can set this new vertex straight by choosing randomly one of the other two outgoing arrows (not the one we just reversed) and reversing that, so that it is pointing inwards. We do this in part (c) of the figure. This fixes one vertex, but creates a new defect at the other end of the reversed arrow, which we can get rid of by reversing still a third arrow (d), and so on. Each reversal sets one vertex to rights but corrupts another. The process ends when our string of arrows finally finds its way back to the point at which we started (e). At this point there is one vertex which has one too few outgoing arrows, and reversing an arrow between this vertex and our

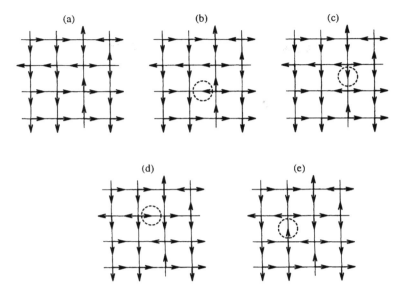

FIGURE 7.6   Flipping arrows one by one along a line across the lattice allows us to change the configuration and still satisfy the ice rules. The only problems are at the ends of the line, but this is fixed if the two ends eventually meet one another forming a closed loop.

wandering defect fixes the arrows around both. The net result is that we have reversed all of the arrows lying around a closed loop on the lattice, and the final configuration will obey the ice rules as long as the initial one did.

In the figure we have illustrated the case of the smallest possible loop, which on the square lattice involves the reversal of just four arrows. However, loops generated by the algorithm above can be much longer than this. By making a random choice at each vertex between the two possible arrows that we could reverse, we generate a species of random walk across the lattice, and even on a finite lattice this walk can take an arbitrarily long time to return to its starting point. For this reason we will refer to this algorithm as the "long loop algorithm". Long loops are not necessarily a bad thing; on the finite lattices we use in our Monte Carlo simulations we are guaranteed that the walk will always return eventually, and although longer loops take longer to generate they also flip a larger number of arrows, which allows the system to decorrelate quicker.

7.2.2   Ergodicity

We have now specified a move that will take us from one correct configuration of the arrows to another, and our proposed Monte Carlo algorithm for square

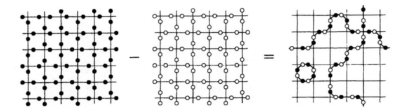

FIGURE 7.7   The difference between any two configurations of an ice
model can be decomposed into a number of loops, which are allowed
to wrap around the periodic boundaries. We can go from one configuration to the other by reversing the arrows around these loops.

ice is simply to carry out a large number of such moves, one after another.
However, we still need to demonstrate that the algorithm satisfies the criteria
of ergodicity and detailed balance.

First, consider ergodicity, whose proof is illustrated in Figure 7.7. The
figure shows how the difference between two configurations of the model
on a finite lattice can be decomposed into the flips of arrows around a finite
number of loops. We can show that this is possible for any two configurations
by the following argument. Each of the vertices in Figure 7.3 differs from each
of the others by the reversal of an even number of arrows. (You can convince
yourself of this simply by looking at the figure, although it follows directly
from the ice rules as well.) Thus, if we take two different configurations of
the model on a particular lattice and imagine drawing lines along the bonds
on which the arrows differ, we are guaranteed that there will be an even
number of such lines meeting at each vertex. Thus these lines form a set of
(possibly intersecting) loops covering a subset of the vertices on the lattice.
It is not difficult to show that these loops can be chosen so that the arrows
around each one all point in the same direction. Since the reversals of the
arrows around these loops are precisely our Monte Carlo moves, and since
there are a finite number of such loops, it follows that we can get from any
configuration to any other in a finite number of steps, and thus that our
algorithm is ergodic. Note that it is important to allow the loops to pass
through the periodic boundary conditions for this to work.[4]

---

[4] It is not too hard to show that the loops which wrap around the periodic boundary
conditions change the polarization, Equation (7.14), of the system, whereas the ones which
don't conserve polarization. Thus, if we didn't allow the loops to wrap around in this way
the polarization would never change and ergodicity would not be satisfied.

7.2.3 Detailed balance

In square ice, all states of the system have the same energy, which means that the condition of detailed balance, Equation (2.14), takes the simple form

$$\frac{P(\mu \to \nu)}{P(\nu \to \mu)} = e^{-\beta(E_\nu - E_\mu)} = 1. \tag{7.4}$$

In other words, the rate for the move from μ to ν should be the same as that from ν to μ. In the long loop algorithm, our Monte Carlo move consists of choosing a starting site S_0 and reversing a loop of arrows starting at that site and ending, m steps later, at the same site $S_m = S_0$. The probability of selecting any particular site as the starting site is $1/N$, where N is the number of sites on the lattice. The probability of making a particular choice from the two possible outgoing arrows at each step around the loop is $\frac{1}{2}$ for each step, so that the probability that we chose a certain sequence of steps is equal to 2^{-m}, and the probability of performing the entire move is $\frac{1}{N}2^{-m}$. For the reverse move, in which the same loop of arrows is flipped back again to take us from state ν back to state μ, the exact same arguments apply, again giving us a probability of $\frac{1}{N}2^{-m}$ for making the move. (The only difference is that for the reverse move we follow the loop in the reverse direction.) Thus the forward and backward rates are indeed the same and detailed balance is observed. This, in combination with the demonstration of ergodicity above, ensures that our algorithm will sample all states of the model with the correct (equal) probabilities.

7.3 An alternative algorithm

As we pointed out in Section 7.2.1, the algorithm presented above can generate arbitrarily long loops even on a finite lattice. It is not clear, as we said, that this is necessarily a problem—we can imagine reasons why longer loops might either help or hinder the rapid equilibration of the system. In Section 7.5, however, we give some simulation results which show clearly that the long loop algorithm is inefficient on large lattices precisely because of the size of the loops it generates. To get around this problem, Rahman and Stillinger (1972) proposed another algorithm which also reverses the arrows around a closed loop of bonds, but which generates shorter loops. For obvious reasons we call this the "short loop algorithm". Here we describe a slightly refined version of the algorithm due to Yanagawa and Nagle (1979).

In the short loop algorithm we choose a starting site S_0 at random from the lattice and reverse one of the outgoing arrows at that vertex, thereby creating two defects, just as we did in the long loop case. We then reverse further arrows so that one of the defects wanders around the lattice randomly, as before. However, rather than waiting until the two defects find

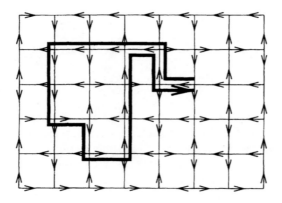

FIGURE 7.8 A typical
path followed by the
wandering defect in the
short loop algorithm.

each other again, we now continue only until the wandering defect encoun-
ters *any* site which it has encountered before in its path around the lattice.
If we call this site S_m, then $S_m = S_l$ with $0 \leq l < m$. From this point, we
retrace our steps *backwards* down the old path of the defect, until we reach
S_0 again, reversing all the arrows along the way. The process is illustrated
in Figure 7.8. The net result is that we reverse all the arrows along the path
from site S_0 to S_l twice (which means that they are the same before and
after the move), and all the arrows in the loop from S_l to S_m once. Since the
wandering defect only needs to find any one of the sites on its previous path
in order to close the loop, we are guaranteed that the length of its walk will
never exceed N steps where N is the number of vertices on the lattice, and
in practice the typical length is much shorter than this. (In fact, the number
of steps tends to a finite limit as the lattice becomes large—see Section 7.5.)

The proof of ergodicity for the short loop algorithm is identical to that
for the long loop case: the difference between any two states on a finite
lattice can be reduced to the reversal of the spins around a finite number of
loops. Since the algorithm has a finite chance of reversing each such loop, it
can connect any two states in a finite number of moves.

The proof of detailed balance is also similar to that for the long loop
algorithm. Consider again a move which takes us from state μ to state ν.
The move consists of choosing a starting site S_0 at random, then a path
$P = \{S_0 \ldots S_l\}$ in which the arrows are left untouched, followed by a loop
$L = \{S_l \ldots S_m\}$ in which we reverse the arrows. (Remember that the last
site in the loop S_m is necessarily the same as the first S_l.) The probability
that we choose S_0 as the starting point is $1/N$, where N is the number of
sites on the lattice. After that we have a choice of two directions at each step
along the starting path and around the loop, so that the probability that we
end up taking the path P is equal to 2^{-l} and the probability that we follow
the loop L is $2^{-(m-l)}$. After the loop reaches site $S_m = S_l$, we do not have
any more free choices. The probability that we move from a configuration μ

to configuration ν by following a particular path P and loop L is thus

$$P(\mu \to \nu) = \frac{1}{N} 2^{-l} 2^{-(m-l)} = \frac{1}{N} 2^{-m}. \tag{7.5}$$

For the reverse move, the probability of starting at S_0 is again $1/N$, and the probability of following the same path P as before to site S_l is 2^{-l} again. However, we cannot now follow the same loop L from S_l to S_m as we did before, since the arrows along the loop are reversed from what they were in state μ. On the other hand, we can follow the loop in the reverse direction, and this again has probability $2^{-(m-l)}$. Thus we have

$$P(\nu \to \mu) = \frac{1}{N} 2^{-l} 2^{-(m-l)} = \frac{1}{N} 2^{-m}, \tag{7.6}$$

exactly as before. This demonstrates detailed balance for the algorithm and, in combination with the demonstration of ergodicity, ensures that all possible states will be sampled with equal probability.

7.4 Algorithms for the three-colour model

We now have two Monte Carlo algorithms which correctly sample the states of the square ice model, and since, as we showed in Section 7.1.3, the states of this model can be mapped onto the states of the three-colour lattice model, we can use the same algorithms to study the three-colour model. In this section however, we explore the other side of the same question: is there a natural Monte Carlo algorithm for the three-colour model which could then be used to sample the states of the ice model? It turns out that there is, and this algorithm provides not only an efficient way of simulating the square ice model in which all states have the same energy, but will also prove useful when we get onto the models of Section 7.6 in which different states are assigned different energies.[5]

In the three-colour representation of square ice, the degrees of freedom— the colours—are located on the plaquets of the lattice, rather than on the bonds, and, as we showed earlier, the ice rules translate into the demand that nearest-neighbour squares have different colours. Just as in the arrow representation, there is no obvious update move which will take us from state to state. Although there are some states in which the colour of a single square can be changed from one value to another without violating the ice rules, there are also states in which no such moves are possible, and therefore

---

[5]The algorithms described in this section are due to Wang *et al.* (1990), although they were not originally conceived as algorithms for ice models. Wang *et al.* were studying anti-ferromagnetic Potts models, but it turns out that the $q = 3$ Potts antiferromagnet at $T = 0$ is identical to the three-colour representation of square ice. This point is explored further in Problem 7.2.

single-plaquet moves of this kind cannot reach these states and so do not lead to an ergodic dynamics. Again then, we must resort to more complex moves. One possibility is to look for clusters of nearest-neighbour plaquets of only two colours, call them A and B, entirely surrounded by plaquets of the third colour C. A move which exchanges the two colours A and B in such a cluster but leaves the rest of the lattice untouched will result in a new configuration which still has an allowed arrangement of colours, and this suggests the following cluster-type algorithm for square ice:

1. We choose a plaquet at random from the lattice as the seed square for the cluster. Suppose this plaquet has colour A.

2. We choose another colour $B \neq A$ at random from the two other possibilities.

3. Starting from our seed square, we form a cluster by adding all nearest-neighbour squares which have either colour A or colour B. We keep doing this until no more such nearest neighbours exist.

4. The colours A and B of all sites in the cluster are exchanged.

There are a couple of points to notice about this algorithm. First, the cluster possesses no nearest neighbours of either colour A or colour B and therefore all its nearest neighbours must be of the third colour, C. In the simplest case, the seed square has no neighbours of colour B at all, in which case the cluster consists of only the one plaquet. It is crucial to the working of the algorithm that such moves should be possible. If we had chosen instead to seed our cluster by picking two neighbouring plaquets and forming a cluster with their colours, single-plaquet moves would not be possible and we would find that the algorithm satisfied neither ergodicity nor detailed balance. Second, notice also that within the boundary of colour C, the cluster of As and Bs must form a checkerboard pattern, since no two As or Bs can be neighbours.

We now want to show that the algorithm above satisfies the conditions of ergodicity and detailed balance. In this case it turns out that detailed balance is the easier to prove. Consider once more a Monte Carlo move which takes us from a state μ to a state ν, and suppose that this move involves a cluster of m squares. The probability of choosing our seed square in this cluster is m/N, where N is the total number of plaquets on the lattice. The probability that we then choose a particular colour B as the other colour for the cluster is $\frac{1}{2}$, and after that there are no more choices: the algorithm specifies exactly how the cluster should be grown from here on. Thus the total probability for the move from μ to ν is $m/(2N)$. Exactly the same argument applies for the reverse move from ν to μ with the same values of m and N, and hence the rates for forward and reverse moves are the same. Thus detailed balance is obeyed.

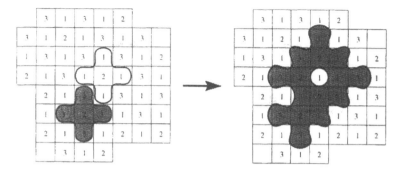

FIGURE 7.9   A checkerboard domain consisting of only two colours (in this case colours 2 and 3) entirely surrounded by the third colour (1 in this case) can be grown by exchanging the two colours in a cluster bordering it (1 and 2).

The proof of ergodicity is a little trickier. It involves two steps. First, we show that from any configuration we can evolve via a finite sequence of reversible moves to a checkerboard colouring (a configuration in which one of the three colours is absent). Then we show that all checkerboard colourings are connected through reversible moves.

Any configuration of the lattice can be broken down into a number of checkerboard regions consisting of only two colours, surrounded by plaquets of the third colour. This is always true since any plaquet which doesn't belong to such a region can be regarded as being the sole member of a checkerboard of one. Under the dynamics of our proposed Monte Carlo algorithm these checkerboard domains can grow or shrink. Consider for example a domain of colours A and B, which is necessarily entirely surrounded by plaquets of the third colour C. We can form a cluster at the boundary of the domain out of plaquets of colour C and one of the other two colours, and by swapping the colours in this cluster we can increase the size of our AB domain. (In order to make this work it may be necessary first to exchange the colours A and B in the AB domain.) By repeating this process we can take a single cluster of one checkerboard pattern and grow it until it covers the entire lattice, leaving the lattice in a checkerboard state. This process is illustrated in Figure 7.9. (As the figure shows, it is possible for moves of this type to leave the occasional odd plaquet in the middle of the checkerboard region. However, these can easily be removed by a Monte Carlo step which takes this single plaquet as its cluster and changes its colour. If such single-plaquet moves were not allowed, then the algorithm would not be ergodic, as we maintained above.)

There are six possible checkerboard colourings in all,[6] and from any one

---

[6]Bear in mind that the checkerboard with colour A on the even sites and B on the odd

of them the others can easily be reached, since on a checkerboard the colour
of any square can be changed on its own without changing any other squares.
Thus for example we can get from a checkerboard of colours A and B to one
of A and C by changing all the Bs to Cs one by one. All other combinations
can be reached by a similar process.

Since we can get from any state to a checkerboard colouring and from
any checkerboard to any other, all via reversible moves, it follows that our
algorithm can get from any state to any other in a finite number of moves.
In other words, it is ergodic.

The algorithm presented above, a single-cluster algorithm, resembles in
spirit the Wolff algorithm for the Ising model which we studied in Section 4.2.
It is also possible to construct a multi-cluster algorithm for the three-colour
model, similar to the Swendsen–Wang algorithm of Section 4.4.1. In this
algorithm we start by choosing at random a pair of colours A and B. Then
we construct all clusters of nearest-neighbour spins made out of these two
colours, and for each cluster we decide at random with 50% probability
whether to exchange the two colours or not. This algorithm satisfies ergodicity for the same reason the single-cluster algorithm did—each move in our
single-cluster algorithm is also a valid move in the multi-cluster version, so
we can apply exactly the same arguments about checkerboard domains to
prove ergodicity. The algorithm also satisfies detailed balance: the probability of selecting a particular two out of the three colours for a move is $\frac{1}{3}$,
and the probability of exchanging the colours in a particular set of clusters
is 2^{-n}, where n is the number of clusters. The probability for the reverse
move is exactly the same, and hence detailed balance is upheld.

7.5   Comparison of algorithms for square ice

In the previous sections, we have described four algorithms for the simulation
of square ice: the long loop algorithm, the short loop algorithm, the single-
cluster three-colour algorithm, and the full-lattice three-colour algorithm.
In this section we consider these algorithms one by one and compare their
efficiency.

The first algorithm we looked at, the long loop algorithm of Section 7.2.1,
involved the creation of a pair of ionic defects, one of which diffuses around
the lattice until it recombines with the first, in the process reversing all
the arrows along the path of its diffusion. To assess the efficiency of this
algorithm, we first measure the average number of steps which the wandering
defect takes before it recombines, as a function of the system size L. For an
ordinary random walker on a square lattice, this number scales as L^2. In the
case of the wandering defect however, we find that it scales instead as $L^{1.67}$—

---

sites is distinct from the one in which the colours are reversed.

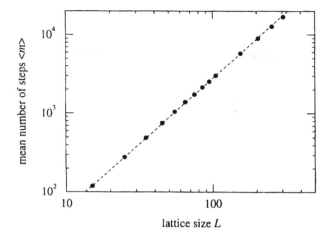

FIGURE 7.10   The mean length $\langle m \rangle$ of loops in the long loop algorithm as a function of system size L. We find that $\langle m \rangle \sim L^{1.665 \pm 0.002}$

see Figure 7.10.[7] The amount of CPU time required per step in our algorithm increases linearly with the size of the loop, and hence we expect the CPU time per Monte Carlo step also to increase with system size as $L^{1.67}$. This is not necessarily a problem; since longer loops reverse more arrows as well as taking more CPU time it is unclear whether longer is better in this case, or worse. To answer this question we need to consider the correlation time τ of the algorithm. This, however, presents a new problem. In order to calculate τ we first need to calculate an autocorrelation (see Section 3.3.1), but what quantity do we calculate the autocorrelation for? In the case of the Ising model in Chapter 3 we calculated the autocorrelation of the magnetization, but the square ice model does not have a magnetization. We might consider calculating the autocorrelation of the internal energy, but all states of the square ice model have the same energy, so the energy autocorrelation would always be zero.

In this case, we have chosen to measure a quantity ρ_{sym} which we define to be the density of the symmetric vertices, numbers 5 and 6 in Figure 7.3. We calculate the autocorrelation of this quantity using Equation (3.21) and then fit the resulting function to an exponential to find the correlation time, as in Figure 3.6. The results are shown in Figure 7.11. As we can see, when we measure time in Monte Carlo steps we find a correlation time which

---

[7]In fact, it is known that the loop size scales exactly as $L^{5/3}$. This result follows from calculations using the so-called Coulomb gas representation of the square ice model. See Saleur (1991).

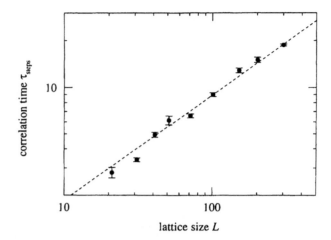

FIGURE 7.11 The correlation time in Monte Carlo steps of the long loop algorithm as a function of system size L. The best fit straight line gives $\tau_{\text{steps}} \sim L^{0.68 \pm 0.03}$.

increases as a power law with the dimension L of the system:

$$\tau_{\text{steps}} \sim L^{0.68 \pm 0.03}. \tag{7.7}$$

As in Section 3.3.1, the quantity we are really interested in is the correlation time in "sweeps" of the lattice, which in this case means arrow flips per bond on the lattice. This is easily calculated however. On average, each Monte Carlo step corresponds to $\langle m \rangle/(2L^2)$ sweeps on our two-dimensional lattice where $\langle m \rangle$ is the mean number of arrows around a loop. This means that the correlation time varies with L as

$$\tau \sim L^{0.68} \frac{L^{1.67}}{L^2} = L^{0.35}. \tag{7.8}$$

The correlation time measures the amount of computer effort we have to invest, per unit area of the lattice, in order to generate an independent configuration of the arrows, and the exponent 0.35 appearing here is a measure of how quickly this correlation time increases with increasing lattice size. The value 0.35 is a reasonably small one—the correlation time is increasing only rather slowly with system size—but as we will shortly see, some of our other algorithms for square ice do better still, possessing exponents not measurably different from zero.[8]

---

[8] It may strike the reader that the power-law variation of the correlation time with system size seen here is reminiscent of the way in which the correlation time of the Ising

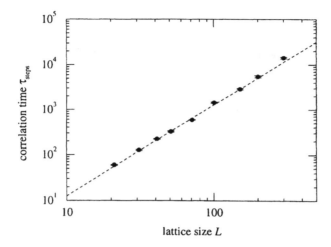

FIGURE 7.12   The correlation time τ_{steps} of the short loop algorithm measured in Monte Carlo steps as a function of system size. The best fit straight line gives $\tau_{\text{steps}} \sim L^{2.00\pm0.01}$

The short loop algorithm of Section 7.3 also involves creating a pair of defects and having one of them diffuse around. Recall however that in this case the wandering defect only has to find *any* of the sites which it has previously visited in order to close the loop and finish the Monte Carlo step. If the diffusion were a normal random walk then this process would generate loops of finite average length. Although the diffusion of defects in square ice is not a true random walk, it turns out once more that the behaviour of the system is qualitatively the same as if it were. From simulations using the short loop algorithm we find that the average number of steps per move is $\langle m \rangle \simeq 13.1$, independent of the lattice size, for a sufficiently large lattice.

The correlation time measured in Monte Carlo steps τ_{steps}, for the same observable ρ_{sym} as above, increases as L^2, as shown in Figure 7.12. Since the mean number of steps in a loop is independent of L, the correlation time per unit area therefore goes as

$$\tau \sim L^2 \frac{L^0}{L^2} = \text{constant.} \tag{7.9}$$

Thus the correlation time of the short loop algorithm does not increase at

---

model varied with system size at the critical temperature in the algorithms studied in Chapter 4. Yet the square ice model has no temperature parameter and therefore presumably cannot have a phase transition. In fact, it turns out that, in a certain technical sense, the square ice model *is* at a phase transition, and the exponent 0.35 measured here is precisely the dynamic exponent z of our algorithm at this transition (see Section 4.1). A discussion of this point is given by Baxter (1982).

all with system size. This is the best behaviour we could hope for in our algorithm, and so the short loop algorithm is in a sense optimal.

Our third algorithm was the single-cluster three-colour algorithm which we developed in Section 7.1.3. For this algorithm each Monte Carlo step corresponds to $\langle n \rangle / L^2$ sweeps of the lattice, where $\langle n \rangle$ is the average cluster size. Like the average loop length in the long loop algorithm, $\langle n \rangle$ scales up with increasing lattice size, and from our simulations we find that

$$\langle n \rangle \sim L^{1.5}. \tag{7.10}$$

The correlation time per Monte Carlo step is measured to go as

$$\tau_{\text{steps}} \sim L^{1.8}, \tag{7.11}$$

and hence the correlation time in steps per site goes as

$$\tau \sim L^{1.8} \frac{L^{1.5}}{L^2} = L^{1.3}. \tag{7.12}$$

The large value of the exponent appearing here indicates that the single-cluster algorithm would be a poor algorithm for studying square ice on large lattices.

Our last algorithm, the full-lattice three-colour algorithm, also described in Section 7.1.3, generates clusters in a way similar to the single-cluster algorithm, but rather than generating only one cluster per Monte Carlo step, it covers the whole lattice with them. For this algorithm we find numerically that the correlation time τ_{steps} measured in Monte Carlo steps is approximately constant as a function of lattice size. Since each Monte Carlo step updates sites over the entire lattice, one step corresponds to a constant number of sweeps of the lattice on average, and hence the correlation time in moves per site goes as

$$\tau \sim L^0 L^0 = L^0. \tag{7.13}$$

Thus, like the short loop algorithm, this algorithm possesses optimal scaling as lattice size increases.

Comparing the four algorithms, clearly the most efficient ones for large systems are the short loop algorithm and the full-lattice three-colour algorithm. In both other algorithms, the computer time required per site to generate an independent configuration of the lattice increases with system size. The larger impact of the larger loops in the long loop algorithm for instance does not compensate for the extra effort invested generating them. Between the short loop algorithm and the full-lattice three-colour algorithm it is harder to decide the winner since both have the same scaling of CPU requirements with system size. The best thing to do in this case is simply try them both out. As far as we can tell, the loop algorithm is slightly faster (maybe 10% or 20%), but on the other hand the three-colour algorithm is considerably more straightforward to program. It's a close race.

7.6 Energetic ice models

An interesting extension of the ice models discussed so far in this chapter is to assign different energies to the different configurations of vertices, in much the same way as we assign different energies to different spin configurations in the Ising models of previous chapters. There are a number of different ways in which the energies can be assigned. One common one was introduced by Slater (1941) who studied a model in which vertices 1 and 2 in Figure 7.3 are favoured by giving them an energy $-\epsilon$, while all the others are given energy zero. Slater suggested that this model might provide an approximate representation of the behaviour of the protons in the compound potassium dihydrogen phosphate (KH_2PO_4), also known as KDP, which we mentioned in a footnote earlier in the chapter. KDP has a four-fold coordinated arrangement of hydrogen bonds as ice does, but appears to favour some arrangements of the hydrogens over others. Because of its use as a model of KDP, Slater's model is usually referred to as the **KDP model**.

Since it is possible to form a domain on a square lattice consisting only of type 1 vertices or only of type 2, the KDP model has two degenerate ground states in which the lattice is entirely covered with vertices of one of these two types, and the model displays a phase transition, similar to that seen in the Ising model (Section 3.7.1), from a high-temperature phase in which the two appear with equal probability to a low-temperature one in which one or the other dominates. A suitable order parameter to describe this transition (see Section 3.7.1) is the **polarization** or average direction of the arrows:

$$\mathbf{P} = \frac{1}{\sqrt{2}N} \sum_i \hat{\mathbf{n}}_i, \qquad (7.14)$$

where the vector $\hat{\mathbf{n}}_i$ is a unit vector in the direction of the ith arrow. In an infinite system the polarization is zero above the critical temperature of the phase transition T_c, and non-zero below it with a direction either upwards and to the right, or downwards and to the left. The factor of $1/\sqrt{2}$ in Equation (7.14) is included so that the magnitude of $\mathbf{P}$ approaches unity as $T \to 0$.

Another widely studied energetic ice model is the so-called **F model** introduced by Rys (1963). In this model the symmetric vertices, numbers 5 and 6 in Figure 7.3, are favoured by giving them a lower energy $-\epsilon$, while all the others are given energy zero. This model has a ground state in which vertices 5 and 6 alternate in a checkerboard pattern across the lattice. There are again two possible such ground states, depending in this case on which type of vertex falls on the even sites of the lattice and which on the odd, and as we lower the temperature there is a phase transition to one or other of these states from a high-temperature phase in which vertices 5 and 6 fall on even and odd sites with equal probability. Since neither symmetric vertex possesses any net polarization, the value of **P**, Equation (7.14), is zero in the thermodynamic limit for the F model, regardless of temperature. One can however define an anti-ferroelectric order parameter which does become non-zero in the low-temperature phase (see Lieb 1967).

The F model and the KDP model are the best known of the energetic ice models, but there are many other possible choices of energies for the six vertex configurations, many of which have been investigated at one time or another. A number of them are discussed by Lieb and Wu (1970). If in addition to the six configurations appearing in Figure 7.3 two more configurations are allowed, the ones with four ingoing or four outgoing arrows, we end up with another popular class of models, known as **eight-vertex models**. The properties of at least one eight-vertex model have been solved exactly (Baxter 1971).

7.6.1 Loop algorithms for energetic ice models

In Section 7.5 we developed a variety of elementary moves for sampling the states of ice models on square lattices, and showed how these could be used to create Monte Carlo algorithms for the square ice model, in which all states have the same energy. We can use the same sets of elementary moves to create Monte Carlo algorithms for energetic ice models as well. Here we give examples of algorithms for the particular case of the F model, but the same ideas can easily be adapted for use with the KDP model or other energetic ice models. The Hamiltonian of the F model is given by

$$H = -\epsilon \sum_i [\delta_{v_i,5} + \delta_{v_i,6}], \qquad (7.15)$$

where v_i is the type of the vertex at site i, represented using the numbering scheme from Figure 7.3. Interest in the F model has primarily focused on its behaviour close to the phase transition, which as Lieb (1967) has shown takes place at $T_c = \epsilon/\log 2$. Most of the developments here will be directed at finding an algorithm which performs well in this region, which means both finding an efficient way of sampling states of the system with their correct Boltzmann probabilities, and also tackling the critical slowing down problems which we typically run into in the vicinity of a phase transition (see Section 3.7.2).

The simplest way to create a correct algorithm for the F model is to generate possible moves, such as the loop or cluster moves of the previous sections, and then employ a Metropolis-type scheme in which instead of accepting every move generated by the algorithm, we accept them with a probability $A(\mu \to \nu)$ which depends on the energy difference $\Delta E = E_\nu - E_\mu$

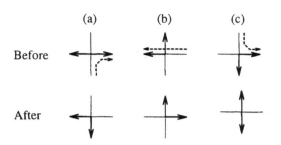

(a) (b) (c)

Before

After

FIGURE 7.13
(a) Symmetric vertices
become non-symmetric if a
loop passes through them.
(b) Non-symmetric vertices
stay non-symmetric if the
loop goes straight through,
but (c) become symmetric
if the loop makes a turn.

between the states μ and ν of the system before and after the move thus:

$$A(\mu \to \nu) = \begin{cases} e^{-\beta \Delta E} & \text{if } \Delta E > 0 \\ 1 & \text{otherwise.} \end{cases} \tag{7.16}$$

(See Section 3.1.)

Let us first consider algorithms in which the proposed moves involve reversing the directions of the arrows around a loop on the lattice, as in the long and short loop algorithms of Sections 7.2.1 and 7.3. To calculate the energy change ΔE resulting from such a move, we could simply measure the total energy of the lattice before and after the move and take the difference of the two. This, however, is wasteful, since most of the vertices on the lattice are unchanged between the two states, so that their contributions to the Hamiltonian, Equation (7.15), cancel out. As we did in Section 3.1.1 with the Ising model, we can use this observation to derive a formula for calculating ΔE which avoids performing sums over the whole lattice.

First, notice that the only vertices which change type (and hence energy) when we perform our Monte Carlo move are those which the loop passes through. Furthermore, as we show in Figure 7.13, a symmetric vertex (type 5 or 6) always becomes non-symmetric if the loop passes through it, thereby increasing the total energy. For non-symmetric vertices the situation is slightly more complicated. If the loop passes straight through a non-symmetric vertex, the vertex remains non-symmetric and its energy is unchanged. On the other hand, if the loop makes a turn as it passes through a non-symmetric vertex, the vertex becomes symmetric and the energy decreases. Thus, given a particular loop, we can calculate ΔE by counting the number m of symmetric vertices which the loop passes through and the number n of non-symmetric vertices in which it makes a 90° turn, and applying the formula

$$\Delta E = \epsilon(m - n). \tag{7.17}$$

The density of symmetric vertices in the F model increases with decreasing temperature, which means that the average number of symmetric vertices through which a loop passes grows as we go to lower temperatures. Since

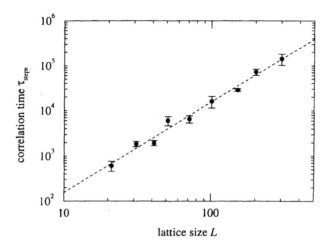

FIGURE 7.14   The correlation time τ_{steps} of the short loop algorithm
for the F model measured in Monte Carlo steps, as a function of system
size. The best fit straight line gives $\tau_{\text{steps}} \sim L^{2.00 \pm 0.09}$.

each such vertex adds an amount ϵ to ΔE, it is clear that the energy cost
of making a move will increase with the length of the loop, especially at low
temperatures. Large values of ΔE mean low acceptance ratios, which are
wasteful of computer time, and this implies that the short loop version of
our F model algorithm should be more efficient than the long loop one.

In Figure 7.14 we show results from a simulation of the F model using
the short loop algorithm at the critical temperature. The figure shows the
correlation time τ_{steps} of the algorithm measured in Monte Carlo steps, and
the best fit to these data gives us

$$\tau_{\text{steps}} \sim L^{2.0}. \tag{7.18}$$

As with square ice, the number of sites updated by a single Monte Carlo
step tends to a constant for large lattices, so that the correlation time in
steps per site is

$$\tau \sim L^{2.0} \frac{L^0}{L^2} = L^0. \tag{7.19}$$

As with the Ising model calculations of Chapter 4, the scaling of τ with the
system size gives us a measure of the dynamic exponent of our algorithm
(see Section 4.2). In the present case, all indications are that our short loop
algorithm has a zero dynamic exponent, at least to the accuracy with which
we can measure it. This is the best value we could hope for. It implies
that the CPU time taken by the algorithm goes up only as fast as the size
of the lattice we are simulating, and no faster. However, it turns out that

the algorithm is still quite inefficient because it has a low acceptance ratio, which means that much of the CPU time is being wasted proposing moves which are never carried out. For example, at T_c the acceptance ratio of the algorithm is 36%, so that nearly two thirds of the computational effort is wasted.

How can we increase the acceptance ratio of our Monte Carlo algorithm? One way is to modify the algorithm to generate moves that are less likely to cost energy and therefore more likely to be accepted. For example, if we could encourage the loop to make turns in non-symmetric vertices, we would on average end up with a lower final energy, since the reversal of the arrows around the loop will create more symmetric vertices. Unfortunately, it turns out to be rather complicated to formulate a correct algorithm along these lines, and the expression for the acceptance ratio becomes quite tedious. There is however an elegant alternative, which is to employ a three-colour algorithm of the type discussed in Section 7.4.

7.6.2 Cluster algorithms for energetic ice models

The equivalent of a symmetric vertex in the three-colour representation of the F model is a group of four squares around a vertex in which both of the diagonally opposing pairs share the same colour. In non-symmetric vertices only one of these two diagonal pairs share the same colour. Making use of this observation we can write the Hamiltonian of the F model, Equation (7.15), in the form

$$H = -\epsilon \sum_{[i,j]} \left(\delta_{c_i c_j} - \tfrac{1}{2} \right) = N\epsilon - \epsilon \sum_{[i,j]} \delta_{c_i c_j}, \qquad (7.20)$$

where the summation runs over all pairs of next-nearest-neighbour squares $[i, j]$, and c_i is the colour of square i. We see therefore that it is energetically favourable to have pairs of next-nearest-neighbour squares with the same colour. We can make use of this observation to create an efficient algorithm for simulating the F model. In this algorithm, as in the algorithms for square ice discussed in Section 7.4, we build clusters of nearest-neighbour plaquets of two colours. But now in addition we also add to the cluster some next-nearest-neighbour plaquets as well. In detail the algorithm is as follows:

1. We choose a plaquet at random from the lattice as the seed square for the cluster. Suppose that this plaquet has colour A.

2. We choose another colour $B \neq A$ at random from the two other possibilities.

3. Starting from our seed square, we form a cluster by adding all nearest-neighbour squares which have either colour A or colour B, and in

addition we now also add to the cluster the squares which are next-nearest neighbours of some square i which is already in the cluster, provided they have the *same* colour as square i. However, we make this latter addition with a temperature-dependent probability $P_{\text{add}} < 1$, whose value we calculate below in order to satisfy the condition of detailed balance. We go on adding squares to the cluster in this way until every possible addition has been considered.

4. The colours A and B of all sites in the cluster are exchanged.

It is straightforward to prove ergodicity for this algorithm. Since our three-colour algorithm for square ice was ergodic (see Section 7.4), and since each move in the square ice algorithm is also a possible move in our F model algorithm (as long as $P_{\text{add}} < 1$), the result follows immediately.

Detailed balance is a little more tricky. As before, we consider two states μ and ν which differ by the exchange of colours in a single cluster of m squares. The probability of choosing the seed square to be in this cluster is m/N and the probability of choosing a particular colour B as the other colour for the cluster is $\frac{1}{2}$, just as in the square ice case. However, we now also have a factor of P_{add} for every square which we add to the cluster which is only a next-nearest neighbour of another and not a nearest neighbour. And we have a factor of $1 - P_{\text{add}}$ for every such site which we could have added but didn't. Thus the overall probability of making the move from μ to ν is

$$P(\mu \to \nu) = \frac{m}{2N} \prod_{[i,j]_{\text{con}}} P_{\text{add}} \prod_{[i,j]_{\text{dis}}} (1 - P_{\text{add}})^{\delta(c_i^\mu, c_j^\mu)}, \qquad (7.21)$$

where the two products run over pairs of next-nearest neighbours which are connected to or disconnected from the cluster respectively. (The notation $\delta(c_i, c_j)$ is just a more readable form of the Kronecker δ of Equation (7.20).) We will find it easier to work with the logarithm of this probability:

$$\log P(\mu \to \nu) = - \log(m/2N) + \log P_{\text{add}} \sum_{[i,j]_{\text{con}}} 1$$
$$+ \log(1 - P_{\text{add}}) \sum_{[i,j]_{\text{dis}}} \delta(c_i^\mu, c_j^\mu). \qquad (7.22)$$

The expression for $\log P(\nu \to \mu)$ is identical except for the exchange of the labels μ and ν. Thus the logarithm of the ratio of the probabilities for the forward and reverse moves is

$$\log \frac{P(\mu \to \nu)}{P(\nu \to \mu)} = \log(1 - P_{\text{add}}) \sum_{[i,j]_{\text{dis}}} \left[\delta(c_i^\mu, c_j^\mu) - \delta(c_i^\nu, c_j^\nu) \right]. \qquad (7.23)$$

The energy difference ΔE between states μ and ν is equal to ϵ times the change in the number of identically coloured next-nearest-neighbour squares

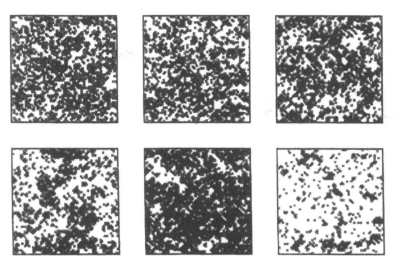

FIGURE 7.15   Sample configurations of the F model for increasing β. White squares denote vertices of type 5 and 6 on even and odd lattice sites respectively, and black vertices the reverse. All other vertices are denoted by grey squares. The temperatures are, on the top row, $\beta/\beta_c = 0.5$, 0.8, and 0.9; and on the bottom row, $\beta/\beta_c = 1.0$, 1.1, and 1.2.

(see Equation (7.20)). The only contribution to this sum comes from next-nearest-neighbour pairs $[i, j]$ such that i belongs to the cluster and j does not, since all other pairs contribute the same amount to the Hamiltonian in state μ as in state ν. Thus

$$\Delta E = E_\nu - E_\mu = -\epsilon \sum_{[i,j]_{\text{dis}}} \left[\delta(c_i^\nu, c_j^\nu) - \delta(c_i^\mu, c_j^\mu) \right]. \qquad (7.24)$$

In order to satisfy the condition of detailed balance we want the ratio of the rates $P(\mu \rightarrow \nu)$ and $P(\nu \rightarrow \mu)$ to be equal to the ratio $\exp(-\beta\Delta E)$ of the Boltzmann weights of the two states. Comparing Equations (7.23) and (7.24), we see that this can be arranged by setting $\log(1 - P_{\text{add}}) = -\beta\epsilon$, or[9]

$$P_{\text{add}} = 1 - e^{-\beta\epsilon}. \qquad (7.25)$$

We can also make a full-lattice version of this algorithm in exactly the same way as we did for the square ice case. We choose two colours A and B at random and create clusters all over the lattice from these two, using the

---

[9] Notice the similarity between this result and Equation (4.13) for the Wolff algorithm. Our cluster algorithm for the three-colour model and the Wolff algorithm for the Ising model have more in common than just a philosophical similarity.

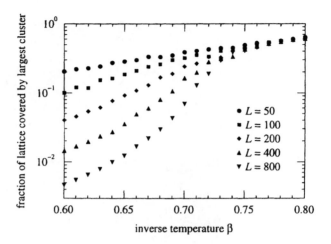

FIGURE 7.16   The average area covered by the largest cluster of symmetric vertices in the F model as a function of inverse temperature β
for a variety of different system sizes.

method above. The proofs of ergodicity and detailed balance for the full-
lattice version of the algorithm follow from the single-cluster version just as
in the case of the square ice model.

So how do these cluster algorithms perform? In Figure 7.15 we show some
results from simulations of the F model with $\epsilon = 1$ and varying temperatures
using the full-lattice version of the algorithm described above. In this figure
we have coloured areas of the two low-energy domains (checkerboards of
symmetric vertices) in black and white—type 5 vertices on even lattice sites
and type 6 vertices on odd lattice sites are black, while type 5 vertices on
odd lattice sites and type 6 vertices on even lattice sites are white. All other
vertices are in grey. The phase transition is clearly visible in the figure as a
change from a state in which black and white appear with equal frequency
to one in which one or the other dominates.

One observable in which we can detect the change in behaviour at the
phase transition is the fraction of the system covered by the largest cluster of
symmetric vertices (the black and white areas in Figure 7.15). In Figure 7.16
we have plotted this fraction as a function of the inverse temperature β for
a number of different lattice sizes L with $\epsilon = 1$. For large β (i.e., low
temperature) the largest cluster percolates (see Section 4.3) and covers a
sizeable fraction of the system. For smaller values the cluster only covers
a small fraction of the system and its size is independent of the size of the
lattice, so that the fraction of the total system size which it covers gets
smaller as L increases. The transition between these two regimes can be

seen in Figure 7.16, although it is only really clear for the larger system sizes ($L = 400$ and above). The critical temperature appears to be somewhere in the range between $\beta = 0.65$ and 0.75, which agrees with Lieb's exact figure of $\beta_c = \log 2 = 0.693$. In Section 8.3 we will show how the technique of finite size scaling can be used to make more accurate estimates of critical temperatures from Monte Carlo data.

More extensive simulations (Barkema and Newman 1998) indicate that the single-cluster three-colour algorithm for the F model is actually rather poor at simulating the model near T_c. The measured dynamic exponent is about $z = 1.3$, making it the worst of the algorithms we have looked at in this chapter. The full-lattice algorithm on the other hand is much better; there is no measurable increase in the correlation time in number of lattice sweeps with system size at T_c. The best estimate of the dynamic exponent is $z = 0.005 \pm 0.022$. Since this algorithm has an acceptance ratio of unity and is in addition relatively straightforward to program, it would clearly be the algorithm of choice for studying the critical properties of the F model.

Problems

7.1 We can extend the square ice model to three dimensions by placing arrows on the bonds of a cubic lattice and requiring that exactly three arrows enter and leave each vertex. Using Pauling's argument (Section 7.1.2), make an estimate of the residual entropy per site of this model.

7.2 In the ground states of the $q = 3$ Potts model with anti-ferromagnetic ($J < 0$) nearest-neighbour interactions, no two adjacent spins may take the same value. This means that the set of ground states is the same as the set of states of the three-colour model described in Section 7.1.3, or equivalently the square ice model. What kind of Potts model is equivalent to the F model of Section 7.6?

7.3 Design a Monte Carlo algorithm to simulate the eight-vertex model described at the beginning of Section 7.6, in the version where all vertices have the same energy.

8

Analysing Monte Carlo data

In the preceding chapters of this book we have looked at Monte Carlo algorithms for measuring the equilibrium properties of a variety of different models. The next few chapters are devoted to an examination of Monte Carlo methods for the study of out-of-equilibrium systems. But before we embark on that particular endeavour, we are going to take a look at some techniques for the analysis of data from equilibrium simulations. There are no new Monte Carlo algorithms in this chapter, but it does describe several methods of data analysis which can significantly improve the quality of the results we extract from the raw data spit out by the algorithms of the previous chapters.

In Chapter 3 we discussed a number of basic techniques for extracting estimates of quantities of interest from Monte Carlo data, and for calculating the errors on these estimates. We covered such topics as the measurement of the equilibration and correlation times, and the calculation of the values of directly measurable quantities such as internal energy, as well as ones which can only be measured indirectly, such as specific heat or entropy. We looked at error estimation techniques like blocking and bootstrapping, and at the calculation of correlation functions and autocorrelations. Most of the methods we discussed are essentially direct applications of ideas common to all data analysis; we deal with our Monte Carlo data very much as an experimentalist would deal with data from an experiment on a real physical system. In this chapter, by contrast, we will examine a number of data analysis techniques which are peculiar to Monte Carlo methods. In Sections 8.1 and 8.2 we look at the "histogram method" of Ferrenberg and Swendsen, in both its single and multiple histogram incarnations. The single histogram method allows us to take a single Monte Carlo simulation performed at some particular temperature and extrapolate the results to give predictions of observable quantities (internal energy, for instance) at other temperatures. The multiple histogram method allows us to interpolate between several different simulations performed at different temperatures. Both methods can also be generalized to allow extrapolation or interpolation in parameters other than the temperature, such as magnetic field in the case of a spin model.

In Sections 8.3 and 8.4 we discuss two techniques which are important for the study of critical phenomena using Monte Carlo methods. In Section 8.3 we look at "finite size scaling" which allows us to calculate critical exponents, critical temperatures and other critical properties by observing the way in which our simulation results vary as we change the size of the system. We have already seen a demonstration of one form of this technique in the calculation of the dynamic exponents for various Ising model algorithms in Chapter 4. In Section 8.4 we look at "Monte Carlo renormalization group" techniques which provide an alternative, and sometimes superior way of measuring critical properties.

8.1 The single histogram method

The **single histogram method** (sometimes just called the "histogram method") is a technique which allows us to perform a single Monte Carlo simulation at some specified temperature and extrapolate the results to other nearby temperatures. Clearly this could save us a great deal of CPU time compared with the alternative of performing a number of different simulations over a range of temperatures to extract the same information. The method is based on an idea put forward by Valleau and Card (1972). The version in common use today, which is the one we present here, is due to Ferrenberg and Swendsen (1988). In essence it is extremely simple.

Equation (2.2), which we repeat here for convenience, defines the way in which we calculate the estimator of a quantity Q from M measurements Q_{μ_i} made during the course of a Monte Carlo simulation:

$$Q_M = \frac{\sum_{i=1}^{M} Q_{\mu_i} p_{\mu_i}^{-1} e^{-\beta E_{\mu_i}}}{\sum_{j=1}^{M} p_{\mu_j}^{-1} e^{-\beta E_{\mu_j}}}. \tag{8.1}$$

In most of the algorithms we have seen so far, the probabilities p_{μ_i} with which the individual states are sampled once the simulation reaches equilibrium were chosen to be the Boltzmann weights for the temperature we are interested in.[1] Suppose however, that the p_{μ_i} are instead the Boltzmann probabilities for *another* temperature $\beta_0 = 1/kT_0$—one close to, but not exactly equal to the temperature we are interested in:

$$p_{\mu_i} = \frac{1}{Z_0} e^{-\beta_0 E_{\mu_i}}, \tag{8.2}$$

---

[1] An exception is the entropic sampling method of Section 6.3, which samples states in inverse proportion to the density of states.

where Z_0 is the partition function at temperature T_0. Substituting this into Equation (8.1), we get

$$Q_M = \frac{\sum_{i=1}^M Q_{\mu_i} e^{-(\beta-\beta_0)E_{\mu_i}}}{\sum_{j=1}^M e^{-(\beta-\beta_0)E_{\mu_j}}}. \tag{8.3}$$

This is the fundamental equation of the histogram method, in its simplest form. It allows us to take a series of measurements Q_{μ_i} of the observable Q, made during a Monte Carlo run at inverse temperature β_0, and extract from them an estimate of the thermal average of Q at inverse temperature β. Note that the energies E_{μ_i} appearing in the equation are the total energies of the states μ_i, and not energies per spin. (If you use energies per spin in this formula, it will give completely incorrect results, a point worth taking note of, since our Monte Carlo codes often calculate energy per spin.)

As an example of the application of the histogram method, we show in Figure 8.1 the internal energy of the two-dimensional normal Ising model on a 32×32 square lattice with $J = 1$ simulated for one million steps of the Swendsen–Wang cluster algorithm (Section 4.4.1). The simulation was performed at the known critical temperature of the 2D Ising model $\beta_c = \frac{1}{2}\log(1 + \sqrt{2})$. We have calculated the internal energy per spin u at β_c directly from the output of this simulation (the large dot in the figure), and then extrapolated to other nearby temperatures using the histogram method (solid line). We have also performed a number of further simulations over a range of temperatures to measure the internal energy directly. These results are shown as the open circles in the figure. As we can see, the extrapolation agrees well with the true result, until we get well away from T_0; extrapolating too far from the temperature of the original simulations gives poor results with the histogram method.

The errors on our extrapolation of the internal energy can be calculated by any of the methods discussed in Section 3.4. We treat the extrapolated values no differently from the way we would treat directly measured ones. However, as Figure 8.1 makes clear, there is another, rather poorly regulated source of error here: the failure of the extrapolation as we move too far away from the temperature at which the original simulation was performed. The obvious way of dealing with this error is to confine ourselves to the region in which the extrapolation is good. But how do we estimate the limits of this region? In Figure 8.1 we have performed a set of extra simulations to check our extrapolation, but this was only for illustrative purposes. In general, it would defeat the point of performing the extrapolation to check it against direct calculations. However, it turns out that there is another, quite straightforward way to calculate the limits of validity of the extrapolation directly from the results of our single simulation at T_0.

Still considering the case of the internal energy for the moment, the measurements Q_{μ_i} in Equation (8.3) are simply measurements E_{μ_i} of the total

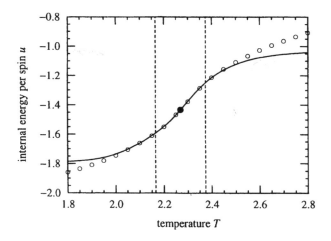

FIGURE 8.1   Illustration of the working of the histogram extrapolation
method for a 32×32 two-dimensional Ising model with $J = 1$. A single
simulation was performed at the critical temperature of the model
(large dot) and the results were extrapolated to nearby temperatures
(solid line). The dashed vertical lines represent the temperature range
over which the extrapolation is expected to remain accurate and the
open circles are the results of extra simulations performed to check the
accuracy of the method.

energy (i.e., the Hamiltonian) of the system. We can rewrite the equation
in the form

$$U = \langle E \rangle = \frac{\sum_E E N(E)\, e^{-(\beta - \beta_0)E}}{\sum_E N(E)\, e^{-(\beta - \beta_0)E}}, \tag{8.4}$$

where the sum now runs over all possible energies E of the states of our
32×32 Ising model (i.e., $E = -2048J, -2044J, \ldots 2048J$) and $N(E)$ is the
number of times that states of energy E occurred during the Monte Carlo
run. $N(E)$ is thus a histogram of the energies of the states sampled, and in
fact in their original exposition Ferrenberg and Swendsen wrote the entire
development of their method in terms of histograms like this, which is where
the name "histogram method" comes from. However, although we will use
the histogram here to show what the limits of the extrapolation should be, it
is not necessary to use it to perform calculations using the histogram method;
Equation (8.3) is all that is necessary. The main reason why one might
want to use a histogram is because the direct application of Equation (8.3)
normally requires us to store all the measurements Q_{μ_i} made during the
simulation. For a long simulation this means storing a large amount of data,
whereas the histogram by contrast is fairly small, and furthermore does not
grow in size with the length of the simulation. However, these days (in

contrast to the situation in 1988 when Ferrenberg and Swendsen published their original paper on the histogram method) disk storage space is cheap, and finding enough to store the results of a simulation is rarely a problem. Furthermore, while it is fine to use a histogram for models such as the Ising model, it is less satisfactory for models which have continuous energy spectra, such as the XY and Heisenberg models of Section 4.5.3. For models such as these, construction of a histogram means approximating our Monte Carlo data by lumping nearby energies into the same bin in the histogram. For models such as these, storing all the data and using Equation (8.3) is clearly a better course.

Returning to the question of the extrapolation range, we can regard Equation (8.4) as a weighted average over the energies E thus:

$$U = \sum_E E W(E), \tag{8.5}$$

with properly normalized weights $W(E)$ which depend on the temperature difference $\beta - \beta_0$:

$$W(E) = \frac{N(E)\, e^{-(\beta-\beta_0)E}}{\sum_E N(E)\, e^{-(\beta-\beta_0)E}}. \tag{8.6}$$

Note that $W(E)$ is proportional to $N(E)$ at the temperature $\beta = \beta_0$ at which the original simulation was performed. In Figure 8.2 we have plotted $W(E)$ as a function of energy for a variety of different values of $\beta - \beta_0$ for our 32×32 Ising model. This figure demonstrates in a graphical fashion how the histogram method works, and also shows why it eventually breaks down when β strays too far from β_0. In effect, the histogram method attempts to estimate what the histogram of energies would be for a simulation carried out at temperature T, based on the histogram measured at T_0, by reweighting the bins of the histogram with exponential factors which depend on the difference of temperatures. However, as we know from Section 1.2.1, the range of energies sampled by a typical Monte Carlo calculation is rather small. If we stray too far from the centre of this range we will find that the bins of the histogram are empty. And in the tails of the histogram the bins will contain very few samples, which means that the fractional sampling error in the number of samples will be large. Thus, attempting to calculate a reweighted histogram too far away from the temperature T_0 of the original simulation will either result in a histogram with large statistical errors, or worse, one with no samples in its bins. This problem is clearly visible in Figure 8.2 as we increase the separation of β and β_0, and is particularly noticeable in the rightmost curve.

The question then is, what is the largest temperature difference $\Delta T = T - T_0$ which we can extrapolate over and still get reliable results from the histogram method? The answer is that the middle of the distribution $W(E)$,

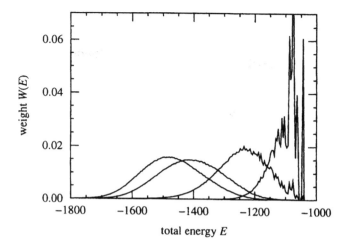

FIGURE 8.2   The weight function $W(E)$ as a function of E for the two-dimensional Ising model on a 32×32 lattice. The original simulation was performed at the critical temperature $T_c = 2.269$ with $J = 1$ and the curves shown are (left to right) for $T = T_c$, 2.3, 2.4 and 2.6.

which is approximately given by the internal energy $U = \langle E \rangle$, Equation (8.5), should fall within the range over which $N(E)$ is significantly greater than 1. If we represent this range by the standard deviation σ_E of $N(E)$, then our criterion is

$$|U(T) - U(T_0)| \leq \sigma_E. \tag{8.7}$$

The square σ_E^2 of the width of the histogram is proportional to the total specific heat C at temperature T_0 (see Equation (1.19)), so we can also write our criterion entirely in terms of the thermodynamic variables U and C thus:

$$[U(T) - U(T_0)]^2 \leq \frac{C(T_0)}{\beta_0^2}. \tag{8.8}$$

(Once again, we have set the Boltzmann constant $k = 1$ for simplicity.) If we further make the approximation that

$$U(T) - U(T_0) \simeq \left. \frac{dU}{dT} \right|_{T_0} (T - T_0) = C(T_0)\, \Delta T, \tag{8.9}$$

then the distance ΔT which we can reliably extrapolate away from T_0 is given by

$$\left[\frac{\Delta T}{T_0} \right]^2 = \frac{1}{C(T_0)}. \tag{8.10}$$

In fact, for simulations in which we take a large number of samples—a million is not atypical—this is quite a conservative estimate. It is quite possible that we will get reasonable results as much as $2\Delta T$ or more away from T_0. However, Equation (8.10) provides a good rule of thumb to guide us in our work.

In the case of our 32×32 Ising model calculation, we performed our simulation at the critical temperature $T_0 = T_c = 2.269$. At this temperature we find that the specific heat is about $C(T_0) = 1.89 \times 10^3$. Applying Equation (8.10), this gives us $\Delta T = 0.052$. In this case our simulation was quite a large one, producing a million samples, so it is probably safe to extrapolate $2\Delta T$ from the central temperature. (Assuming normally distributed samples, going two standard deviations σ_E away from the mean sample energy still leaves 5% of the samples in the region around the peak of the reweighted histogram. For a million sample run 5% is 50 000 samples, which should be plenty to give a reasonable estimate of u.) We have marked these limits with the vertical dashed lines in Figure 8.1, and, as we can see, these do indeed represent a fair estimate of the region in which the extrapolated results are in agreement with the direct measurements.

The case we have discussed here of the measurement of the internal energy is a particularly simple one. Both the histogram and the Boltzmann weight are functions of the same variable E, which makes the equations especially simple. However, the application of the histogram method to other variables is not difficult: the fundamental Equation (8.3) applies perfectly well to any other observable, such as magnetization for example, and performing the extrapolation just involves the application of this equation. Furthermore, the arguments given here for the limits of the extrapolation still apply: the width of the energy distribution of the sampled states still gives an estimate of the temperature range over which we can reliably extrapolate. So Equation (8.10) still applies. The only change occurs if we wish to perform the extrapolation using an actual histogram, as in Equation (8.4). In this case, the correct generalization is

$$Q_M = \frac{\sum_{E,Q} Q N(E,Q) \, e^{-(\beta - \beta_0)E}}{\sum_{E,Q} N(E,Q) \, e^{-(\beta - \beta_0)E}}, \qquad (8.11)$$

where $N(E,Q)$ is a two-dimensional histogram of energy and the observable Q. In the case of an Ising model, for example, $N(E,M)$ would be the number of times during the Monte Carlo run that a state with energy E and magnetization M appeared. Unfortunately, this two-dimensional histogram may have many millions of bins for a large system, making it impractical to calculate. What's more, a separate histogram has to be constructed for each observable we wish to measure. For these reasons, as well as for those given earlier, it is normally better just to apply Equation (8.3) directly, and this is the course of action we would usually recommend.

8.1.1 Implementation

The single histogram method is usually very simple to implement, as the example given in the last section demonstrates. However, there is one problem which crops up, particularly with simulations of large systems, which can make direct implementation of the method problematic. Recall that the energies appearing in Equation (8.3) are *total* energies of the system simulated, which means that they increase as the volume of the simulated system. For larger systems therefore, it is not unusual for the exponentials to overflow or underflow the range of reals which the computer can represent (which is about $10^{\pm 300}$ on most modern computers). There are a couple of tricks we can use in this situation.

Since the properties of our system are independent of where we set the origin of measured energies, we can add or subtract any constant E_0 from all energies E_μ and still get the same answer for the expectation of any observable quantity. (Quantities which depend directly on the energy will of course be shifted by an amount corresponding to E_0, but this is easy to correct for.) Thus we can reduce the probability of the exponentials causing overflows or underflows by subtracting the mean energy $E_0 = \langle E \rangle$ from all energies before performing the histogram calculation. The range of energies sampled by the calculation increases only as $\sqrt{N}$, where N is the size of the system, so by using this trick we should be able to simulate significantly larger systems before overflow occurs. Often this is sufficient to remove the problem for all system sizes of interest. For very large systems however, it may not be enough, in which case we can use another trick, similar to the one we used for the density of states in the entropic sampling method (see Section 6.3.3). Instead of calculating the terms in Equation (8.3) directly, we calculate their logarithms, which have a much smaller dynamic range and will never overflow the limits of the computer's accuracy. Evaluating the equation then requires us to calculate a sum of quantities for which we know only the logarithms. Moreover, we want to do this without taking the exponential of these logarithms, since we are using the logarithms precisely because the exponentials are expected to overflow the numerical limits of our computer. We can get around this problem using the following trick. Suppose ℓ_1 and ℓ_2 are the logarithms of two numbers x_1 and x_2, and that $\ell_1 \geq \ell_2$. Then the logarithm of the sum of x_1 and x_2 is

$$
\begin{aligned}
\log(x_1 + x_2) &= \log(e^{\ell_1} + e^{\ell_2}) = \log(e^{\ell_1}(1 + e^{\ell_2 - \ell_1})) \\
&= \ell_1 + \log(1 + e^{\ell_2 - \ell_1}).
\end{aligned} \tag{8.12}
$$

The exponential inside the logarithm is, by hypothesis, less than one, so the expression can be safely evaluated without risk of overflow. In the case where $\ell_1 < \ell_2$ we can use the same expression but with ℓ_1 and ℓ_2 swapped around. Most computers provide a library function which will evaluate $\log(1+x)$ accurately, even for very small x (which is often the case with Equation (8.12)), and if such a function is available we advise you to use it. (In C, for example, it is usually called `log1p()`.)

Using Equation (8.12) requires more computational effort than performing the simple sums involved in Equation (8.3) and it can slow down the calculation considerably. However, the time taken is usually still small compared with the time invested in performing the Monte Carlo simulation itself, so this is not normally an issue of any importance.

Another slight problem is that some observables, such as energy or magnetization, may have negative values, which means that we cannot take the logarithms of the terms in the denominator of (8.3). Normally, however, there is a way around this problem. In the case of the energy for instance, since we are at liberty to shift its value by any constant amount E_0, we can simply shift the energies by an amount big enough to make them all positive. In the case of the magnetization, we normally work with the absolute value anyway, so sign problems do not arise. If there is no other alternative, we can always represent the absolute magnitude of a quantity using a logarithm and store the sign of the quantity in a separate variable.

8.1.2   Extrapolating in other variables

We can also use the histogram method to extrapolate in variables other than the temperature. In particular, if we can write the Hamiltonian in the form

$$H = \sum_k J^{(k)} E^{(k)}, \tag{8.13}$$

then we can perform an extrapolation in any of the variables $J^{(k)}$. Here, the $J^{(k)}$ are "coupling constants" which give a scale to the energies in the Hamiltonian (such as the parameter J appearing in the Ising Hamiltonian), and the $E^{(k)}$ are dimensionless functions of the spins or other degrees of freedom in the model we are studying. For example, in a uniform external magnetic field, the Hamiltonian of the Ising model is

$$H = -J \sum_{\langle ij \rangle} s_i s_j - B \sum_i s_i, \tag{8.14}$$

which is of the form given in Equation (8.13) with

$$J^{(1)} = -J, \qquad E^{(1)} = \sum_{\langle ij \rangle} s_i s_j, \tag{8.15}$$

$$J^{(2)} = -B, \qquad E^{(2)} = \sum_i s_i. \tag{8.16}$$

We can extrapolate in any of the variables $J^{(k)}$ using a simple general-ization of Equation (8.3). First, we rewrite Equation (8.1) for the estimator of Q as

$$Q_M = \frac{\sum_{i=1}^{M} Q_{\mu_i} p_{\mu_i}^{-1} \exp\left(-\beta \sum_k J^{(k)} E_{\mu_i}^{(k)}\right)}{\sum_{j=1}^{M} p_{\mu_j}^{-1} \exp\left(-\beta \sum_k J^{(k)} E_{\mu_j}^{(k)}\right)}. \tag{8.17}$$

Suppose we wish to calculate Q_M at some point in the parameter space represented by the values $\{J^{(k)}\}$, but our Monte Carlo run was actually performed at some nearby point $\{J_0^{(k)}\}$. Then the sampling probabilities p_{μ_i} are given by

$$p_{\mu_i} = \frac{1}{Z_0} \exp\left(-\beta \sum_k J_0^{(k)} E_{\mu_i}^{(k)}\right). \tag{8.18}$$

Substituting this into Equation (8.17), we then get

$$Q_M = \frac{\sum_{i=1}^{M} Q_{\mu_i} \exp\left[-\beta \sum_k \left(J^{(k)} - J_0^{(k)}\right) E_{\mu_i}^{(k)}\right]}{\sum_{j=1}^{M} \exp\left[-\beta \sum_k \left(J^{(k)} - J_0^{(k)}\right) E_{\mu_j}^{(k)}\right]}. \tag{8.19}$$

This is the appropriate generalization of Equation (8.3) for extrapolation in a number of variables. In order to use this equation, we need to perform a Monte Carlo simulation of our system at a particular point $\{J_0^{(k)}\}$ in the parameter space and record the value of each of the energy terms $E^{(k)}$ for each sample we take. Note that it is not adequate to record only the total energy E_{μ_i} of each state as we did before.

8.2 The multiple histogram method

If we want to estimate our observable Q over a larger range of temperatures than the range allowed by the single histogram method (see Equation (8.10)), then we need to perform simulations at more than one temperature. (Again, we will for the moment restrict our discussion to variations in temperature, but the techniques we introduce in this section can be generalized to other parameters in a straightforward fashion.) If these simulations sample sets of states with overlapping ranges of energy, then the ranges over which we can extrapolate the results of each individual simulation will overlap, and so we should in theory be able to employ the extrapolation method intro-duced in the last section to calculate the value of any observable over the entire temperature range spanned by our simulations. However, we can do better than this. It is clear, for instance, that at the extreme ends of the extrapolation range of the single histogram method, the accuracy of the ex-trapolation decreases. In the case of several simulations with overlapping ranges however, the upper end of one simulation's range is the lower end of another's. Although both extrapolations may be rather poor in this region, one feels that it ought to be possible to combine the estimates from the two simulations to give a better estimate of $\langle Q \rangle$. Indeed, since every simulation gives an estimate (however poor) of the value of $\langle Q \rangle$ at every temperature, we should be able to combine all of these estimates in some fashion (presumably giving greater weight to ones which are more accurate) to give the best possible figure for $\langle Q \rangle$ given our several simulations. This in essence is the idea behind the **multiple histogram method**, first proposed by Ferrenberg and Swendsen in 1989. The multiple histogram method is in fact not particularly closely related to the single histogram method of Section 8.1, and it is probably better to regard it as an entirely new technique, rather than an extension of the ideas of the last section.

There are a variety of ways in which we might reasonably combine estimates of an observable derived from several different Monte Carlo calculations. The most obvious approach is to take some kind of weighted average of the single histogram extrapolations from each of our simulations. It turns out, however, that this approach is rather error-prone and does not give good results except when the quality of our Monte Carlo data is very high. Here we describe an alternative prescription which concentrates instead on the evaluation of a best estimate for the density of states. From this estimate, we can then derive the value of any other observables we are interested in. As we did in the last section, we will start off by considering the observable E—the internal energy—which is the simplest case. In Section 8.2.2 we give the appropriate generalization to other quantities.

For a Monte Carlo simulation with normal Boltzmann importance sampling, the probability $p(E)$ of generating a state with energy E on any one time-step is

$$p(E) = \rho(E)\frac{e^{-\beta E}}{Z}, \tag{8.20}$$

where $\rho(E)$ is the (energy) density of states—the number of states of the system[2] with energy E—and Z is the partition function (neither of which we usually know). If we perform a Monte Carlo simulation at inverse temperature β and make n independent measurements of the total energy E of the system, then we can make an estimate of $p(E)$ thus:

$$p(E) = \frac{N(E)}{n}, \tag{8.21}$$

where the histogram $N(E)$ is the number of times out of n that the system was measured to have energy E. Equation (8.20) then tells us that the

---

[2]We are assuming here that the system has discrete energies, as the Ising model for example does. If it has a continuous spectrum of energies, as in the Heisenberg model for example, then $\rho(E)$ is replaced by $\rho(E)\,dE$, the number of states in an interval dE, and the following equations are modified appropriately. As it turns out however, the final equations for the multiple histogram interpolation in the continuous case are exactly the same as for the discrete case considered here.

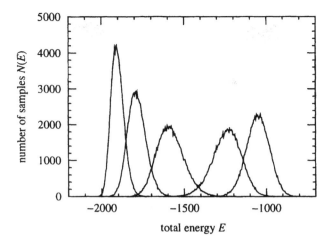

FIGURE 8.3 Histograms $N(E)$ of energies measured for a two-dimensional Ising model on a 32×32 square lattice with $J = 1$ in five runs of 100 000 Monte Carlo steps each using the Swendsen–Wang algorithm. The temperatures of the runs are (left to right) $T = 1.8$, 2.0, 2.2, 2.4 and 2.6.

density of states is given by

$$\rho(E) = \frac{N(E)}{n} \frac{Z}{e^{-\beta E}}. \tag{8.22}$$

Now suppose we perform a number of different simulations at a number of inverse temperatures β_i, giving a number of different estimates of the density of states:

$$\rho_i(E) = \frac{N_i(E)}{n_i} \frac{Z_i}{e^{-\beta_i E}}. \tag{8.23}$$

Since $\rho(E)$ is a single function which depends only on the system we are studying, and not on the temperature, each of these estimates $\rho_i(E)$ is an estimate of the same function. Now we ask ourselves, "What is the best estimate we can make of the true density of states, given these many different estimates from our different simulations?" To make the point clearer, look at Figure 8.3, in which we have plotted the histograms $N_i(E)$ of energies measured in five simulations of a two-dimensional Ising model at five different temperatures. As we can see, the simulations sample different but overlapping ranges of energy. Clearly Equation (8.23) will give a fairly good measure of the density of states in a region where the corresponding histogram $N_i(E)$ has many samples. But in other regions where it has few (or none at all) it will give a very poor estimate. What we want to do is form a weighted average over the estimates $\rho_i(E)$ to get a good estimate over the whole range of energies covered by our histograms. This average should give more weight to individual estimates in regions where the corresponding histograms have more samples.

The standard way to perform such a weighted average is as follows. If we have a number of measurements x_i of a quantity x, each of which has some associated standard error σ_i, then the best estimate of x is[3]

$$\bar{x} = \frac{\sum_i x_i / \sigma_i^2}{\sum_j 1/\sigma_j^2}. \tag{8.24}$$

In other words, we take a weighted average in which we weight each measurement with the inverse of its variance about the true value.[4] It turns out that the variances $\sigma_i^2(E)$ on our estimates $\rho_i(E)$ of the density of states are proportional to $N_i^{-1}(E)$, so that to get the best estimate of $\rho(E)$ we simply weight each individual estimate according to the number of samples in the corresponding histogram at that energy. Although this result is satisfyingly intuitive, its proof is quite subtle.

Assuming that our measurements of the energy of the system are independent, the error $\Delta N_i(E)$ on the number of samples in each bin of a histogram should be simply Poissonian. In other words

$$\Delta N_i(E) = \sqrt{N_i(E)}. \tag{8.25}$$

In fact, and this turns out to be an important point, the true error $\Delta N_i(E)$ is a function of the *average* histogram $\overline{N_i(E)}$ at temperature β_i, taken over many runs.[5]

$$\Delta N_i(E) = \sqrt{\overline{N_i(E)}}. \tag{8.26}$$

---

[3] This assumes that the errors are normally distributed. As we will see, the errors on our histograms are Poissonian, but it is a reasonable assumption to approximate Poissonian errors as Gaussian in this case.

[4] It is in fact not too difficult to prove this formula. Writing $\bar{x}$ as a weighted average over the measurements, with weights w_i thus:

$$\bar{x} = \frac{\sum_i w_i x_i}{\sum_j w_j},$$

we can then calculate the difference between this estimate and the true value x, which we don't know. Then we square this quantity and minimize, so as to make $\bar{x}$ as close as possible to the true value. Making the assumption that the errors σ_i in the different measurements are all independent random variables, we can then show that the best estimate is achieved when $w_i = 1/\sigma_i^2$.

[5] Similarly, we can make an estimate of the standard deviation of a quantity x by making a number of measurements x_i and calculating the standard deviation of the sample. However, the true standard deviation is only given by taking an infinite number of such measurements.

To clarify what we mean here, imagine making a very large number of simulations all at inverse temperature β_i, taking n_i measurements of E in each, forming a histogram of each run, and then averaging the histograms bin by bin. The square root of these average bins would then give us the correct estimate of the error $\Delta N_i(E)$ on any one histogram. The reason that this is important is that $\overline{N_i(E)}$ is related to $\rho(E)$ thus (see Equation (8.23)):

$$\rho(E) = \frac{\overline{N_i(E)}}{n_i} \frac{Z_i}{e^{-\beta_i E}}, \qquad (8.27)$$

where $\rho(E)$ is the *exact* density of states. In other words, if we were to do an infinite number of simulations at β_i, then we could in principle calculate the true density of states for all energies from the results. Of course, we cannot do this, but Equation (8.27) will allow us to simplify our equations for the multiple histogram method considerably.

The Poissonian variation in $N_i(E)$ is the only source of error in Equation (8.23), so the error σ_i on $\rho_i(E)$ is

$$\sigma_i = \frac{\Delta N_i(E)}{n_i} \frac{Z_i}{e^{-\beta_i E}} = \frac{\sqrt{N_i(E)}}{n_i} \frac{Z_i}{e^{-\beta_i E}}, \qquad (8.28)$$

and the variance is

$$\sigma_i^2 = \frac{\overline{N_i(E)}}{n_i^2} \left[\frac{Z_i}{e^{-\beta_i E}} \right]^2 = \frac{\rho^2(E)}{N_i(E)}, \qquad (8.29)$$

where we have used Equation (8.27). Recalling (Equation (8.24)) that the weights in our weighted average over the $\rho_i(E)$ are $1/\sigma_i^2$, this shows, as we contended above, that the best estimate of the true $\rho(E)$ is arrived at by weighting each $\rho_i(E)$ in proportion to the number of samples in the corresponding histogram at that energy.[6] Performing the weighted average, our best estimate of $\rho(E)$ is thus

$$\rho(E) = \frac{\sum_i \overline{N_i(E)}\,[N_i(E)/n_i][Z_i/e^{-\beta_i E}]}{\sum_j \overline{N_j(E)}} = \frac{\sum_i N_i(E)}{\sum_j n_j Z_j^{-1} e^{-\beta_j E}}, \qquad (8.30)$$

where we have used Equation (8.27) again to get rid of the quantities $\overline{N_i(E)}$, whose values we don't know.

Unfortunately however, we are not finished yet. This expression is still of little use to us, because it contains the partition functions Z_j of the system at each of the simulated temperatures β_j on the right-hand side. These are

----

[6]Strictly, it's the *ideal* histogram $\overline{N_i(E)}$, not the measured one.

also quantities which we don't know. We get around this by noting that the partition function itself is given by

$$Z_k = \sum_E \rho(E)\, e^{-\beta_k E} = \sum_E \frac{\sum_i N_i(E)}{\sum_j n_j Z_j^{-1} e^{(\beta_k - \beta_j)E}}. \qquad (8.31)$$

Implementing the multiple histogram method involves solving this equation iteratively for the partition functions Z_k at each of the inverse temperatures β_k used in the simulations. We make some initial estimate of the values of the Z_k, plug them into this equation and arrive at a different, better estimate. Repeating this process many times, we eventually converge on a "fixed point" of the equations, which gives the correct values for the partition functions. Given these values, we can then calculate the partition function at any other temperature from a simple generalization of (8.31):

$$Z(\beta) = \sum_E \frac{\sum_i N_i(E)}{\sum_j n_j Z_j^{-1} e^{(\beta - \beta_j)E}}. \qquad (8.32)$$

Combining this with Equation (8.30) we can then get an estimate of, for instance, the internal energy:

$$\begin{aligned}
U(\beta) &= \sum_E E\rho(E)\frac{e^{-\beta E}}{Z(\beta)} \\
&= \frac{1}{Z(\beta)} \sum_E E \frac{\sum_i N_i(E)}{\sum_j n_j Z_j^{-1} e^{(\beta - \beta_j)E}}.
\end{aligned} \qquad (8.33)$$

Note that, just as with the single histogram method, the energies E appearing in all of these formulae are total energies, not energies per spin. If we use energies per spin the method will give erroneous results.

The multiple histogram method is undoubtedly quite a complicated technique, but in practice it gives excellent results. As an example, consider Figure 8.4. Here we have taken the five Ising simulations which produced the histograms in Figure 8.3 and calculated the internal energy for each one by a direct average over the measured energies. These five measurements are represented by the five large dots. Then we have used the five histograms to evaluate the partition function at each of the five temperatures by iterating Equation (8.31) until it converges to a stable fixed point. Then we have calculated the internal energy over the entire range of temperatures using Equations (8.32) and (8.33). The results are shown as the solid line in the figure. As we can see, the line passes neatly through each of the large dots, and interpolates smoothly between them. Finally, we have performed a number of further simulations at temperatures intermediate between the first simulations, as a check of the histogram interpolation. From these we

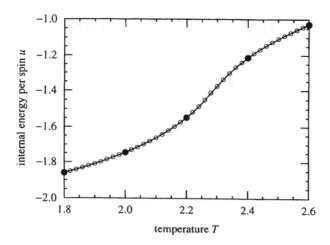

FIGURE 8.4 Illustration of the working of the multiple histogram method for our 32×32 two-dimensional Ising model. The five large dots are direct measurements of the internal energy at five temperatures. The solid line is the histogram interpolation between these five values. The open circles are the results of extra simulations performed to check the accuracy of the interpolation.

have made direct measurements of the internal energy which we show as the open circles on the figure.

We can also calculate the errors on our interpolated results by applying any of the methods described in Section 3.4. For example, we could repeat the entire histogram procedure several times over with bootstrap resampled data sets (Section 3.4.3) drawn from the data produced by the simulations, and calculate the variance of the interpolated results over all repetitions. In fact, in the case of the calculation shown in Figure 8.4, the error on u is smaller than the width of the line, so there isn't much point in our including it on the graph.

As the figure shows, the histogram interpolation approximates extremely well the true value of the internal energy in the regions between the five initial simulations. This is the power of the multiple histogram method: it allows us to perform a small number of simulations at different temperatures and interpolate between them to get a good estimate of an observable over the entire range,[7] thereby saving us the effort of performing more simulations at intermediate temperatures. The amount of CPU time involved in performing

---

[7] Actually, the method also allows us to extrapolate a short distance outside the range sampled by the simulations—it extrapolates well over the same range as the single histogram method (see Equation (8.10)).

the iteration of Equation (8.31) is small compared with the amount needed to perform most Monte Carlo simulations, especially on large systems, so the method can save us a lot of time when we want to measure some quantity over a large range of temperatures.

8.2.1   Implementation

Before moving on to other topics, we should say a few words on how you actually implement the multiple histogram method. First, how widely spaced should our Monte Carlo simulations be to allow accurate interpolation using the multiple histogram method? Well, the range over which any one of our estimates $\rho_i(E)$ of the density of states is valid is the range over which the corresponding histogram $N_i(E)$ is significantly greater than one. Thus, the linear combination used to estimate $\rho(E)$ will only give good results over the entire temperature range if the histograms from the different simulations overlap, just as we suggested they should earlier. As in the case of the single histogram method, we can write a criterion for this overlap in terms of the standard deviations of the samples in the histograms, or in terms of the specific heat. A simple version, which works perfectly well in most cases, is just to double the temperature range allowed by Equation (8.10) and make this the temperature separation of adjacent simulations.

The complicated part of the histogram calculation is the iteration of Equation (8.31) to calculate the partition functions Z_k. There are a number of tricks which are useful here. The main problem is that the values of the partition functions can become very large or very small—recall that the partition function is exponentially dependent on the total energy E of the states of the system. We can alleviate this problem somewhat by normalizing the partition functions. Notice that Equation (8.31) remains unchanged if we multiply all the Z_k by the same normalization factor A. Furthermore, any such normalization will cancel out of Equation (8.33) when we come to calculate U. It makes sense to normalize the partition functions by some factor which reduces the chances of their either over- or under-flowing the numerical range of the computer (about $10^{\pm 300}$). A simple way to achieve this is to set A equal to one over the geometric mean of the largest and smallest partition functions at each iteration:

$$A = \frac{1}{\sqrt{Z_{\text{large}} Z_{\text{small}}}}. \tag{8.34}$$

This makes the normalized values of Z_{large} and Z_{small} reciprocals of one another, and therefore equally far away from the limits of precision of the machine.

It is not unusual however, for the ratio $Z_{\text{large}}/Z_{\text{small}}$ to exceed even the 600 orders of magnitude over which a typical computer can represent real numbers accurately, in which case the trick above will not be sufficient and at least one of our partition functions will pass outside the allowed range, causing the histogram method to fail. This is particularly common with large system sizes, since the value of the partition function increases exponentially with the volume of the system simulated. In this case, we can use the same trick as we did in the single histogram method and work with the logarithms of the partition functions. (The partition function is always positive, so there is no danger of our trying to take the logarithm of a negative number.) To evaluate the sums in Equations (8.31), (8.32) and (8.33) we use Equation (8.12).

Another issue is the question of what the starting values of the Z_k should be, and how many iterations we need to perform to calculate their final value. The starting values are in fact not particularly important. The only crucial point is that they should be greater than zero. In the calculations for Figure 8.4 we used starting values of $Z_k = 1$ for all k. If we were performing the calculation using the logarithms of the partition functions, then the equivalent starting values of the logarithms would be zero. As for the number of iterations we should use, we can estimate this from the amount by which the partition functions change on each iteration. In the calculations presented here, we gauged this by calculating the quantity

$$\Delta^2 = \sum_k \left[\frac{Z_k^{(m)} - Z_k^{(m-1)}}{Z_k^{(m)}} \right]^2, \tag{8.35}$$

where $Z_k^{(m)}$ is the value of Z_k at the mth iteration. When this quantity falls below some predefined target value ϵ^2 we know that the fractional change per iteration in the most quickly changing of the Z_k is less than ϵ. In our calculations we used a value of $\epsilon = 10^{-7}$. (Strictly, this gives us no guarantees about the accuracy of our results, since we have no information about the speed with which the iteration converges. In practice however, it converges very fast—exponentially in fact—and the criterion above gives perfectly good results.)

A further point worth noting is that, as with the single histogram method, it is not necessary to express all the formulae in terms of actual histograms $N(E)$. In fact, it is often more convenient to work with the raw data themselves. We can rewrite the fundamental iterative relation, Equation (8.31), as follows:

$$\begin{aligned} Z_k &= \sum_E \frac{\sum_i N_i(E)}{\sum_j n_j Z_j^{-1} e^{(\beta_k - \beta_j)E}} = \sum_{i,E} \frac{N_i(E)}{\sum_j n_j Z_j^{-1} e^{(\beta_k - \beta_j)E}} \\ &= \sum_{i,s} \frac{1}{\sum_j n_j Z_j^{-1} e^{(\beta_k - \beta_j)E_{is}}}, \end{aligned} \tag{8.36}$$

where the sum over s is over all states sampled during the ith simulation, and E_{is} is the total energy of such a state. Equation (8.32) can similarly be rewritten

$$Z(\beta) = \sum_{i,s} \frac{1}{\sum_j n_j Z_j^{-1} \mathrm{e}^{(\beta - \beta_j) E_{is}}}, \tag{8.37}$$

and Equation (8.33) as

$$U(\beta) = \frac{1}{Z(\beta)} \sum_{i,s} \frac{E_{is}}{\sum_j n_j Z_j^{-1} \mathrm{e}^{(\beta - \beta_j) E_{is}}}. \tag{8.38}$$

As with the single histogram method, these forms of the equations are also better if we are studying systems with continuous energy spectra, since in that case the construction of a histogram necessarily entails throwing away some of the information from our Monte Carlo simulation. The equations above by contrast make use of all the available information and so will in general give a more accurate answer for U.

8.2.2 Interpolating other variables

The multiple histogram method can easily be extended to provide interpolation of quantities other than the internal energy. If we wish to interpolate between measurements of an observable Q, for instance, we would still iterate Equation (8.36) (or equivalently Equation (8.31)) to calculate the partition functions for each of the temperatures simulated and then use Equation (8.37) (or (8.32)) to calculate Z for all intervening temperatures. Then the expectation of Q is given by the appropriate generalization of (8.38):

$$\langle Q \rangle = \frac{1}{Z(\beta)} \sum_{i,s} \frac{Q_{is}}{\sum_j n_j Z_j^{-1} \mathrm{e}^{(\beta - \beta_j) E_{is}}}. \tag{8.39}$$

It is possible to make a similar generalization of Equation (8.33) for U to other quantities, but, as with the single histogram method, this involves constructing a two-dimensional histogram, which takes up a lot of space, as well as being inappropriate for systems with continuous energy spectra. For these reasons we recommend using (8.39) directly in most cases.

It is also possible to extend the multiple histogram method to give interpolation in other variables, such as the external magnetic field in an Ising model for example. The developments are straightforward, and we leave the details to the interested reader.

8.3   Finite size scaling

A surprisingly large fraction of the published papers on Monte Carlo calculations, and indeed statistical physics papers in general, deal with critical phenomena. As discussed in Section 3.7.1, critical phenomena are phenomena seen when a system is close to a phase transition, in particular, a continuous phase transition. Although it might seem that the critical region is only one small portion of a system's parameter space and not deserving of all the attention it has received, there are a number of reasons why we should be interested in it.

First, critical phenomena are intrinsically interesting; some systems display quite unusual behaviour in the vicinity of a phase transition. Examples are the onset of ferromagnetism at the Curie point of a ferromagnet, or the optical effect known as **critical opalescence**, which causes large fluctuations in the refractive index of critical mixtures of certain liquids such as methanol and hexane.

Second, phase transitions are interesting because of the phenomenon of universality, which we discussed briefly in Section 4.1. Universality means that the values of critical exponents for a model (as well as a number of other measurable quantities) are independent of most of the parameters of the model, such as the interaction energy J between spins in the Ising model for example, or the topology of the lattice. Only fairly gross properties of the system, such as the dimensionality d of the lattice, or the number of dimensions of the order parameter, affect the critical exponents. This means that the values of the exponents fall into a number of discrete **universality classes**, and every model which falls in a particular universality class will have the same values for these exponents. This in turn means that we can measure their values using some simple model for which the calculation is relatively easy, and still get an answer that applies to a far more complicated real-world system which just happens to fall in the same class. As an example, the phase transition of a liquid/vapour system at its tri-critical point is believed to be in the same universality class as the ordinary Ising model, and indeed the measured critical exponents of the two agree quite well.

A third reason for the large amount of interest which critical phenomena have attracted, and one which should not be ignored, is that studying them, either experimentally or theoretically, has proved to be quite difficult. And if there's one thing scientists like, it's a challenge.

All of these points and more have been discussed at great length elsewhere and it is outside the scope of this book to discuss critical phenomena much more thoroughly than we already have. The interested reader is referred to one of the many texts on the subject. However, one aspect of the area which does concern us is the calculation of critical exponents using Monte Carlo techniques, and it is to this subject that the remainder of this chapter is devoted. In this section we study the most widely used technique for extracting the values of exponents, **finite size scaling**. In Section 8.4 we study "Monte Carlo renormalization group" techniques, which are more complex but can in some cases give better answers.

8.3.1   Direct measurement of critical exponents

In Section 4.1 we defined a number of critical exponents which govern the singular behaviour of various quantities close to the critical temperature T_c. For convenience, we repeat those definitions here. If we define the reduced temperature t as a measure of our distance from the critical temperature:

$$t = \frac{T - T_c}{T_c}, \tag{8.40}$$

then in the thermodynamic limit (i.e., infinite system size) the behaviour of the correlation length ξ in the critical region is given by

$$\xi \sim |t|^{-\nu}, \tag{8.41}$$

where ν is a critical exponent. For specific models we can usually define other exponents also. For example, in the normal Ising model we can define exponents γ and α which govern singularities in the magnetic susceptibility and specific heat thus:

$$\chi \sim |t|^{-\gamma}, \tag{8.42}$$
$$c \sim |t|^{-\alpha}. \tag{8.43}$$

Also, although we didn't mention it in Chapter 4, we can define an exponent β which governs the behaviour of the magnetization:

$$m \sim |t|^{\beta}. \tag{8.44}$$

This equation is only useful below the critical temperature however, since m is zero for $T \geq T_c$.

How should we go about measuring these critical exponents? The most obvious approach is the direct one, simply to fit the data from our Monte Carlo simulations to the suggested asymptotic forms. For example, in Figure 8.5 we show a number of measurements of the magnetic susceptibility of a 1000×1000 two-dimensional Ising model below its critical temperature, made using the Swendsen–Wang cluster algorithm of Section 4.4.1. We have plotted them on logarithmic scales against the values of the reduced temperature t and, as the figure shows, they tend to what appears to be a straight line as we approach the critical temperature at $t = 0$. The slope of this line should give us the value of the critical exponent γ defined in Equation (8.42).

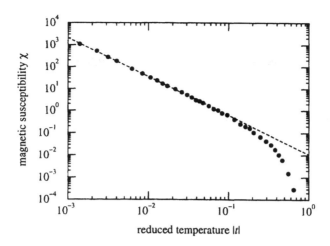

reduced temperature |t|

FIGURE 8.5 Monte Carlo results for the magnetic susceptibility χ of a 1000×1000 Ising model on a square lattice as a function of the absolute reduced temperature $|t|$. Fitting the results to a straight line gives a figure of $\gamma = 1.76 \pm 0.02$ for the corresponding critical exponent (see Equation (8.42)).

Fitting a straight line to the last twenty or so points, as shown in the figure, gives us a value of $\gamma = 1.76 \pm 0.02$, which compares very well with the known exact result $\gamma = \frac{7}{4}$. We could probably improve on this value still further by using the multiple histogram method of Section 8.2 to interpolate between the points in the figure. However, there is in fact little point in doing this, because this direct method of measuring exponents turns out to have a number of problems which make it unsuitable for general use. One problem is already evident in Figure 8.5. The simple power-law form of the susceptibility is only followed closely when we are very near the critical temperature. Away from T_c the line in the figure is no longer straight. In order to perform a fit to the data, we need to estimate where the critical region ends, and the value we get for the exponent γ varies depending on this estimate. This makes the error on γ difficult to estimate. Furthermore, some systems, such as the random-field Ising model of Section 6.1.1, show two different values of their critical exponents. The real one is only seen if you get sufficiently close to the critical temperature. Further away the system appears to behave according to Equation (8.42), or the equivalent for the exponent of interest, but with the wrong value of the exponent. Such "cross-over" effects are very difficult to allow for if we don't know where the cross-over actually takes place.

Another problem is that in order to calculate the reduced temperature t we need to know the value of the critical temperature (see Equation (8.40)).

In general we don't know this value, which makes it difficult to perform the fit. It is possible to guess T_c and then vary the guess to make the line in Figure 8.5 as close to straight as possible. However, this process is highly susceptible to error. The curvature which is already present in the line as we move away from T_c can be particularly misleading when we are trying to gauge the straightness of the line, and it also turns out that rather small miscalculations in the value of T_c can lead to large errors in the measured critical exponent.

There are also other reasons for avoiding the direct method of measuring exponents. It requires simulations of rather high accuracy to give reasonable results, and it requires us to work with large systems to avoid problems with "finite size effects" (see below). In short, the method is to be avoided. In its place another technique has been developed which circumvents all of these problems: finite size scaling.

8.3.2 The finite size scaling method

The finite size scaling method is a way of extracting values for critical exponents by observing how measured quantities vary as the size L of the system studied changes. In fact, we saw a simple version of finite size scaling in action in Chapter 4, where we used it to calculate the dynamic exponent of a number of different algorithms for simulating the Ising model by measuring the correlation time τ for systems of a variety of different sizes. However, the techniques we used for those calculations required that we perform simulations exactly at the critical temperature of the model, which in turn requires us to know T_c. In the case of the Ising model we know T_c in two dimensions from the exact solution of the model. In some other cases we may be able to apply a method such as the invaded cluster algorithm of Section 4.4.4 to calculate T_c. However, in most cases we do not know T_c, which means that the simple method of Chapter 4 will not work. We now describe a more sophisticated version of the finite size scaling technique which not only does not require us to know T_c, but in fact returns a value of T_c itself, as well as giving us estimates of the critical exponents. We illustrate the technique for the case of the susceptibility exponent γ.

We begin, as we did in Section 4.1, by expressing the quantity of interest in terms of the correlation length ξ of the system. In this case the quantity of interest is the magnetic susceptibility. Eliminating $|t|$ from Equations (8.41) and (8.42) we get

$$\chi \sim \xi^{\gamma/\nu} \tag{8.45}$$

in the vicinity of the phase transition. Now, consider what happens in a system of finite size L, such as we use in our Monte Carlo simulations. In finite systems, as depicted in Figure 4.1, the correlation length is cut off as it approaches the system size, so that the susceptibility χ will also be cut off; in systems of finite size the susceptibility never actually diverges. We can express this cut-off mathematically as follows. If we continue to denote by ξ the value which the correlation length *would* have in an infinite system at temperature t, then the cut-off takes place when $\xi > L$. As long as $\xi \ll L$ the value of χ should be the same as that for the infinite system. We can express this by writing

$$\chi = \xi^{\gamma/\nu}\chi_0(L/\xi), \qquad (8.46)$$

where χ_0 is a dimensionless function of a single variable which has the properties

$$\chi_0(x) = \text{constant} \qquad \text{for } x \gg 1 \qquad (8.47)$$

and

$$\chi_0(x) \sim x^{\gamma/\nu} \qquad \text{as } x \to 0. \qquad (8.48)$$

The precise way in which the susceptibility gets cut off close to T_c is contained in the functional form of χ_0. It is this function which we will measure in our Monte Carlo simulations.

Equation (8.46) in fact contains all the information we need about the behaviour of our system with varying system size. However, it is not in a very useful form, since it still contains the variable ξ, the correlation length at temperature t in the infinite system, which we don't know. For this reason it is both conventional and convenient to reorganize the equation a little. Defining a new dimensionless function $\tilde{\chi}$ thus:

$$\tilde{\chi}(x) = x^{-\gamma}\chi_0(x^\nu). \qquad (8.49)$$

and making use of Equation (8.41), we get

$$\chi = L^{\gamma/\nu}\tilde{\chi}(L^{1/\nu}|t|). \qquad (8.50)$$

In fact, to be strictly correct, we should have two equations such as this, one for positive and one for negative values of t with different functions $\tilde{\chi}$, since the behaviour of χ is not symmetric on the two sides of the phase transition. However, we can easily combine these two equations into one by extending the definition of $\tilde{\chi}(x)$ to negative values of x. Then we can write

$$\chi = L^{\gamma/\nu}\tilde{\chi}(L^{1/\nu}t). \qquad (8.51)$$

This is the basic equation for the finite size behaviour of the magnetic susceptibility. It tells us how the susceptibility should vary with system size L for finite systems close to the critical temperature. Note that we have derived this equation for the susceptibility *per spin* as defined in Section 1.2.2. If we were to use the extensive susceptibility, the leading power of L would be $L^{\gamma/\nu+d}$ instead of just $L^{\gamma/\nu}$, where d is the dimensionality of the system. It is very important to recognize this distinction if you want to get the correct answers for exponents using the finite size scaling method. All the equations given in this section are correct for intensive quantities but need to be modified if you are going to use extensive ones.[8]

Equation (8.51) contains the unknown function $\tilde{\chi}(x)$ which we call the **scaling function** for the susceptibility. Although the scaling function is unknown, there are certain things we do know about it. Equation (8.48) tells us that

$$\tilde{\chi}(x) \to x^{-\gamma}(x^{\nu})^{\gamma/\nu} = \text{constant} \qquad \text{as } x \to 0. \qquad (8.52)$$

In other words, $\tilde{\chi}$ is finite at the origin, which in this case means close to the critical temperature. Another important point is that, by design, all the L-dependence of χ is displayed explicitly in Equation (8.51); the scaling function does not contain any extra hidden dependence on L which is not accounted for. In other words, if we measure $\tilde{\chi}(x)$ we should get the same result regardless of the size of the system. It is this last fact that allows us to use Equation (8.51) to calculate the exponents γ and ν and the value of the critical temperature.

Suppose we perform a set of Monte Carlo calculations of the system of interest for a variety of different system sizes L over a range of temperatures close to where we believe the critical temperature to be. (With this method we do not need to be exactly at the critical temperature, only in the rough vicinity. We can estimate where this region is by, for example, looking for the tell-tale peak in the magnetic susceptibility or the specific heat—see Figure 3.10.) For each system size, we measure the magnetic susceptibility $\chi_L(t)$ at a set of temperatures t. We can now rearrange Equation (8.51) thus

$$\tilde{\chi}(L^{1/\nu}t) = L^{-\gamma/\nu}\chi_L(t), \qquad (8.53)$$

to get an estimate of the scaling function $\tilde{\chi}$ for several different values of the scaling variable

$$x = L^{1/\nu}t \qquad (8.54)$$

for each system size. Since the scaling function is supposed to be the same for all system sizes, these estimates should coincide with one another—they should all fall on the same curve if we plot them together on one graph. However—and this is the crucial point—this will only happen if we use *the correct values* of the exponents γ and ν in Equation (8.53). Also, although it's not immediately obvious from the equation, we must use the correct value of the critical temperature T_c, which enters in the calculation of the reduced temperature t through Equation (8.40). The idea behind finite size scaling therefore is to calculate $\tilde{\chi}(x)$ for each of our different system sizes,

---

[8]The reader might be interested to work out where in the preceding derivation the extra powers of L would come in if we were to use the extensive susceptibility.

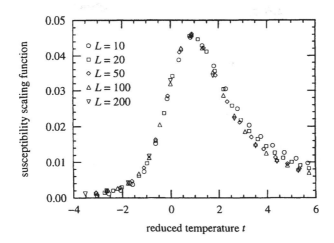

FIGURE 8.6 Data collapse of magnetic susceptibility data for the two dimensional Ising model. The points are taken from Monte Carlo measurements of the susceptibility for five different sizes of system as indicated. From this collapse we find $\gamma = 1.76$, $\nu = 1.00$ and $T_c = 2.27J$. Notice that the collapse fails once we get sufficiently far away from the critical temperature ($t = 0$).

and then vary the exponents γ and ν and the critical temperature until the resulting curves all fall or **collapse** on top of one another. An example of such a calculation is shown for the two-dimensional Ising model in Figure 8.6. Here we performed simulations around the critical temperature for square systems of size $L = 10$, 20, 50, 100 and 200. The best data collapse is obtained when $\gamma = 1.76 \pm 0.01$, $\nu = 1.00 \pm 0.05$ and $T_c/J = 2.27 \pm 0.01$. These values are in good agreement with the exact known values of $\gamma = \frac{7}{4}$, $\nu = 1$ and $T_c = 2.269J$. (The error bars are quite rough here, calculated by estimating the region over which the collapse appears optimal. Error estimation is discussed in more detail in the following section.)

This method can easily be extended to quantities other than the susceptibility. For the Ising model for example, we can derive scaling equations similar to (8.51) for the specific heat and the magnetization by arguments closely similar to the ones given above. The results are:

$$c = L^{\alpha/\nu}\tilde{c}(L^{1/\nu}t), \tag{8.55}$$

$$m = L^{-\beta/\nu}\tilde{m}(L^{1/\nu}t). \tag{8.56}$$

Performing data collapses using these equations yields values for α and β, as well as values for ν and T_c again. (If we perform collapses for a number of different quantities, we can use the several values of ν and T_c which we get as a consistency check on the different calculations.) Note the different sign in (8.56), which arises from the definition of the exponent β in Equation (8.44).

8.3.3   Difficulties with the finite size scaling method

Although the basic finite size scaling method outlined in the last section often works extremely well, there are nonetheless a number of problems with the method as described. How, for instance, are we to assess the errors on our critical exponents? In the example above, we gauged by eye when the various curves best fit one another and took our values for the critical exponents and critical temperature at that point. The errors were estimated by judging (again by eye) the range of values of the exponents over which the collapse appears equally good. When the collapse begins to look significantly poorer, we have reached the limits of the acceptable values of the exponents. A more thorough method of error estimation is to use one of the techniques discussed in Section 3.4, such as bootstrap resampling. By resampling our data a number of times and repeating the data collapse for each resampled data set, we can estimate the errors from the variation in the values of the exponent from one set to another. In order to do this however, it helps greatly if we have a quicker way of performing the data collapse than simply fiddling with values of the exponents until it looks good. This brings us on to another problem. Performing the collapse by eye is, after all, hardly a very scientific way to proceed. It is much better if we can construct a quantitative measure of the quality of our data collapse. One such measure is the variance of the set of curves, integrated over a range of values of x close to criticality. In the case of the susceptibility for instance, we could calculate

$$\sigma^2 = \frac{1}{x_{\max} - x_{\min}} \int_{x_{\min}}^{x_{\max}} \sum_L \tilde{\chi}_L^2(x) - \left[\sum_L \tilde{\chi}_L(x)\right]^2 \mathrm{d}x. \qquad (8.57)$$

This however is a little difficult to estimate, since the points at which the scaling function is evaluated are different for each system size. What we need is some way of interpolating between these points, and the perfect technique is provided by the multiple histogram method of Section 8.2. If we use this method, then we can directly evaluate (8.57) by using for example a simple trapezium rule integration (or any other form of numerical integration that we happen to favour) and then minimize it to give an estimate of the critical exponents and T_c. Bootstrapping then gives an estimate of our statistical errors.

There is still a problem however, which is evident in Figure 8.6: if we stray too far from the critical temperature, our scaling equation (8.51) is no longer correct, simply because we are no longer in the critical region.

Furthermore, the range of values of the scaling variable, Equation (8.54), over which the critical region extends varies as a function of L, so that the region over which it is safe to perform our data collapse depends on the size of the systems studied. How are we to estimate the extent of this region? There are a couple of commonly used methods. One is to perform a series of collapses over smaller and smaller ranges Δx and extrapolate the results to the limit $\Delta x \to 0$. In order to use this method, we need to choose a point x_0 about which to centre our data collapse. The standard choice is to set x_0 to be the point at which the scaling function (or equivalently the susceptibility or other quantity of interest) is a maximum. (For scaling functions such as $\widetilde{m}$ which are monotonic and therefore do not have a maximum, we choose x_0 to be the point at which the scaling function has its steepest gradient.) Note that the maximum in the scaling function is not at the critical temperature $t = 0$ (see Figure 8.6). This is a reflection of the fact that the maximum of the susceptibility (or specific heat, or whatever) is also not precisely at the critical temperature, except in an infinite system. However, the maximum of the susceptibility (and hence the scaling function) does correspond approximately to the maximum value of the correlation length in the finite system and hence marks roughly the centre point of the critical region, which makes it a good choice for x_0.

Now if we perform our data collapse, Equation (8.57), over a range from $x_{\min} = x_0 - \Delta x$ to $x_{\max} = x_0 + \Delta x$, make an estimate with errors of the critical exponents, and then repeat the calculation for successively smaller values of Δx, we can extrapolate the result to the limit $\Delta x \to 0$ by making a weighted fit through the resulting data points. The authors used this technique, for example, in a calculation of the critical exponents for the random-field Ising model of Section 6.1.1 (Newman and Barkema 1996). In Figure 8.7 we have reproduced the results for the exponents ν and γ. The quantities measured directly by the data collapse are $1/\nu$ and γ/ν. The extrapolation to $\Delta x = 0$ was done in this case by the weighted fit of a quadratic to the data points, giving $1/\nu = 0.98 \pm 0.06$ and $\gamma/\nu = 1.85 \pm 0.07$. This in turn gives us values of $\nu = 1.02 \pm 0.06$ and $\gamma = 1.89 \pm 0.13$ for the exponents themselves.

Another common method for circumventing the question of the size of the critical region is to perform the data collapse at only one point, the point x_0 at which the scaling function is a maximum (or has maximum gradient in the case of the magnetization). This is clearly less accurate than the method described above, but in many cases, especially when the accuracy of our Monte Carlo data is good, it gives perfectly adequate results. With this method we can use either the single or multiple histogram method to calculate the temperature T_0 at which the quantity of interest (susceptibility for example) is greatest, which is also the point at which the corresponding scaling function is greatest. We repeat this for each system size L. Since

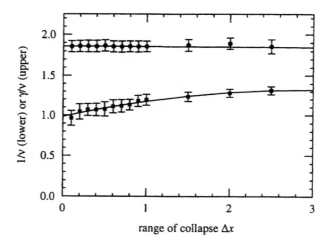

FIGURE 8.7   Calculation of critical exponents for the random-field Ising model by collapsing data over successively smaller ranges Δx. The limit as $\Delta x \to 0$ gives us the values of the exponents. In this case the method measures the quantities $1/\nu$ and γ/ν. The end result is $\nu = 1.02 \pm 0.06$ and $\gamma = 1.89 \pm 0.13$. After Newman and Barkema (1996).

the scaling function is the same for each system size, the values x_0 of the scaling variable at which we find these maxima should be the same for all system sizes. From Equations (8.40) and (8.54) we see that the temperature T_0 corresponding to x_0 is given by

$$T_0 = T_c(1 + x_0 L^{-1/\nu}). \tag{8.58}$$

Thus, if we plot T_0 against $L^{-1/\nu}$, the resulting points should lie on a straight line, provided we use the correct value of ν. And the intercept of this line with the vertical axis gives us an estimate of T_c. So, for example, we can fit our data points to a straight line using a least-squares fit, and minimize the variance on the fit to give the best straight line. The result is an estimate of both ν and the critical temperature. As before, we can use any of our standard methods of error estimation to calculate the errors on these values.

Once we have estimated ν and T_c in this fashion, we can use them to make estimates of the values of the other exponents. Since the scaling function is the same for all system sizes, its value $\tilde{\chi}(x_0)$ at x_0 should be independent of L. The value of the susceptibility at its maximum should therefore take the particularly simple form

$$\chi_L(T_0) \propto L^{\gamma/\nu}, \tag{8.59}$$

with $\tilde{\chi}(x_0)$ being the constant of proportionality. Thus if we plot the maximum value of χ as a function of L on logarithmic scales we should again get

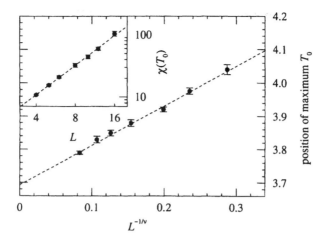

FIGURE 8.8   An illustration of finite size scaling in the random-field
Ising model using only one point on the scaling function. The main
figure shows the calculation of the exponent ν. The inset shows the
calculation of γ. The results in this case were $\nu = 1.1 \pm 0.2$, $\gamma =$
1.7 ± 0.2. After Rieger (1995).

a straight line. The slope of this line gives us a measure of γ/ν, and hence
γ, since we already know ν. We can calculate an error on our figure just as
before.

In Figure 8.8 we show an example of the application of this method,
again to the random-field Ising model, taken from Rieger (1995). The main
figure shows the calculation of ν and T_c from Equation (8.58) and the inset
shows the calculation of γ. The results for the exponents, $\nu = 1.1 \pm 0.2$
and $\gamma = 1.7 \pm 0.2$, are in respectable agreement with those from the other
method.

A third problem with the finite size scaling method as we have described it
here is that scaling equations such as Equation (8.51) are only approximate,
even close to the critical point. In particular, the arguments leading to the
scaling equations are only valid for sufficiently large system sizes. If the
system size L becomes small—how small depends on the particular model
being studied—there are correction terms in the equations which become
important. These terms can lead to systematic errors in the values of our
exponents if they are not taken into account in the analysis. In fact, except
for very high-resolution studies, such corrections are usually not important.
Furthermore, a detailed discussion of them would take us some way away
from the Monte Carlo simulations which are the principal topic of this book.
For more information on corrections to the scaling forms discussed here, we
therefore refer the reader to other sources. For the case of the normal Ising

model a good discussion has been given by Ferrenberg and Landau (1991). Binder and Heermann (1992) also cover the subject in some detail.

8.4 Monte Carlo renormalization group

A completely different approach to the calculation of critical exponents and critical temperatures is the **Monte Carlo renormalization group** method pioneered by Swendsen (1979, 1982) and others, based on the renormalization group ideas introduced in the early 1970s by Wilson and Fisher (1972, Wilson and Kogut 1974, Fisher 1974). This method is often more accurate than the finite size scaling method of the last section for the same expenditure of computational effort, and furthermore it only requires us to perform simulations for one system size, or at most two. Its main disadvantage is that it contains some rather poorly controlled approximations which make it difficult to estimate accurately the errors in our values for the critical exponents. Nevertheless, for many problems it is sufficiently superior to the finite size scaling method that we would recommend it as the method of choice for the calculation of critical properties.

The developments of this section will require us to learn a little about the ideas behind the renormalization group. We will not go into the subject in great depth however, and we would encourage those interested in learning more to consult one of the many books devoted to the topic.

8.4.1 Real-space renormalization

Despite its formidable-sounding name, real-space renormalization is really quite simple. Here we first describe how it can be used to calculate the critical temperature of a system. In Section 8.4.2 we describe the calculation of critical exponents.

The fundamental concept behind the real-space renormalization method is that of **blocking**,[9] whereby the spins or other degrees of freedom on a lattice are grouped into blocks to form a coarser lattice. Consider the configurations of the ordinary two-dimensional Ising model shown in Figure 8.9. In the first frame we show a state of the model on a 16×16 square lattice, with the black and white squares representing up- and down-spins respectively. The frame next to it shows the result of performing a blocking procedure on this state: we group the spins into blocks of four and in the blocked system we represent each such block by just one bigger **block spin**. In this case the value of this spin is assigned by majority voting amongst the spins making up the block. That is, we set it to point up or down depending on whether

---

[9] Not to be confused with the error estimation technique of the same name introduced in Section 3.4.2.

FIGURE 8.9 The "majority rule" blocking procedure applied to the Ising model. The top two frames show the procedure in detail for a 16 × 16 system. The bottom two show the same procedure for a 200 × 200 system. The simulations were performed at a temperature of $T = 2.29J$, which is slightly above the critical temperature of $T_c = 2.269J$.

there were more ups or downs in the original block of four spins. If there were exactly two ups and two downs then we choose the value of the new block spin at random. As we can see, the blocked system preserves the gross features of the spin configuration of the original system, but only has one quarter as many spins. This is made clearer in the remaining two frames of the figure, in which we have performed the same procedure on a larger 200 × 200 Ising model, reducing it by blocking to a 100 × 100 one. It is clear from these frames that the blocked system does indeed preserve the large-scale features of the configuration of the system.

The blocking procedure used here shrinks the size L of the system (measured in lattice spacings) by a factor of 2 in each direction. This **rescaling factor** is normally denoted b, and the number of spins in the system decreases on blocking by b^d, where d is the dimensionality of the system. The choice $b = 2$ is a common one for square or cubic lattices. For other lattices other choices are appropriate. For a triangular lattice, for example, the most common blocking scheme groups the spins in blocks of three, giving a rescaling factor of $b = \sqrt{3}$.

The "majority rule" described here is not the only way of performing the blocking. Many others have been suggested, with different ones being appropriate for different systems. For Ising models another common choice is the **decimation** procedure, in which we discard all but one spin from each block and set the new block spin equal to that one. In the example of Figure 8.9 for instance, we might discard all but the top left spin of each block of four and use the value of this one spin for our new block spin. The decimation method also works well with Potts models (see Section 4.5.1). For models with a continuum of spin values a simple additive rule may be appropriate, in which the renormalized spin is a (suitably normalized) sum of the spins in the original block. For the moment, however, let us concentrate on the simple case of the Ising model and the majority rule.

We now come to the crucial assumption of the renormalization group method. We assume that the blocked spin configuration is a typical spin configuration of another Ising model on a lattice of dimension $L' = L/b$. In other words, imagine sampling a set of states of our original Ising system, each with the correct Boltzmann probability for some temperature T, for example by performing a Monte Carlo simulation. For each state we perform the blocking procedure described above and generate a blocked state of the smaller system. Our assumption is that the series of blocked states we generate *also appear with their correct Boltzmann probabilities*, if we calculate their energies using the same Ising Hamiltonian as we used for the original system. In fact, as we might well imagine, this assumption is not normally correct and this is the primary source of the uncontrolled errors in the method, to which we alluded earlier. However, for the moment let us accept this assumption and see where it leads us.

It is clear that the blocked states cannot normally appear with the Boltzmann probabilities appropriate to the same temperature T as the original set of states which we started with. To see this, consider the correlation length ξ of the system. (See Section 3.7.1 for a definition of the correlation length.) When we block the system, the average correlation between two spins which are far apart must remain approximately the same, since, as we pointed out, the large-scale features of the system do not change on blocking. This means that the correlation length of the system should also stay the same, except that the number of spins in the system decreases by a factor of b in each direction. This in turn means that, when measured in terms of the lattice spacing, the correlation length of the blocked system is

$$\xi' = \xi/b. \tag{8.60}$$

(In general, we will use primed variables to denote the values of quantities in the blocked system.) In the case of the systems illustrated in Figure 8.9 we have $b = 2$, so that the rescaled system has a correlation length a half that of the original system, when measured in terms of the lattice parameter. As we know, the correlation length varies with temperature, so that the blocked states of the system must be typical of a different temperature from the original ones. We assume then that the blocked states are typical of an Ising model at a temperature T' for which the correlation length is a half that of the original system.

Now suppose that we calculate some measurable property of our system, such as, for example, the internal energy per spin u, and average it over our complete sequence of states. Since internal energy is also a function of temperature, we will presumably get two different answers for the average internal energies u and u' of the original and rescaled systems. To the extent that the blocked states appear with the correct Boltzmann weights for a system at temperature T', the internal energy u' of the rescaled system is presumably the appropriate internal energy for the system at that temperature.

Here, then, is the clever bit: there is one temperature at which the correlation length of the system is the same for the original and rescaled systems and that is the critical temperature. At this temperature, as discussed in Section 3.7.1, the correlation length becomes infinite, which means that the rescaled correlation length, Equation (8.60), is also infinite. At this one point, therefore, our original and rescaled systems have the same correlation length and hence the same temperature $T = T' = T_c$. They also therefore have the same values of all other (intensive) quantities such as the internal energy per spin. So here is our scheme for calculating the critical temperature:

1. We perform a Monte Carlo simulation at temperature T and calculate the internal energy u.

2. We take each of the states generated by our simulation and block them, using for example the majority rule blocking scheme described above.

3. We calculate the internal energy u' for the blocked system, averaged over all of these blocked states.

4. Now we vary the temperature T until we find the point at which $u' = u$. This is the critical temperature of the system.

As a practical consideration, the part about varying the temperature can be done most efficiently using the single or multiple histogram techniques of Sections 8.1 and 8.2, rather than performing a separate simulation for every temperature we are interested in. In order to extrapolate the value of the internal energy of the rescaled system, we treat it as an observable quantity in the *original system* (albeit one with a slightly unusual definition), and reweight with the Boltzmann factors appropriate to the original system.

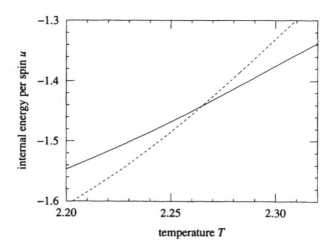

FIGURE 8.10   The internal energy per spin *u* as a function of temperature for a 32 × 32 Ising model (solid line) and for the same system
rescaled to 16 × 16 (dotted line). The two lines cross at 2.265*J* which
gives us an estimate of the critical temperature of the model.

In Figure 8.10 we show the results of a Monte Carlo renormalization
group calculation of the critical temperature of the two-dimensional Ising
model. In this calculation we performed a simulation of a 32 × 32 system
at a single temperature and calculated the values of u and u' over a range
of temperatures using the single histogram method. The figure shows the
two resulting curves, and their intersection gives an estimate of the critical
temperature of the model. In this case we find $T_c = 2.265J$, which compares
favourably with the exact figure of 2.269J, especially considering it was
derived from a short (about five minute) simulation on quite a small system.
The finite size scaling method of Section 8.3.2 by contrast gave a less accurate
result for the same quantity even though we simulated systems up to 200 ×
200 in size.

   In fact, we cheated a little in this calculation, since our single simulation was performed at the known critical temperature of the infinite two-
dimensional Ising model on a square lattice. Clearly in real life we would
not be performing a simulation to measure the critical temperature of a system for which we already knew the answer. In a real calculation therefore,
we would probably make a series of simulations, each one at the value of T_c
estimated from the previous one, so as to converge on the correct answer.
Typically we might have to perform three or four iterations of the procedure to find a good answer. Still, however, the Monte Carlo renormalization
group method is far quicker and more accurate than finite size scaling.

There are, however, a number of problems with the renormalization group method as we have described it. The main one is that we have no method of estimating the error in our calculation. We could certainly perform a bootstrap resampling (Section 3.4.3) or similar analysis to make an estimate of the statistical error in our figure for T_c. However, this would not tell us very much since there are other sources of error in the calculation which we have not taken care of, and it turns out that some of these are very important. Ignoring them can give highly inaccurate answers for calculations with some models.

One source of error is finite size effects in our simulation. The renormalization group method reduces the error arising from finite size effects considerably, as the accuracy of the above calculation shows. However, we will still get more accurate answers if we perform a simulation on a larger system. Finite size effects can also cause problems because the sizes of the original and rescaled systems in our simulation are different. (One is half the linear dimension of the other in the square lattice example given here.) We would get a more accurate answer for the critical temperature if we compared the internal energies of two systems which had the same size. A simple way to achieve this is to perform an extra simulation of a smaller system of the same size as the rescaled one, for comparison purposes. In the example above, for instance, we would perform a direct simulation of a 16×16 Ising system and compare its internal energy with that of our 16×16 rescaled system. (If we do this, then we no longer need to worry about whether we are working with the intensive or extensive internal energy. Either will work if the systems compared are the same size.) This takes somewhat more computer time than the simple comparison above, but usually improves the results. In fact, for the case of the internal energy of the two-dimensional Ising model, it does not make much difference, but for some other quantities it is very important. One example is the magnetization. In Figure 8.11 we show three curves for the magnetization per spin of our Ising system, one for the magnetization of the 32×32 system, one for the rescaled 16×16 system, and one for the magnetization of a separate simulation of a 16×16 system. As the figure shows, the finite size effects are such that the curves for the original 32×32 and rescaled systems do not cross at all, so this calculation provides no estimate of the critical temperature. On the other hand, when the curves describe two systems of the same 16×16 size, they cross neatly at $T = 2.268J$, which is a very respectable estimate of the critical temperature. Since performing the extra simulation on the smaller system only takes a fraction of the time spent on the larger simulation, we recommend taking this extra precautionary step in all such calculations.[10]

---

[10] In fact, in the case of the internal energy, doing this gives a somewhat *worse* answer of $T_c = 2.252J$ for the critical temperature. This is most likely because of cancellation of different sources of error. In Section 8.4.5 we will discuss how the accuracy of the method

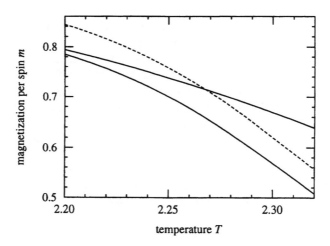

FIGURE 8.11   The magnetization per spin m as a function of temperature for 16×16 and 32×32 Ising models (upper and lower solid lines) and for the rescaled 16×16 system (dotted line). The 32×32 and rescaled lines do not cross at all, as a result of finite size effects due to the different system sizes. However, the crossing point of the two 16×16 systems gives a good estimate of $2.268J$ for the critical temperature.

Another important source of error in our calculations is the approximation inherent in the renormalization group method, whereby we assume that the sequence of blocked states generated by our simulation is typical of an Ising model at some temperature T', which in general is different from the temperature T of the original system. This is actually not such a bad approximation in the case of the two-dimensional Ising model, but in other models it can be extremely poor. In Section 8.4.5 we discuss a technique for getting around this problem, but first we look into the issue of calculating critical exponents.

8.4.2   Calculating critical exponents: the exponent ν

In the last section we saw how the Monte Carlo renormalization group approach can be used to calculate the critical temperature of a model, using only a single simulation, or perhaps two, on a comparatively small system. In this section, we look at the calculation of critical exponents, which will require us to delve a little deeper into the theory of renormalization group calculations.

can be improved to remove errors such as this.

First, we need to introduce the idea of a **renormalization group transformation**. Consider again the Ising model calculation of the last section. In this calculation we simulated a 32×32 Ising model and calculated its internal energy per spin u as a function of temperature using the single histogram method. We also calculated the internal energy per spin for the blocked system which had a quarter as many spins. As we pointed out, the blocked states approximate reasonably well to the states that would be generated if we directly simulated a 16×16 Ising system, but at a different temperature T'.

We can calculate what this temperature T' is by looking at the two internal energy curves in Figure 8.10. For a given temperature T, the temperature T' is that temperature at which

$$u'(T) = u(T'). \tag{8.61}$$

(Notice the arrangement of the primes on the variables here—it might not be exactly what you expect, but it is correct.) T' is thus given in terms of T by

$$T' = u^{-1}(u'(T)). \tag{8.62}$$

This mapping of T onto T' is a simple example of a renormalization group transformation, and it is by the study of this mapping that we can extract values for the critical exponents of our system. In Figure 8.12 we show a graph of T' against T calculated from our measurements of the internal energy of the rescaled 16×16 Ising model, and the direct simulation of another system of the same size. The critical temperature is a fixed point of the transformation, by definition. We refer to it as the **critical fixed point**. Above the critical temperature the value of the rescaled temperature T' is greater than T, and below it T' is less than T. In other words, the renormalization group transformation "pushes" the temperature in the direction away from the fixed point. It is common to regard the renormalization group transformation as giving rise to a "flow" through the parameter space of our model. In this case the parameter space is one-dimensional, the temperature being our only parameter. The temperature flows away from the critical fixed point under the transformation.

It is simple enough to see the physical reason why the temperature should flow away from the critical fixed point in this way. If we are above the critical temperature then the spins are arranged in clusters of typical size ξ, which is finite. When we apply the blocking procedure, the typical clusters of blocked spins become a factor of b smaller, producing a configuration with a smaller ξ, which is typical of a higher temperature $T' > T$. Below the critical temperature there is a spontaneous magnetization (see Section 3.7.1) so that a majority of the spins are pointing in the same direction. Let us suppose that this majority direction is the up direction. This means that there will be

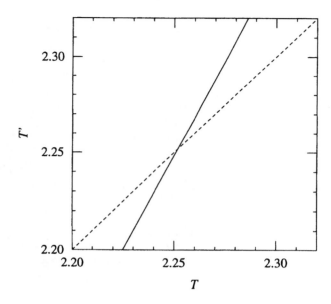

FIGURE 8.12 The temperature of the rescaled system T' as a function of T. The dotted line is the line $T' = T$, and the point at which the two cross is the critical fixed point of the renormalization group transformation.

many blocks in which three or four spins are pointing up, and the majority rule then transforms these all into upward-pointing blocked spins. In other words, a minority of downward-pointing spins tends to get "washed out" in the blocking process, leaving a larger majority of up-spins in the rescaled system, a configuration typical of a lower temperature $T' < T$.

So how do we use our knowledge of the renormalization group transformation to calculate a critical exponent? Consider the exponent ν, defined in Equation (4.2), which we repeat here for convenience:

$$\xi \sim |t|^{-\nu}. \tag{8.63}$$

Recall that the variable t is the reduced temperature, which measures our distance from the critical point:

$$t = \frac{T - T_c}{T_c}. \tag{8.64}$$

In our blocked system, the rescaled correlation length ξ' is also given by Equation (8.63), except that we must substitute in the transformed temperature T', giving

$$\xi' \sim |t'|^{-\nu}. \tag{8.65}$$

Dividing (8.63) by (8.65) and using Equation (8.60), we then get

$$\left(\frac{t}{t'}\right)^{-\nu} = b. \tag{8.66}$$

If we know the functional form of the transformation from T to T' then we can employ this equation to calculate ν. In fact, since Equations (8.63) and (8.65) are only valid close to the critical temperature, what we actually need to know is the relationship between T' and T close to T_c. The standard way of looking at this is to **linearize** the renormalization group transformation about the critical fixed point by performing a Taylor series expansion to leading order thus:

$$T' - T_c = (T - T_c)\frac{\mathrm{d}T'}{\mathrm{d}T}\bigg|_{T_c} \tag{8.67}$$

Substituting into Equation (8.66) using (8.64), and rearranging, we then get

$$\nu = \frac{\log b}{\log \frac{\mathrm{d}T'}{\mathrm{d}T}\big|_{T_c}}. \tag{8.68}$$

Performing a numerical derivative on the transformation shown in Figure 8.12 at the value of the critical temperature which we measured earlier, this equation gives us a figure of $\nu = 1.02$ for our two-dimensional Ising model. A similar calculation using the magnetization data from Figure 8.11 gives $\nu = 1.03$. The correct value for this exponent is $\nu = 1$, so these figures are quite accurate despite the small system size used in the calculation. Furthermore, we note that the gradient of the transformation function in Figure 8.12 varies very little in the vicinity of the critical point. This is one of the merits of the Monte Carlo renormalization group method. Even if we do not know the position of the fixed point very accurately, it does not normally affect our estimate of the critical exponents to any significant degree. The method however still suffers from the problems mentioned at the end of Section 8.4.1; we cannot estimate the error on our value for ν because there are systematic errors which we have not taken account of. Although these do not introduce a very large bias in the present calculation, they can in some cases generate considerable deviations from the true result. In Section 8.4.5 we address this problem when we discuss how the Monte Carlo renormalization group method can be improved by the introduction of longer-range interactions into the Hamiltonian. First, however, we discuss the calculation of other exponents such as the exponents α, β and γ defined in Sections 4.1 and 8.3.1.

8.4.3   Calculating other exponents

As we will shortly see, it is often not necessary to calculate directly all
the exponents which we are interested in: we can calculate them instead
using "scaling relations" which relate one exponent to another. However,
some Monte Carlo calculations are performed precisely to test these scaling
relations, and in this case we need to be able to measure all exponents. Also
in some circumstances it is more accurate to make a direct measurement of
an exponent of interest than to use a scaling relation.

To illustrate the calculation of other critical exponents, let us take the
example of the exponent β, defined in Equation (8.44), which governs the
singularity in the magnetization m close to the critical temperature:

$$m \sim |t|^{\beta}. \tag{8.69}$$

(We use the magnetization per site here again, although if, as we recommend,
you compare the magnetization of the blocked system with a separate simulation of a system of the same size, it makes no difference whether you use the
total magnetization or the magnetization per site.) Using Equation (8.63),
we can rewrite this as

$$m \sim \xi^{-\beta/\nu}. \tag{8.70}$$

If we now rescale our system using a blocking transformation such as the
majority rule introduced at the beginning of Section 8.4.1, we can calculate
the magnetization m' per site of the blocked system, and write

$$m' \sim \xi'^{-\beta/\nu}. \tag{8.71}$$

Dividing these two equations one by the other, and using Equation (8.60),
we get

$$\frac{m'}{m} = b^{\beta/\nu}, \tag{8.72}$$

and hence

$$\frac{\beta}{\nu} = \frac{\log \frac{m'}{m}}{\log b}. \tag{8.73}$$

The trouble with this equation is that it is strictly only true in an infinite
system, since Equation (8.69) is only true in an infinite system. On the other
hand, if we did know the magnetization for the infinite system, we wouldn't
be able to calculate m'/m at T_c, because both m and m' are zero at this
point. Instead however we can use l'Hôpital's rule, which says that as m and
m' go to zero at T_c, the limiting value of their ratio is

$$\frac{m'}{m} = \frac{\mathrm{d}m'}{\mathrm{d}T} \bigg/ \frac{\mathrm{d}m}{\mathrm{d}T} = \frac{\mathrm{d}m'}{\mathrm{d}m}. \tag{8.74}$$

And thus

$$\frac{\beta}{\nu} = \frac{\log \frac{dm'}{dm}\big|_{T_c}}{\log b}. \tag{8.75}$$

The clever thing about this equation is that the derivative $\frac{dm'}{dm}$ does not vary much as we vary the size of the system simulated, so that the equation gives good results for finite systems and not just infinite ones. This makes Equation (8.75) a better way to calculate β in a Monte Carlo simulation than Equation (8.73). Normally, in a finite system we should interpret m as meaning the absolute value of the magnetization, as we discussed in Section 3.7.1.

Note the similarity of form between Equation (8.75) and Equation (8.68) for the exponent ν. We can also, by a very similar line of argument, derive equivalent equations for other exponents. For example, the exponents α and γ are given by

$$\frac{\alpha}{\nu} = -\frac{\log \frac{dc'}{dc}\big|_{T_c}}{\log b}, \tag{8.76}$$

$$\frac{\gamma}{\nu} = -\frac{\log \frac{d\chi'}{d\chi}\big|_{T_c}}{\log b}. \tag{8.77}$$

Note the minus signs, which arise from the slight difference between the definitions of the exponents α and γ, and that of β.

In the simplest case, we can evaluate the derivatives numerically, as we did for the calculation of ν, although we will shortly see that there is a better way of evaluating these exponents which avoids the calculation of a derivative altogether.[11]

8.4.4 The exponents δ and θ

One exponent which we cannot calculate by exactly the method above is the exponent δ, which we have not encountered previously in this book. This exponent governs the way in which the magnetization of our model varies with an applied external field at the critical temperature:

$$m \sim B^{1/\delta}. \tag{8.78}$$

To calculate this exponent we use a method analogous to that for the calculation of β. First we look at the way in which the correlation length diverges

---

[11]In fact, we can in this case do better by noting that dm'/dT and dm/dT can both be calculated by analytically differentiating the single histogram formula, Equation (8.3), with respect to T and then using Equation (8.74). We can do a similar thing for the derivatives in Equations (8.76) and (8.77). Unfortunately, there is no equivalent trick for the derivative in Equation (8.68).

as $B \to 0$ at T_c. We define another exponent θ to describe this:

$$\xi \sim |B|^{-\theta}.$$
(8.79)

The equivalent of Equation (8.68) for θ is then

$$\theta = \frac{\log b}{\log \frac{dB'}{dB}\big|_{B=0}},$$
(8.80)

and the equivalent of (8.75) is

$$\frac{1}{\theta\delta} = \frac{\log \frac{dm'}{dm}\big|_{B=0}}{\log b}.$$
(8.81)

Combining the last two equations we then get

$$\delta = \frac{\log \frac{dB'}{dB}\big|_{B=0}}{\log \frac{dm'}{dm}\big|_{B=0}}.$$
(8.82)

The exponent θ itself is actually an interesting quantity. As it will turn out, the exponents ν and θ which govern the divergence of the correlation length with temperature and magnetic field as we approach the critical point are directly associated with the two "relevant variables" in the renormalization group transformation for the Ising model. In some sense, these two exponents are "fundamental" exponents of the model and all the other exponents studied here can be calculated from them via scaling relations.

8.4.5   More accurate renormalization group transformations

In this section we extend the approach introduced in the last few sections, and show how it can systematically be made more accurate. At the same time, we will develop a more complete picture of the Monte Carlo renormalization group technique, and show how the critical exponents can be calculated without resorting to the numerical derivatives of the last section.

The primary source of uncontrolled errors in the renormalization group methods described in the previous sections is the approximation discussed at the end of Section 8.4.1. In order to perform the calculation, we had to assume that the blocked states generated by our simulation were typical of the states that would have been sampled by a direct simulation of the model on a smaller lattice at some other temperature. Whether this is a good approximation or not depends on the particular model we are working with, and on the details of the blocking transformation which we use. Unfortunately, there is no easy way to tell in advance whether a particular transformation will work well for a particular model. As we saw, the majority rule transformation of Section 8.4.1 worked well for the two-dimensional Ising model, but other transformations might not have been so successful, and the majority rule might not work so well with other models.

There is however a way to improve the accuracy of our renormalization group transformation, regardless of the blocking scheme we use, although some schemes will still give better results than others. The trick is to introduce longer-range interactions into our Hamiltonian.

Consider once more the example of the two-dimensional Ising model. For this model we perform a simulation at temperature T of some system of size $L \times L$, and block each state generated to give a sequence of states of a system a factor of b smaller along each side. These states will to some approximation be typical of an Ising model at another temperature T', but this is only an approximation. We can make this approximation better by regarding the blocked states instead as the states of a system which has extra interaction parameters in its Hamiltonian. For example, we might introduce a next-nearest-neighbour interaction, or longer interactions still. If we are working with zero magnetic field, then symmetry dictates that all of these interactions involve an even number of spins on the lattice. Thus three-spin interactions are not allowed, but we might include perhaps the interactions of groups of four spins in a square. It is clear that if we introduce enough such interactions, we can approximate the behaviour of the blocked system to any desired degree of accuracy, since the more parameters we add, the more freedom we have to tune the Hamiltonian to take the desired value in any particular state. It turns out in practice, however, that quite good results can be achieved by including only a small number of interactions—usually no more than five or six—if they are carefully chosen.

Let us write a general expression for the Hamiltonian of this generalized Ising model thus:

$$\beta H = \sum_i K_i S_i. \qquad (8.83)$$

Here we have absorbed the factor of the inverse temperature β which appears in the Boltzmann weight into the coupling constants K_i. The quantities S_i represent sums over all sets of similar groups of spins on the lattice. For example, one possible set is the set of all nearest neighbours:

$$S_{\mathrm{nn}} = \sum_{\langle ij \rangle} s_i s_j. \qquad (8.84)$$

Other possibilities are the set of next-nearest neighbours, and the set of all groups of four spins in a square, as described above.

Our normal Ising model with nearest-neighbour interactions only is defined by setting all the K_i to zero except for the one which multiplies S_{nn}.

However, even if we start off simulating this model, the blocked states generated in the simulation will usually be best represented using a model in which some of the longer-range interactions are non-zero (although the nearest-neighbour interaction is probably still the largest of the parameters K_i). In other words, the renormalization group flows (see Section 8.4.2) move away from the line in K-space representing the normal Ising model into the larger space of the generalized model. However, and this is a crucial result, the critical exponents of this generalized model are the same as those of the original model which did not possess any longer-range interactions. We will not actually prove this result here. It is another consequence of the phenomenon of universality which we discussed in Section 4.1. It is proved in, for example, Binney *et al.* (1992). Here, we will just make use of the result. If we can find the values of the coupling constants K'_i for the blocked system near the fixed point of the renormalization group transformation in our new higher-dimensional space of parameters K_i, then we can calculate the critical exponents of the generalized model, which are the same as those for the normal Ising model. However, the renormalization group transformation for the generalized model is more accurate than the one we found for the normal Ising model, since we have more parameters to play with, and this improves the accuracy of our results. By introducing longer and longer range interactions we can, in a moderately systematic way, improve the accuracy of the calculation, as well as making a rough estimate of the error introduced by truncating the Hamiltonian at interactions of some finite range.

We calculate the critical exponents for our generalized model as follows. Denoting by $\mathbf{K}$ and $\mathbf{K}'$ two vectors whose elements are the couplings K_i and K'_i for the original and rescaled systems, we consider $\mathbf{K}'$ to be a function of $\mathbf{K}$ via a renormalization group transformation, just as we did in the simpler one-parameter case of Section 8.4.2. Then, by analogy with Equation (8.67), we linearize this transformation about the critical fixed point, which is traditionally denoted $\mathbf{K}^*$. This gives

$$K'_i - K^*_i = \sum_j (K_j - K^*_j) \frac{\partial K'_i}{\partial K_j}\bigg|_{\mathbf{K}^*} \tag{8.85}$$

Or, in matrix form

$$\mathbf{K}' - \mathbf{K}^* = \mathbf{M} \cdot (\mathbf{K} - \mathbf{K}^*), \tag{8.86}$$

where the elements of the matrix $\mathbf{M}$ are

$$M_{ij} = \frac{\partial K'_i}{\partial K_j}\bigg|_{\mathbf{K}^*}. \tag{8.87}$$

$\mathbf{M}$ defines the renormalization group transformation in the vicinity of the fixed point.

Now we change variables to a new set $\{x_i\}$ which are linear combinations of the components $K_i - K_i^*$:

$$x_i = \sum_j Q_{ij}(K_j - K_j^*), \qquad (8.88)$$

so that

$$\mathbf{x}' = \mathbf{QMQ}^{-1}\mathbf{x}. \qquad (8.89)$$

We choose the matrix $\mathbf{Q}$ to make $\mathbf{QMQ}^{-1}$ diagonal, and thus end up with

$$x_i' = \lambda_i x_i, \qquad (8.90)$$

where the λ_i are the eigenvalues of $\mathbf{M}$. Given that the correlation length diverges at the critical fixed point, we can define an exponent ν_i governing that divergence along the direction of any of the eigenvectors of $\mathbf{M}$ thus:

$$\xi \sim |x_i|^{-\nu_i}. \qquad (8.91)$$

Now, using the same arguments which led us to Equation (8.68), we can show that

$$\nu_i = \frac{\log b}{\log \frac{\mathrm{d}x_i'}{\mathrm{d}x}} = \frac{\log b}{\log \lambda_i}. \qquad (8.92)$$

Note that this gives a negative value of ν_i for any eigenvalues λ_i which are less than one. There are some standard terms used to describe this situation. **Relevant variables** are the variables x_i which grow under the renormalization group flows defined by the matrix $\mathbf{M}$ because the corresponding eigenvalue in Equation (8.90) is greater than one. **Irrelevant variables** are those which get smaller under the renormalization group flows because the corresponding eigenvalue is less than one. In the special case where $\lambda = 1$ the corresponding variable x_i is said to be **marginally relevant**. Only the relevant variables contribute to the power-law divergence in the correlation length as we approach the critical fixed point. Marginally relevant ones can produce logarithmic divergences, though we have to examine each case separately to determine whether they in fact do.

The number of relevant variables is also the number of independent critical exponents in the model. It turns out that all the other exponents studied earlier, such as α, β, γ and δ in the Ising model, can be calculated from the values of the exponents ν_i defined in Equation (8.92). In the case of the Ising model, it turns out that there are just two independent critical exponents, the exponents ν and θ encountered earlier. We should emphasize that it is by no means obvious that there should only be two. There are many models which have only one independent exponent, or more than two, and many more for which we do not yet know how many there are. The number of independent exponents is one of the crucial defining properties of a model which can be extracted from a renormalization group calculation.

In terms of ν and θ, the rest of the exponents of the Ising model can be calculated from the following **scaling relations**:

$$\alpha = 2 - \nu d, \tag{8.93}$$

$$\beta = \nu \left[d - \frac{1}{\theta} \right], \tag{8.94}$$

$$\gamma = \nu \left[\frac{2}{\theta} - d \right], \tag{8.95}$$

$$\delta = \frac{1}{d\theta - 1}, \tag{8.96}$$

where d is the dimensionality of the system. We will not prove these relations here; we merely state them for completeness. The interested reader can find a proof in any good book on critical phenomena.

8.4.6 Measuring the exponents

The only question remaining then is how we calculate the elements of the matrix $\mathbf{M}$ so that we can diagonalize it and find the relevant variables of the renormalization transformation. The standard approach employs a clever trick introduced by Swendsen (1979), which allows us to avoid the calculation of the derivative in Equation (8.87).

The Hamiltonian for the rescaled system can be written as

$$\beta H' = \sum_i K_i' S_i', \tag{8.97}$$

where the quantities S_i' are the same as those appearing in Equation (8.83) except that they are evaluated using the blocked spins, rather than the original spins. Let us denote by $\langle S_i \rangle$ and $\langle S_i' \rangle$ our best estimates of the values of S_i and S_i' averaged over the states sampled in our Monte Carlo simulation. Now we write

$$\frac{\partial \langle S_i' \rangle}{\partial K_k} = \sum_j \frac{\partial \langle S_i' \rangle}{\partial K_j'} \frac{\partial K_j'}{\partial K_k}. \tag{8.98}$$

This equation contains no physics; it is simply an application of the chain rule of differential calculus. We can rewrite it in matrix form as

$$\mathbf{A} = \mathbf{BM}, \tag{8.99}$$

where $\mathbf{M}$ is the matrix defined in Equation (8.87). The elements of $\mathbf{A}$ and $\mathbf{B}$ are

$$A_{ij} = \frac{\partial \langle S_i' \rangle}{\partial K_j} = \langle S_i' S_j \rangle - \langle S_i' \rangle \langle S_j \rangle, \tag{8.100}$$

$$B_{ij} = \frac{\partial \langle S_i' \rangle}{\partial K_j'} = \langle S_i' S_j' \rangle - \langle S_i' \rangle \langle S_j' \rangle, \qquad (8.101)$$

where we have made use of the linear response theorem, Equation (1.22), to calculate the expressions on the right-hand side. (The factor of β in Equation (1.22) has disappeared because we absorbed a factor of β into our definition of the quantities K_i and K_i' in Equations (8.83) and (8.97).)

Our calculation then proceeds as follows: we evaluate the two matrices $\mathbf{A}$ and $\mathbf{B}$ by calculating the correlation functions, Equations (8.100) and (8.101), and from these we calculate $\mathbf{M} = \mathbf{B}^{-1} \mathbf{A}$. Then we diagonalize this matrix to find the eigenvalues λ_i, and for each $\lambda_i > 1$ we calculate one independent exponent using Equation (8.92). From these, all the other exponents can then be calculated, using scaling relations such as Equations (8.93–8.96).

The only remaining issue which we have not addressed is the question of how we find the position of the critical fixed point itself. In fact, as we saw in Figure 8.12, the values of the critical exponents are not very sensitive to the position at which we evaluate our eigenvalues, which is one of the nice things about this method. However, we still need to make a rough estimate of the position of the fixed point. In the case where we explicitly calculated the renormalization group transformation for the temperature (see Equation (8.62)), this simply meant finding the fixed point of that transformation. In the present case however, where we don't explicitly calculate the transformation, it is not so obvious what we should do. One solution has been given by Swendsen (1979). In his calculations, he performed two consecutive iterations of the blocking procedure, creating systems of size b and b^2 smaller than the original system simulated. He then calculated the exponent ν for each of these systems using the methods outlined above. Exactly at the critical point, these calculations should give the same answer, apart from finite size effects, but away from the critical point, where the parameters K_i drift with the renormalization group flows, the two will give slightly different answers. By minimizing the difference between the two values for ν, Swendsen was able to make an estimate of the position of the critical fixed point with sufficient accuracy to give good measurements of the elements of the matrix $\mathbf{M}$.

As an example of this technique, Blöte and Swendsen (1979) calculated critical exponents for the three-dimensional Ising model on a cubic lattice. In their calculation they included in the Hamiltonian all two-spin interactions up to fourth-nearest neighbours, and also all four-spins-in-a-square interactions. Since they only included interactions which are even in the number of spins[12] their calculation only gives information about the exponent ν and

---

[12] In order to study θ we need to introduce an external magnetic field, which couples to an odd number of spins. The up–down symmetry of the Ising model then ensures that

not about θ. They found a figure of $\nu = 0.637$, which compares reasonably well with the best available figure of $\nu = 0.627 \pm 0.002$ from the simulations of Ferrenberg and Landau (1991), who used finite size scaling. Since Blöte and Swendsen gave no error for their figure it is difficult to say exactly how good the agreement is, but one should bear in mind that this calculation predated that of Ferrenberg and Landau by more than a decade, which is a long time in the development of computers. The renormalization group calculation used far less computer power than the later calculation and yet produced results of very reasonable accuracy, an indication of the power of the Monte Carlo renormalization group method.

It is worth noting that, in theory at least, the technique of introducing longer-range interactions into the Hamiltonian allows us to make an estimate of our errors, since we know that in theory we can get a perfect result (except for statistical fluctuations due to finite simulation length) by including an infinite number of such interactions. Thus the change in our values for the critical exponents as we add in more interactions can give us a crude estimate of the accuracy of those values. (It is only crude since we have no real way of knowing which interactions are most important, or how the result will converge to the true value as the number of interactions included is increased.)

Problems

8.1 Equation (8.10) gives a criterion for estimating the temperature range over which the single histogram extrapolation is reliable. For a simulation performed at the critical temperature, how does this scale with system size? (Hint: notice that Equation (8.10) involves the *total* specific heat, not the specific heat per site.)

8.2 The criterion given in Equation (8.10) is only a rough one. In particular, it does not, as we pointed out, take into account the improvement in the extrapolation range on increasing the length of our Monte Carlo simulation. Assuming that the distribution of energies sampled by a simulation is roughly Gaussian, show that the extrapolation range increases as $\sqrt{\log n}$, where n is the number of independent energy measurements made during the simulation.

8.3 Suppose we want to use the multiple histogram method to calculate the value of a quantity over a certain temperature range. How does the number of different simulations needed vary with the size of the system?

---

the K_i for the even and odd interactions transform separately. In a calculation such as that of Blöte and Swendsen, in which no odd interactions were introduced to begin with, there can thus never be any odd ones appearing.

8.4 Why does the peak of the scaling function in Figure 8.6 fall slightly above the true critical point $t = 0$?

8.5 Derive the equivalent of Equation (8.51) for the scaling of the correlation time τ.

8.6 Write a program to perform a Metropolis Monte Carlo simulation of the four-dimensional Ising model, and use it to estimate the missing number in Table 4.1, the dynamic exponent of the Metropolis algorithm in four dimensions, by finite size scaling using the formula derived in the previous problem. (Hint: you might want to take the Metropolis program given in Appendix B as a starting point. Further hint: we don't know the answer to this problem. As we said in Chapter 4, this calculation has never been done, as far as we know. You probably shouldn't attempt this problem unless you have a lot of spare time.)

Part II

Out-of-equilibrium simulations

9

The principles of out-of-equilibrium Monte Carlo simulation

In Part I of this book we looked at ways of measuring the equilibrium properties of model systems using Monte Carlo simulations. The next few chapters of the book are devoted to systems which are out of equilibrium. In studying these out-of-equilibrium systems we are commonly interested in one of two things. Either we want to know how the system relaxes from an initial state to equilibrium at a particular temperature, or we are studying a system which never reaches equilibrium because it has a driving force which continually pushes it into states which are far from equilibrium. We will see examples of both these types of simulations in the following pages.

The statistical mechanics of out-of-equilibrium systems is a less well-developed field of study than that for equilibrium systems, and there is no one general framework to guide our calculation such as we had in the equilibrium case. As the subject stands at the moment at least, each system must be considered independently. In many cases the mathematical hurdles to formulating an accurate analytic theory of a system's behaviour are formidable. In this part of the book we will look at a number of out-of-equilibrium problems and see how they may be studied using Monte Carlo methods instead. In so doing, we will introduce some of the more important techniques that are used for these calculations, including different types of dynamics, energy barrier calculations, continuous time algorithms and the mapping of models onto other models. First though we take a few pages to give a brief overview of the difficulties occurring in studies in this area.

9.1   Dynamics

The fundamental difficulty which makes out-of-equilibrium Monte Carlo sim-ulations harder than their equilibrium counterparts is that we are not at liberty to choose the dynamics of the Monte Carlo algorithm, as we were in the equilibrium case. Consider the single-spin-flip algorithms we looked at in Chapters 3 and 4 for Ising, Potts and continuous spin models. In the case of the Ising model for example, we suggested a number of different choices for the acceptance probability of a single-spin-flip algorithm, all of which produce the same Boltzmann distribution of states once the simulation has reached equilibrium. We showed that one of them—the Metropolis algo-rithm of Section 3.1—is the most efficient such algorithm, and, since the results produced are in all other respects the same for all the algorithms, it makes sense to use the Metropolis algorithm rather than any other.

If we are interested in the behaviour of the system away from equilib-rium however, this is no longer the case. All we required before was that the equilibrium distribution of states sampled by an algorithm be the correct Boltzmann distribution, and we showed that the conditions of ergodicity and detailed balance were between them sufficient to ensure this. These conditions say nothing however about the way in which the system comes to equilibrium, and it is clear that different choices of the dynamics of the algorithm will make a difference here. In Figure 9.1 we show the magnetiza-tion of the two-dimensional Ising model just below its critical temperature at $T = 2.1J$ simulated using the Metropolis algorithm and the heat-bath algorithm of Section 4.5.1. The results are averaged over a number of runs and, as the figure shows, the two algorithms come to equilibrium at different rates, even though the final results for the equilibrium magnetization must necessarily agree. If we were interested in measuring the rate of equilibration then, it would be crucial to specify which dynamics we were using. And if it were the heat-bath dynamics that we wanted to study, we could not then perform our simulation using the faster Metropolis algorithm; we would be stuck using the relatively inefficient heat-bath algorithm.

The situation is not always quite this bad however. There are some properties which do not vary when we make certain kinds of changes to the dynamics of our model. For example, the dynamic exponent z of the Ising model (see Section 4.1) appears to be the same for the Metropolis and heat-bath algorithms. Such algorithms are said to fall in the same **dynamic universality class**. If we are only interested in calculating the dynamic exponent, or one of a small number of other quantities which are the same for all members of such a class, then we are at liberty at least to choose the most efficient algorithm in that class to perform our simulation.

Some of the best algorithms we saw for the Ising model however achieved their efficiency precisely by falling in a different dynamic universality class.

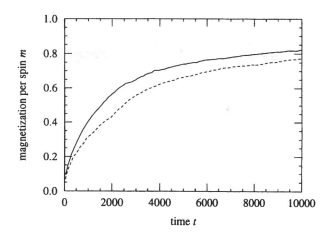

FIGURE 9.1   The absolute value of the magnetization of a 100×100 Ising model averaged over 200 runs for the Metropolis algorithm (solid line) and the heat-bath algorithm (dotted line). Although the final equilibrium magnetization is guaranteed to be the same in each case, the way in which the two algorithms come to equilibrium is clearly different.

The Wolff algorithm of Section 4.2 for example is particularly good for simulations near the critical temperature because it has a dynamic exponent close to zero. If we are interested in the way in which the Ising model comes to equilibrium under a single-spin-flip dynamics such as the heat-bath algorithm, then under no circumstances can we perform a simulation using the Wolff algorithm to study the problem. This is a great shame, since it means that many of the cleverest ideas which we saw in Part I of this book are inapplicable to simulations of out-of-equilibrium systems. Cluster algorithms, non-local algorithms for conserved-order-parameter models, simulated tempering and entropic sampling algorithms for glassy systems, and any number of other ingenious methods cannot be applied because they will give completely erroneous answers. What then can we do to improve the efficiency of these calculations?

There are a number of tricks which can be used to speed up our calculations without altering the dynamics. One of the most important we have seen already. The continuous time Monte Carlo method, which was introduced in Section 2.4 and which we applied to the conserved-order-parameter Ising model in Section 5.2.1, is a way of improving the efficiency of a Monte Carlo algorithm, particularly at low temperatures, without altering the effective dynamics of the algorithm. In Section 10.3.2 we apply the continuous time method to the problem of domain growth in an equilibrating conserved-order-parameter model, and show that one can achieve a speed-up of better than a factor of 10^8 over the normal Metropolis method under certain circumstances.

Another way of speeding up out-of-equilibrium Monte Carlo simulations is to look for mappings which turn a model of interest into another model which may be simulated more efficiently. In Chapter 12 we study the "repton model" of DNA gel electrophoresis. This is a polymer model which is normally defined in two or three dimensions. We show how the model can be mapped onto a much simpler model in one dimension, which can then be simulated efficiently using a relatively straightforward Monte Carlo algorithm.

A third method for speeding up our calculations, and perhaps the one which has been exploited most thoroughly, is to use any of a variety of programming tricks to improve the efficiency of our code. Some of these tricks we have seen before, such as the calculation of look-up tables of quantities needed by the algorithm. Others include using parallel computers, which we discuss in Chapter 14, and "multispin coding", which is discussed in Chapter 15.

9.1.1   Choosing the dynamics

So, if it is so important that we use the right dynamics in our out-of-equilibrium simulations, how do we know exactly what the correct dynamics is? In our equilibrium simulations it is clear that the dynamics of our algorithms is not usually a good approximation to the way in which real systems work. We can, for example, use the Metropolis algorithm to simulate an Ising or Heisenberg model, but in real Ising or Heisenberg magnets (and they do exist) the spin-flip dynamics may be completely different from the Metropolis algorithm. Even if the true dynamics of the system is a single-spin-flip dynamics, the wrong choice of acceptance ratio could mean that our simulation would give very different results from those obtained in an experiment on a real system when we examine the way in which the system comes to equilibrium.

In some cases this is not a problem. Quite often we are interested merely in the dynamics of the model system, and not in emulating some real material. For example, many people have studied the dynamical properties of the ordinary Ising model under the Metropolis algorithm, without regard to the dynamics of real Ising magnets. This may seem like a pointless exercise to some. Why study a model system which is not actually a model of anything? However, the approach is not entirely without merit. We would like to learn how systems come to equilibrium in general, to improve our currently rather meagre understanding of non-equilibrium statistical mechanics. And it makes sense to start off with a simple system with a simple dynamics, rather than jumping in at the deep end with a realistic simulation of a real material.

In other cases, where we are concerned with simulating a real material, it may be possible to use our understanding of the physics of that material to work out what dynamics we should use in our Monte Carlo algorithm. In Chapter 11, for instance, we look at the simulation of adatoms on metal surfaces. In this case quite accurate methods are available for studying the dynamics of individual surface atoms, including a variety of semiclassical techniques for calculating the interaction energies of the atoms. These methods allow us to make an estimate of the correct dynamics, so as to make our Monte Carlo algorithm mimic the true system as closely as possible.

The simulation of adatom diffusion however highlights another problem—the existence of energy barriers between states. When we are interested only in equilibrium properties of a system then the condition of detailed balance, Equation (2.2.3), tells us that we only need to know the energies of the states themselves in order to construct a correct algorithm. If we are interested in the out-of-equilibrium behaviour however, we need to know the way in which the energy varies "between" the states of the simulation. In going from one state μ to another ν (for example, when an adatom moves from one hollow on a metal surface to another), the system must pass through a succession of intermediate states with higher energy, and the rate for the transition $\mu \rightarrow \nu$ depends on the energies of these intermediates. In effect, there is an energy barrier between the two states, and we expect the rate for the system to cross this barrier to go exponentially with its height (the so-called Arrhenius law—see Section 11.1.1).

There is no reason why any system we study should not have energy barriers between states. In the Ising model, for instance, we can introduce energy barriers for the flipping of individual spins by putting in extra intermediate states between up and down spin states which have higher energy. Such barriers can have a profound effect on the dynamics of the model, and in fact a number of authors have looked into this very problem in some detail.

So our focus in the next few chapters will be to look at the simulation of out-of-equilibrium systems, giving a number of examples of algorithms for particular models which illustrate some of the techniques in use in this area. First, we start with a chapter on the Ising model, which is simple and familiar, but nonetheless illustrates a number of important points.

10

Non-equilibrium simulations of the Ising model

As our first example of the ideas introduced in Chapter 9 we now look at how the Ising model comes to equilibrium. As well as being a simple demonstration of most of the important ideas of out-of-equilibrium Monte Carlo simulation, our studies of the Ising model also tell us about real physics. As we explain in the next section, the equilibration of the Ising model below its equilibrium phase transition temperature sheds light on the physical phenomenon known as "spinodal decomposition".

10.1 Phase separation and the Ising model

Demixing is a phenomenon of current interest in non-equilibrium statistical physics. If two immiscible fluids, such as oil and water, are mixed together then, as common experience tells us, they have a tendency to separate again. Typically such fluids demix by first forming small bubbles or domains of the two substances which then coalesce into larger domains and so on, until the substances have completely separated. (We will ignore the effects of gravity, which make a heavier fluid like water sink below a lighter one like oil. We assume that we are in zero gravity, or that the two fluids in question have the same density.)

The growth of domains in this fashion can be modelled using the lattice gases of Chapter 5 which we mapped in turn onto the conserved-order-parameter (COP) Ising model. As we pointed out at the beginning of Chapter 5, the COP Ising model undergoes phase separation at sufficiently low temperatures into domains of up- and down-pointing spins. This phase separation is qualitatively similar to the demixing of fluids, and we can mimic demixing by suddenly cooling or **quenching** an Ising model from a high temperature, one above the critical point, to one below it and watching the formation of these domains and the way in which they grow. The formation and growth of domains of this kind is often called **spinodal decomposition**. (The **spinodal** is the line on the phase diagram between the regions in which phase separation takes place and the regions in which it does not.[1])

You might well ask what use it could be to simulate the Ising model when, as we pointed out in Chapter 5, its properties are only rather loosely related to the properties of real fluids. Previously we got around this problem by noticing that the COP Ising model actually gives a good estimate of some equilibrium properties of a real gas because the two systems fall in the same universality class. This means that quantities like critical exponents (see Section 4.1) take the same values for both systems, despite the obvious differences in the physics involved. As discussed in Section 9.1 however, this argument does not apply to the out-of-equilibrium properties of the two models. Even models in the same universality class can give completely different results when they are not in equilibrium. So what makes us think that the Ising model can tell us anything about the non-equilibrium properties of real fluids?

The answer is that certain properties of domain growth in spinodal decomposition fall into dynamic universality classes of the type discussed in Section 9.1. In particular, the average dimension R of the domains grows as a power law $R \sim t^\alpha$ with time t, where the value of α depends only on the coarse properties of the model and the dynamics used to simulate it, but not on the details, in a way reminiscent of equilibrium universality. This result was first demonstrated by Lifshitz and Slyozov (1961), and independently also by Wagner (1961). The argument goes as follows.

When a system phase separates, the domains in the system coalesce and grow because by doing so they reduce the total area of domain walls in the system; domain walls cost energy, so reducing them lowers the energy. The energy cost of having a domain wall is characterized by a surface tension σ, i.e., the energy per unit area of the wall. The surface energy of a domain thus scales as σR^{d-1}, where d is the dimensionality of the system. The volume of a domain scales as R^d. So the energy density u—the cost of the domain walls per unit volume of the system—goes down as R increases according to $u \sim \sigma/R$, and the gradient of the energy density scales according to $du/dR \sim \sigma/R^2$. If we assume that the domain walls move diffusively,[2] then

---

[1] In practice, the point at which separation takes place does not coincide precisely with the phase boundary depicted in Figure 5.1. There is a metastable region below the boundary in which the system remains mixed on laboratory time-scales, even though the demixed state has lower energy. For this reason the location of the spinodal is not well defined; it occurs closer to the phase boundary the longer one is prepared to wait for the nucleation of phase separated domains.

[2] We have argued previously that the value $z = 2$ of the dynamic exponent for single-

their velocity is proportional to this energy gradient:

$$\frac{dR}{dt} = C\frac{\sigma}{R^2}, \tag{10.1}$$

where C is a constant. The solution to this equation tells us that

$$R \sim t^{1/3}, \tag{10.2}$$

which is of the form postulated above, with $\alpha = \frac{1}{3}$. Note that this result is independent of the dimensionality d and so should apply for systems in either two or three dimensions, both of which we look at in this chapter.

The crucial assumption in this argument is that the movement of the domain walls obeys a diffusion equation—their velocity is proportional to the energy gradient they are moving across. For some systems this is clearly not true. However, it does turn out to be correct for systems in which (a) the motion of the domain walls is driven by thermal fluctuations, as is the case for normal fluids and for our Ising models, and (b) the dynamics by which the molecules or spins in the system rearrange is a local one. In the language of our spin models for example, a dynamics like that of the Kawasaki algorithm of Section 5.1 in which up- and down-pointing spins on adjacent sites swap values is fine. However, a dynamics like that of the non-local algorithm of Section 5.2 would not be acceptable. For the non-local algorithm the argument given above breaks down and a different formula for R applies. However, as long as we stick to Monte Carlo algorithms with local dynamics, both (a) and (b) above are satisfied. They are also satisfied for real fluid systems, and so we expect both the real system and our simulations to display the same growth law for domains.

A similar scaling argument can be made for the growth of domains in the ordinary Ising model in which the order parameter is not conserved. In this case we find that the domain size increases somewhat faster than in the COP case, as

$$R \sim t^{1/2}. \tag{10.3}$$

Again this result only applies if the dynamics is local, as in the single-spin-flip Metropolis and heat-bath algorithms, for example. Non-local algorithms such as the Wolff cluster algorithm of Section 4.2 show completely different behaviour.

Beyond scaling arguments of this sort there is no reliable analytical theory of phase separation to date, and most of what is known comes from computer simulations, usually of very simple models such as the ordinary Ising model or the COP Ising model. The problem with these simulations is that we are restricted to using only local dynamics, such as the Metropolis algorithm.

---

spin-flip algorithms indicates that the domain walls do indeed move diffusively (see Problem 4.1).

This prevents us from using most of the clever methods we developed in the first part of this book to speed up our simulations. However, as we will see, there are still some tricks we can play to make things faster. In the next two sections we will illustrate a number of techniques for simulating these kinds of systems by developing algorithms to test our scaling hypotheses for domain growth, Equations (10.2) and (10.3). We will look at the problem first for the ordinary (non-conserved-order-parameter) Ising model, which is the simpler example, and then for the conserved-order-parameter version.

10.1.1 Phase separation in the ordinary Ising model

The obvious way of simulating the phase separation of the ordinary Ising model is to use the single-spin-flip Metropolis algorithm. This algorithm satisfies the requirement discussed above that the dynamics be local and, as discussed on Section 3.1, is the most efficient single-spin-flip algorithm for the Ising model.

In Figure 10.1 we show a number of snapshots of a simulation of a two-dimensional Ising system on a 500×500 square lattice using the Metropolis algorithm. In this simulation we started the system off in an infinite temperature state, i.e., one in which the values of all the spins are chosen at random (see Section 3.1.1). The simulation itself however was run at a finite temperature $T = 2J$, which is some way below the critical temperature $T_c = 2.269J$ of the model. In effect, the system was quenched instantaneously from $T = \infty$ to $T < T_c$ at time $t = 0$. The figure clearly shows the formation of domains of up- and down-pointing spins in the system, which grow larger as time progresses. The frames in the figure show the state of the system after $t = 1, 10, 100$ and 1000 Monte Carlo steps per site, or "sweeps" of the lattice.

We are of course at liberty to use another algorithm for this simulation if we want to, as long as the dynamics is still a local one. For example, we could use a heat-bath algorithm, Equation (4.44). Although this would not be as efficient a way of simulating the model as the Metropolis algorithm, it is a useful way of checking the dynamical scaling law, Equation (10.3). If it really is true that all types of local dynamics give the same functional form for the growth of the domains over time, then we can check this by comparing the results of simulations using different algorithms. In Section 10.2, we will see how to extract the typical size of the domains from these simulations to check the scaling law.

10.1.2 Phase separation in the COP Ising model

For the case of domain growth in demixing fluids discussed at the beginning of this chapter the ordinary Ising model is not appropriate. As they demix,

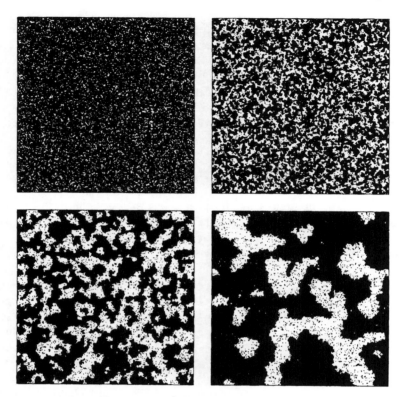

FIGURE 10.1 Snapshots of a computer simulation of phase separation in the two-dimensional normal Ising model using the Metropolis algorithm. The system was quenched from $T = \infty$ to $T = 2J$ at time $t = 0$. From top left, the frames show the configuration of the system after 1, 10, 100 and 1000 sweeps of the lattice, with black and white dots representing up- and down-pointing spins respectively.

fluids preserve the volume occupied by each fluid in the system, and in this case the conserved-order-parameter Ising model is the correct model to use.

As with the ordinary Ising model there are a variety of possible choices for the dynamics of a COP simulation, always with the constraint that the moves in the algorithm should be local. The simplest and commonest choice is the Kawasaki algorithm of Section 5.1 in which randomly chosen pairs of nearest-neighbour spins are exchanged with an acceptance probability which depends on the resultant change ΔE in the energy of the system according to

$$A = \begin{cases} e^{-\beta \Delta E} & \text{if } \Delta E > 0 \\ 1 & \text{otherwise.} \end{cases} \qquad (10.4)$$

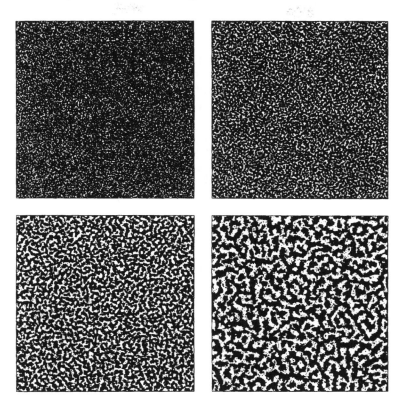

FIGURE 10.2 Snapshots of a computer simulation of phase separation in the two-dimensional conserved-order-parameter Ising model using the Kawasaki algorithm. The system was quenched from $T = \infty$ to $T = 2J$ at time $t = 0$. From top left, the frames show the configuration of the system after 10, 100, 1000 and 10 000 sweeps of the lattice.

(See Equation 5.1.) This algorithm ensures that the number of up- and down-pointing spins is preserved at every step of the simulation.

In Figure 10.2 we show the results of a simulation of a two-dimensional COP Ising model on a 500×500 square lattice using the Kawasaki algorithm. The frames in this figure are taken at times $t = 10$, 100, 1000 and 10 000 sweeps of the lattice—ten times as long as the ones in Figure 10.1. Even so, the domains of up- and down-spins are much smaller than in the non-conserved-order-parameter case. This we might have anticipated. The scaling form for the growth of domains in the COP model, Equation (10.2), goes as $t^{1/3}$, which increases more slowly than the $t^{1/2}$ of the normal Ising model. In the next section we show how to estimate the typical domain size R so that we can check these hypotheses quantitatively.

10.2   Measuring domain size

How do we measure the typical size R of the domains in our simulations?
A simple approach might be to measure the density of domain walls by
counting the number of times the direction of the spins switches from up
to down or *vice versa* along some line through the system. The typical
domain size could then be estimated by dividing the length of the line by
this number. This however would be a poor way of estimating R, because
although the direction of the spins does change at every domain wall, it can
also change temporarily just because a random spin gets flipped over in the
middle of a domain by our Monte Carlo algorithm. This means that we
will almost certainly underestimate the domain size if we use this simple
counting method, possibly very badly. A much better way to calculate the
domain size is to use a correlation function.

10.2.1   Correlation functions

A reliable way of estimating the typical domain size R is to calculate the
two-point spin–spin connected correlation function $G_c^{(2)}(\mathbf{r})$ defined in Section 1.2.1. We can do this for example by making use of the Fourier transform result given in Section 3.6:

$$\widetilde{G}_c^{(2)}(\mathbf{k}) = \frac{1}{N}\langle |\tilde{s}'(\mathbf{k})|^2 \rangle, \qquad (10.5)$$

where $\tilde{s}'(\mathbf{k})$ is the Fourier transform of $s_i' \equiv s_i - m$. Then we invert the
Fourier transform to get the correlation function. If the model being studied
is isotropic, then we can either calculate the correlation function along just
one direction or we can average it over all directions to improve our statistics.
For anisotropic models we may be interested in one particular direction, or
we may wish to calculate the domain size in several different directions, in
which case we should calculate separate correlations functions in each such
direction.

In Figures 10.3 and 10.4 we show the results of calculations of $G_c^{(2)}(\mathbf{r})$
along one axis for our ordinary and COP Ising model simulations. For spins
which are only a short distance apart the correlation function is large because
the spins are very likely to be pointing in the same direction. As the magnitude r of the displacement becomes larger however, the correlation should
fall off, and we see precisely this behaviour in the figures. The correlation
function tends to one at short distances and to zero at large distances in
both the non-conserved and conserved-order-parameter cases. The different
curves in each figure show the correlation function at a succession of different
times throughout the simulation. (In fact, we used the same times as for the
snapshots in our earlier figures, except for the last two curves in Figure 10.4

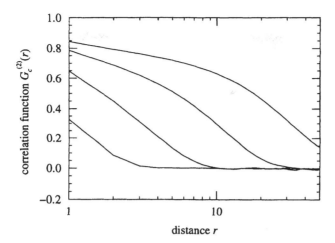

FIGURE 10.3   The two-point connected correlation function $G_c^{(2)}(r)$ of the two-dimensional normal Ising model as a function of displacement r along one axis for the spin states shown in the snapshots of Figure 10.1.

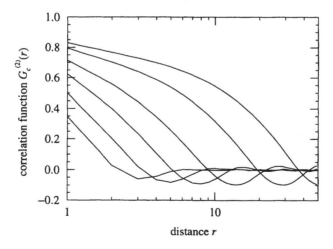

FIGURE 10.4   The two-point connected correlation function $G_c^{(2)}(r)$ of the two-dimensional conserved-order-parameter Ising model as a function of displacement r along one axis for the spin states shown in the snapshots of Figures 10.2 and 10.6.

which were for longer times and made use of a more sophisticated algorithm which we describe in Section 10.3.1.)

There is one important difference between the forms of the correlation functions for the two different models. In the case of the ordinary Ising model the correlation function is, apart from small statistical fluctuations, always positive, whereas in the COP case it is oscillatory, falling quickly to zero and becoming negative before rising again. The reason for this is that in order to form a domain of up-pointing spins in the COP model we have to take up-pointing spins from surrounding regions. Thus the area surrounding such a domain tends to be depleted of up-pointing spins, giving rise to a negative value for the correlation function.

We can now use the correlation function to extract an estimate of the typical domain size. In the case of the ordinary Ising model, the conventional choice is to define the domain size R to be the distance over which the correlation function falls from 1 to a value of $1/e = 0.368$. In the COP case it is defined as the distance to the point at which the correlation function first becomes negative. With these definitions R is roughly the distance from the centre of a domain to the surrounding domain wall; the domain diameter is $2R$.

Clearly these definitions are arbitrary to a degree, as they must necessarily be since a domain is not a very well-defined object in these models. However, if we look at Figures 10.3 and 10.4 again, we notice an important point. As time passes by, the correlations of spins in the system become longer in range, but the functional form of the correlation function stays the same; the curve moves to the right on our figures—corresponding to multiplication by a constant on the logarithmic scales we have used—but its shape remains roughly constant. This means that, as far as the study of domain growth goes, it does not matter precisely what definition we choose for the domain size R. Different definitions will give answers which differ by a multiplicative constant, but universal scaling laws like (10.2) and (10.3) will not be affected.

Using the definitions of domain size above, we have extracted values of R from the correlation functions shown in Figures 10.3 and 10.4 and plotted the results in Figure 10.5. We also show the conjectured scaling laws, Equations (10.2) and (10.3), which appear as straight lines on the logarithmic scales used in the figure. As we can see, the results for the ordinary Ising model (the circles in the figure) follow the $t^{1/2}$ scaling quite convincingly, even for the rather short simulation we have performed here. The results for the conserved-order-parameter model (the squares) are not such a good fit, but for longer times the match between the simulation results and the theory is reasonable.

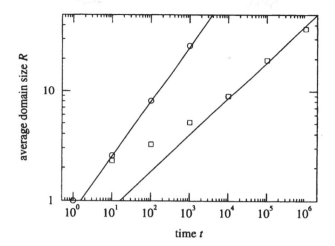

FIGURE 10.5   The typical domain size R as a function of time for our simulations of the ordinary Ising model (circles) and the conserved-order-parameter Ising model (squares). The solid lines show the scaling laws for these quantities, Equations (10.2) and (10.3).

10.2.2   Structure factors

An alternative and frequently more convenient way to estimate the domain size makes direct use of the Fourier transform (10.5) of the two-point correlation function, often called the **structure factor** for the system.[3] As the wave vector **k** tends either to zero or to infinity, the structure factor tends to zero. Somewhere between these two extremes there is a peak in the structure factor at the wave vector corresponding to the typical domain size. By measuring the position of this peak we can determine R.

The usual way to estimate the position of the peak is to average the structure factor over all directions to get the **circularly averaged structure factor** $S(k)$ which depends only on the magnitude of the wave vector k. Then we calculate the first moment k^* of this function thus:

$$k^* = \frac{\int_0^\infty k\,S(k)\,\mathrm{d}k}{\int_0^\infty S(k)\,\mathrm{d}k}. \tag{10.6}$$

The typical radius of a domain is then given by

$$R = \frac{2\pi}{k^*}. \tag{10.7}$$

---

[3]This name comes from the theory of X-ray scattering, in which the structure factor is the quantity directly measured by the scattering experiment.

In fact, this method does not give exactly the same estimate of R as the one discussed in the last section, but again the two differ only by a multiplicative constant so that either is fine for investigating the scaling behaviour of R with time.

The structure factor contains more information about the structure of domains than simply their average size. As $k \to \infty$ we normally find that the structure factor tends to zero as $k^{-(d+1)}$ where d is the dimensionality of the system, a behaviour referred to as **Porod's law**. In the language of scattering experiments, Porod's law is the result of scattering from the domain walls and only persists up to values of k corresponding to the typical width of such a wall (Shinozaki and Oono 1991). The behaviour of the structure factor in the limit of small k is not so well understood, and is still a matter of some debate.

10.3 Phase separation in the three-dimensional Ising model

We have shown that it is possible to perform quite accurate simulations of spinodal decomposition in the two-dimensional Ising model using simple Monte Carlo algorithms like the Metropolis and Kawasaki algorithms. In the case of the ordinary Ising model in particular, where the order parameter is not conserved, we can get results for the rate of increase of the average domain size which are in very reasonable agreement with the predicted scaling laws. And even for the conserved-order-parameter model, in which we expect slower domain growth (the domains grow as $t^{1/3}$, as opposed to $t^{1/2}$ in the non-conserved case), we can still see the beginnings of the expected scaling behaviour towards the end of a long simulation. When we look at three-dimensional models however, this is no longer the case. For the ordinary Ising model, three-dimensional simulations take so much longer than two-dimensional ones that even at the end of an extremely long run it is difficult to convince oneself that one is seeing any more than the very beginnings of the expected $t^{1/2}$ scaling. And for the COP model the situation is entirely hopeless; there is no trace of the $t^{1/3}$ scaling law in even the longest simulations. For three-dimensional models we therefore need to look for more efficient simulation techniques.

In the following sections we look at two different approaches to this problem. First we look at ways in which we can make our current algorithms more efficient without changing the dynamics. Then in Section 10.4 we look at how we can change the actual physics of the problem in such a way as to preserve the expected scaling laws but at the same time speed up the growth of the domains. In these discussions we will concentrate on the conserved-order-parameter Ising model, since this model is more closely related to real phase separation experiments, and also because it is the harder of the two models to get good results from. If we can find an efficient way of simulating the conserved-order-parameter model in three dimensions, then we can surely do the easier non-conserved-order-parameter model as well.

10.3.1 A more efficient algorithm

Taking the example of the conserved-order-parameter Ising model then, we want to find a way of improving the efficiency of our Monte Carlo simulation so that we can simulate the model for long enough times to verify the scaling hypothesis, Equation (10.2). In Section 5.2 we demonstrated an algorithm which made use of non-local spin exchange moves to reach equilibrium faster. We cannot take this approach in the present case however, since, as we pointed out in Section 10.1, only a local diffusive dynamics gives the correct kind of domain growth. Still, even if we stick to Kawasaki-type Monte Carlo moves in which the values of pairs of adjacent spins are exchanged, we can still improve a lot on the simple algorithm of Section 10.1.2.

One problem with the simple Kawasaki algorithm is that we frequently pick a pair of spins which are pointing in the same direction. Swapping the values of such a pair is pointless because it doesn't change the state of either of them. So we can improve our algorithm by only picking pairs of spins which are pointing in opposite directions. As long as the pairs we pick are still nearest neighbours our dynamics is still a local one, and the improvement in efficiency by making this small change in the algorithm can be quite large. If we take the example of a COP Ising system with half the spins pointing up and half pointing down, the fraction of adjacent pairs of spins which are anti-aligned starts off at 50% for the initial $T = \infty$ state. This means we get a improvement in efficiency of a factor two right from the start, and this factor increases as the simulation progresses because as the spins in the system coalesce into domains, more and more of them have neighbours which are pointing in the same direction. The only regions of the system in which spins with anti-aligned neighbours are common are the domain boundaries, which constitute a smaller and smaller fraction of the total volume of the system as time goes by. In the simulation shown in Figure 10.2, for instance, we find that by the end of the run only 14% of the nearest-neighbour spin pairs in the system are anti-aligned, so that an algorithm which only considers these pairs will be about seven times faster than the simple Kawasaki algorithm.

We have to be a little careful about how we implement our algorithm, however. As we pointed out in Section 5.1, the number m of anti-aligned spin pairs is not a constant—it can change when we flip a pair—and this means that a naive implementation of the algorithm will not satisfy the condition of detailed balance. We can fix this however by a trick similar to that used in continuous time Monte Carlo algorithms: we make our Monte Carlo steps correspond to varying amounts of real time. In detail here is how the algorithm works.

First we make a list of all the anti-aligned pairs of nearest-neighbour spins for the initial configuration of the system. Then at each step of the algorithm we perform the following operations.

1. We select a pair of spins at random from our list and calculate the change in energy which would result if their values were swapped over. We then perform the swap (or not) with the Kawasaki acceptance probability, Equation (10.4).

2. We add an amount Δt to the time, where

$$\Delta t = \frac{1}{m}. \tag{10.8}$$

3. If the values of the spin were in fact exchanged, we must update our list of spin pairs to reflect the change. The pair that we exchanged can just stay where they are in the list, since they are still anti-aligned after the move. All the neighbouring pairs will have to be changed however. All the ones which previously were anti-aligned are now aligned and should be removed from the list and all the ones which previously were aligned are now anti-aligned and should be added.

To see that this algorithm satisfies detailed balance, notice that the probability per Monte Carlo step of any anti-aligned spin pair being chosen is just $1/m$, and hence that the probability per unit simulated time of it being chosen is $1/(m \, \Delta t)$ which equals 1, independent of m. Given the form of the acceptance probability for the move, this then ensures detailed balance. Ergodicity is also obeyed, for exactly the same reasons that it is in the normal Kawasaki algorithm.

Using this algorithm (and with a bit more patience) we have been able to simulate the two-dimensional COP Ising model for two decades of time longer than we previously could with the simple Kawasaki algorithm. The results are shown in Figure 10.6.

10.3.2   A continuous time algorithm

The algorithm we proposed in the last section is a good deal more efficient than the simple Kawasaki algorithm, especially at late times in the phase separation process, when the number of anti-aligned spin pairs becomes small. However, there is still one source of inefficiency in the algorithm: many of the Monte Carlo moves we propose are rejected simply because of the acceptance ratio, so we waste CPU time by generating moves which are never

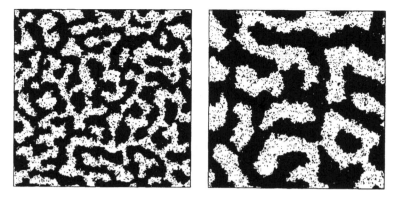

FIGURE 10.6 Snapshots of a simulation of the two-dimensional
conserved-order-parameter Ising model using the algorithm described
in Section 10.3.1, after 10^5 and 10^6 sweeps of the lattice.

performed. In Section 5.2.1 we saw how to get around this problem in the
case of an equilibrium simulation by using a continuous time Monte Carlo
method. We can use exactly the same trick with the current non-equilibrium
problem to produce an algorithm in which no moves are ever rejected. This
algorithm is a very straightforward implementation of the continuous time
idea as set it out in Section 2.4. It goes like this.

First we take the initial $T = \infty$ configuration of the lattice and construct
separate lists of anti-aligned nearest-neighbour spin pairs, grouped according
to the change in energy ΔE which would result if their values were to be
swapped over. On a cubic lattice in three dimensions, for example, the value
of ΔE ranges from $-20J$ to $+20J$ in steps of $4J$, so there would be 11 lists
in all. Suppose that the number of entries in the ith list is m_i. Each Monte
Carlo step then consists of performing the following operations:

1. We choose one of our lists at random, with the probability of choosing
   the ith list being proportional to $m_i A_i$, where A_i is the value of the
   acceptance ratio, Equation (10.4), for the moves in list i.

2. We select one of the pairs of spins in the chosen list uniformly at
   random.

3. We exchange the values of the two spins in the selected pair.

4. We add an amount Δt to the time, equal to the reciprocal of the sum
   of the rates for all possible moves, which in this case means

$$\Delta t = \frac{1}{\sum_i m_i A_i}. \tag{10.9}$$

5. We update our lists to reflect the move we have made.

In fact, it is this last step which takes all the time in this algorithm. The business of updating all the lists after every move is quite involved and can take a lot of work. For this reason, it turns out that it only really pays to use a continuous time algorithm at very low temperatures. At the temperature $T = 2J$ which we used in our Ising model simulations earlier in this chapter, for instance, the lowest values of the Kawasaki acceptance ratio occurred towards the end of our simulations and averaged about 34%. This means that our continuous time algorithm, which has an effective acceptance ratio of 100%, should be about three times as fast. However, the extra work necessary to keep the lists of spin pairs up to date means that each Monte Carlo step of the continuous time algorithm takes a good deal more than three times as much CPU time as a step of the simpler algorithm, and so the simpler algorithm is more efficient in the long run. As the simulation temperature is lowered, the acceptance ratio in Equation (10.4) becomes exponentially small, so there will always be a temperature low enough that using the continuous time algorithm will pay off. For our purposes in this chapter however, the continuous time method does not help.[4]

So what are we to do if we want to probe longer times still? In the next section we investigate a completely different approach to speeding up our simulation. Instead of changing the Monte Carlo algorithm, we look at changing the actual physics of the system simulated.

10.4   An alternative dynamics

Up until this point, we have striven to find an efficient algorithm for simulating phase separation in the conserved-order-parameter Ising model without changing the dynamics of the model itself. This is a sensible approach since, as discussed in Section 10.1, we are not at liberty to toy around with the dynamics in the same way as we were with the equilibrium algorithms in the first half of this book. We have however been rather more strict than we need to be. As we have pointed out, the crucial restriction on the dynamics is only that our Monte Carlo moves should be local. In this section, we show how we can make use of our knowledge of the physics of spinodal decomposition to find a different dynamics for our simulation which still employs local moves, but gives much faster domain growth.

---

[4]We didn't know that this was going to be the case before we tried it. Often finding the best Monte Carlo algorithm for a problem is just a question of trial and error.

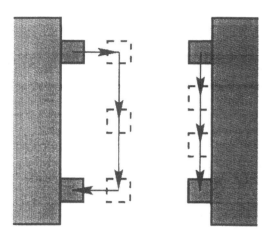

FIGURE 10.7   The two
different mechanisms by
which domains grow during
phase separation: bulk
diffusion (left) and surface
diffusion (right).

10.4.1   Bulk diffusion and surface diffusion

In order to understand how to speed up our simulation, we need to know a
little more about the physics of spinodal decomposition. As we showed in
Section 10.1, the typical domain size R in the COP Ising model is expected
to grow as $t^{1/3}$ with time. In fact, as we can see in Figure 10.5, this scaling
behaviour is only followed closely when t becomes quite large. At shorter
times the curve of R against t is flatter than $t^{1/3}$. Why is this? It turns out
that there are two important processes going on in the system, only one of
which we have taken into account so far.

The movement of particles (or spins, if you prefer) in the conserved-
order-parameter model can take place in two different ways, as illustrated in
Figure 10.7. **Bulk diffusion**, shown on the left of the figure, is the name we
give to the movement of particles from one domain to another, or between
parts of the same domain, by diffusion through an intervening domain of
vacancies. In the language of spins, bulk diffusion takes place when, for
example, an up-pointing spin breaks off from the surface of a domain and
travels through a domain of down-pointing ones to join up with another
up-pointing domain elsewhere. This process contributes to domain growth
because the activation energy needed for a spin to break off from a small
domain—one with high average surface curvature—is less than that for a
large domain, so that on average more spins move from smaller domains to
larger ones than the other way round.

There is however another mechanism by which spins move: **surface diffusion**, which is illustrated on the right-hand side of Figure 10.7. Surface
diffusion refers to the sliding of particles (or spins) along the surface of a domain, without their ever detaching from the domain. Obviously this process
cannot move particles from one domain to another. However, it can move

whole domains around the lattice and in this way cause domain growth by joining pairs of smaller domains into larger ones.

Surface diffusion takes place much more easily than bulk diffusion because there is no activation energy involved. It costs energy to break a spin off from a domain, but it costs nothing to simply slide one along the surface. For this reason surface diffusion tends to dominate the dynamics of domain growth at early times. As time draws on, however, it becomes less and less important, because it can only take place on the surfaces of the domains. As we argued in Section 10.1, the ratio of domain surface to volume decreases as $1/R$, so the opportunity for surface diffusion dwindles as R grows. Thus, it is only the bulk diffusion which contributes to the long-time scaling of domain sizes which we see in our simulations.

In fact, surface diffusion turns out not only to be irrelevant to the results we are looking for in our simulations, it is also the main reason why our simulation is slow. Using scaling arguments similar to the ones we gave in Section 10.1 we can show that the growth of domains as a result of surface diffusion should go as $t^{1/4}$. Given that we expect surface diffusion to dominate at early times, this implies that we should see R growing slowly as $t^{1/4}$ initially, and then switching over to a $t^{1/3}$ growth law later on. This is precisely the behaviour which we remarked on earlier in Figure 10.5. So if we could find an algorithm in which surface diffusion was suppressed in favour of bulk diffusion, it should show the $t^{1/3}$ behaviour much earlier in the simulation. We now show how to create precisely such an algorithm.

10.4.2 A bulk diffusion algorithm

The trick behind creating an algorithm which favours bulk diffusion over surface diffusion is to find an alternative form for the acceptance ratio, Equation (10.4), which gives the motion of a particle along the surface of a domain a lower acceptance ratio than in our previous algorithms, thereby suppressing it relative to bulk motion. It turns out that it is actually rather simple to do this. Our algorithm makes use of the concept of the "spin coordination number" which we introduced in Section 5.2.1. The spin coordination number n_i of a spin i is the number of nearest-neighbour spins which are pointing in the same direction as spin i. If we exchange the values of two nearest-neighbour spins, i and j, we can write the resultant change ΔE in the energy of the system in terms of the spin coordination numbers n_i and n_j before the exchange thus:

$$\Delta E = 4J(n_i + n_j - z + 1), \tag{10.10}$$

where z is the *lattice* coordination number, i.e., the total number of nearest neighbours of each site on the lattice.

Our proposed Monte Carlo algorithm can also be expressed in terms of the spin coordination numbers. The acceptance ratio for making the transition from a state μ to a state ν by exchanging the values of two adjacent spins is

$$A(\mu \to \nu) = \frac{e^{-4\beta Jn_i^\mu} + e^{-4\beta Jn_j^\mu}}{2}, \tag{10.11}$$

where n_i^μ and n_j^μ are the coordination numbers in the initial state μ. As we will shortly demonstrate, this choice does indeed suppress surface diffusion relative to bulk diffusion, and it is also very simple to implement. First however, we have to prove that it satisfies the conditions of ergodicity and detailed balance.

Ergodicity is easy. The individual moves in this algorithm are the same exchanges of adjacent spin values as for the Kawasaki algorithm and so the proof of ergodicity which we gave in Section 5.1 applies here as well. To prove detailed balance the crucial step lies in noticing that the spin coordination numbers of sites in the final state ν are related to those in the initial state μ by

$$n_i^\nu = z - 1 - n_i^\mu, \tag{10.12}$$

$$n_j^\nu = z - 1 - n_j^\mu. \tag{10.13}$$

Using these relations we then find that

$$\begin{aligned}
\frac{A(\mu \to \nu)}{A(\nu \to \mu)} &= \frac{e^{-4\beta Jn_i^\mu} + e^{-4\beta Jn_j^\mu}}{e^{-4\beta Jn_i^\nu} + e^{-4\beta Jn_j^\nu}} = e^{4\beta J(n_i^\mu + n_j^\mu - z + 1)} \\
&= e^{-\beta \Delta E},
\end{aligned} \tag{10.14}$$

as required.

So how does this algorithm help us? Consider again the two scenarios shown in Figure 10.7. With the Kawasaki acceptance ratio, Equation (10.4), the bulk diffusion process on the left is slow because the move which splits the spin off from its domain has an acceptance ratio which is less than one and becomes exponentially small as temperature falls. For the surface diffusion process on the right, by contrast, the spin never splits off from the domain and all moves have an acceptance ratio of one. In the new algorithm, the bulk diffusion process still has an acceptance ratio which is less than one. So, however, does the surface diffusion process, since the spin coordination numbers n_i and n_j of a pair of spins on the surface are both greater than zero. This has the effect that fewer surface moves will be accepted relative to bulk ones in the new algorithm. As a result the $t^{1/4}$ domain growth produced by the surface diffusion should be less strong and we should see the late-time $t^{1/3}$ setting in much earlier.

The acceptance ratio, Equation (10.11) can be used either with a simple accept/reject algorithm of the type described in Section 10.3.1, which would

FIGURE 10.8 Two-dimensional slices of simulations of the three-dimensional COP Ising model, using the surface-diffusion-suppressing algorithm described in the text. The four frames on the left are from a simulation at temperature $T = 0.1T_c$ at times $t = 10^{11.5}$, $10^{12.5}$, $10^{13.5}$ and $10^{14.5}$ (top to bottom). The frames on the right are at $T = 0.6T_c$ at times $t = 10^{2.5}$, $10^{3.5}$, $10^{4.5}$ and $10^{5.5}$.

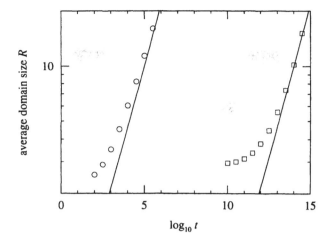

FIGURE 10.9   The average domain radius R as a function of time for the simulations shown in Figure 10.8. The circles are for $T = 0.6T_c$ and the squares for $T = 0.1T_c$. The expected scaling of $t^{1/3}$ is shown as the two solid lines.

be appropriate for simulations at relatively high temperatures, or with a continuous time algorithm similar to the one given in Section 10.3.2, which would work well at lower temperatures. In Figure 10.8 we show the results of simulations using a continuous time algorithm on a $172 \times 172 \times 172$ lattice. The frames show two-dimensional slices of the three-dimensional system at successive decades of time throughout the run. The frames on the left are for $T = 0.1T_c$ and the ones on the right are for $T = 0.6T_c$. Especially for the lower temperature simulation, the algorithm is very efficient and allowed us to reach total simulation times of 3×10^{14} Monte Carlo steps per site. Using the simple Kawasaki algorithm which we started out with at the beginning of this chapter, the same simulation would have taken us about 50 million years. In combination with the enhancement of bulk diffusion, this algorithm allows us finally to make out the $t^{1/3}$ scaling behaviour in a three-dimensional system. Figure 10.9 shows the average domain sizes at the two different temperatures as a function of time, measured using the correlation function technique described in Section 10.2.1. In the last few decades of each run we can see the results tending asymptotically to the expected $t^{1/3}$.

Problems

10.1 What is the state of a phase-separating COP Ising model on an $L \times L$ square lattice in the limit $t \to \infty$ when the numbers of up- and down-pointing spins are equal?

10.2 The implementation of the bulk diffusion algorithm described in Section 10.4.2 as a continuous time algorithm turns out to be charmingly simple. We maintain lists of spins, as we did for the equilibrium algorithm of Section 5.2.1, divided according to their spin coordination numbers n_i. Then we choose a spin i at random from one of these lists with probability proportional to $(z - n_i) \exp(-4\beta n_i)$ and exchange its value with that of one of its anti-aligned neighbours, chosen at random. Show that this algorithm does indeed result in a transition probability proportional to the acceptance ratio of Equation (10.11).

11

Monte Carlo simulations in surface science

In the last chapter we used the results of simulations of the Ising model to study the growth of domains under spinodal decomposition. In this chapter we look at another use of the Ising model in quite a different field: surface science.

Out-of-equilibrium Monte Carlo calculations have become an important tool in surface science. They provide an efficient way of simulating the movement of atoms on crystalline surfaces. Most often we are interested in metals, the majority of which have a face-centred cubic (fcc) crystalline structure. The surfaces of such a crystal provide a natural lattice for our Monte Carlo simulation. For example, the atoms which make up the (001) facet of an fcc crystal form a square lattice, as shown in Figure 11.1. The atoms of a (111) facet form a triangular lattice.

Now consider what happens if we add a few extra atoms to our surface. These **adatoms** might be atoms of the same element as the crystal, or of a different element. In either case, there will be attractive forces between the adatoms and the atoms of the surface which mean that it is energetically favourable for the adatoms to be next to as many surface atoms as possible. On a (001) surface, for instance, they will prefer to sit in the so-called **four-fold hollow sites** where they are in contact with four surface atoms simultaneously, as shown in Figure 11.1. Such four-fold hollow sites also form a square lattice on the surface.

On a (111) surface the best an adatom can do is to sit in a **three-fold hollow site**, in contact with three surface atoms. The three-fold hollow sites divide into two distinct types depending on their position relative to the atoms in lower layers of the metal (see Figure 11.1 again). An **hcp site**, named after the hexagonal-close-packed lattice to which it belongs, is

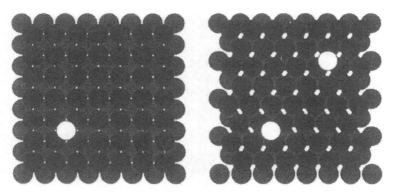

FIGURE 11.1   Left: if a metal with an fcc crystal structure is cut in the (001) direction, the atoms in the resulting facet are arranged in a square lattice. Right: a cut in the (111) direction results in a triangular lattice; the upper of the two adatoms is in an hcp site and the lower one is in an fcc site.

a three-fold hollow site directly over an atom in the layer immediately below the surface. An **fcc site** is one which lies directly over one of the atoms two layers below the surface. Both the set of hcp sites and the set of fcc sites form triangular lattices on the surface of the crystal. The combined set of all three-fold hollow sites forms a honeycomb lattice. (See Section 13.1.2 for a description of the honeycomb lattice.)

Many different types of crystalline surfaces are of interest in surface science. In this chapter we will illustrate our Monte Carlo techniques using the square lattice of a (001) surface, since the square lattice is the easiest to work with. All the ideas and algorithms we will describe, however, are equally applicable to other lattices. Techniques for performing simulations on lattices such as triangular and honeycomb lattices are discussed in Chapter 13.

Consider then a collection of adatoms in the four-fold hollow sites of the (001) surface of an fcc metal. (Actually, it could just as well be a body-centred cubic (bcc) metal. For the (001) surface it makes no difference.) Each adatom is bound to the metal surface with some binding energy E_s, which is typically a few electron volts (eV). This binding energy is the same for all sites. However, two adatoms can lower their energy further by being in adjacent sites; there is an extra binding energy E_b between adjacent adatoms which is typically on the order of a few tenths of an eV. This extends to adatoms which are adjacent to more than one other adatom: we gain an extra E_b in energy for every pair of adjacent adatoms on the surface. For most metals it is a pretty good approximation to neglect binding energies between adatoms which are any further apart than adjacent four-fold hollow sites. Calculations for copper surfaces using semi-empirical potentials, for instance, indicate that the error introduced by making this approximation is less than one per cent (Breeman *et al.* 1995).

Perhaps you can already see that the energy of a set of adatoms on the surface of a metal is closely related to the conserved-order-parameter Ising model which we studied in the last chapter. We now make this connection explicit. The arguments given here follow quite closely those used at the beginning of Chapter 5 to make the connection between lattice gases and the Ising model.

Suppose we have a lattice of N sites on which there are n adatoms, including p pairs of nearest neighbours. The total binding energy H of these atoms can be written as

$$H = -nE_s - pE_b. \tag{11.1}$$

The first term $-nE_s$ here is a constant as long as the number n of adatoms does not change. Only the second term depends on the configuration of the atoms. Let us represent this configuration as we did in Chapter 5 using a set of variables $\{\sigma_i\}$ such that σ_i is 1 if site i is occupied by an adatom and 0 if the site is empty. In terms of these variables we can write the number of nearest-neighbour pairs of atoms as

$$p = \sum_{\langle ij \rangle} \sigma_i \sigma_j, \tag{11.2}$$

where $\langle ij \rangle$ indicates that sites i and j are nearest neighbours. Making a change of variables to the Ising spins

$$s_i = 2\sigma_i - 1, \tag{11.3}$$

this becomes

$$p = \frac{1}{4} \sum_{\langle ij \rangle} (s_i + 1)(s_j + 1)$$

$$= \frac{1}{4} \sum_{\langle ij \rangle} s_i s_j + \frac{1}{2} z \sum_i s_i + \frac{1}{4} zN, \tag{11.4}$$

where z is the lattice coordination number introduced in Section 3.1 (equal to the number of nearest neighbour sites of each site). Using Equation (11.3), the second term here can be written as

$$\sum_i s_i = 2 \sum_i \sigma_i - 1 = 2n - 1, \tag{11.5}$$

which is a constant as long as the number of occupied sites remains fixed. The third term in (11.4) is also a constant, since both z and N are constants.

Substituting for p in Equation (11.1) we thus arrive at

$$H = -\frac{1}{4}E_b \sum_{\langle ij \rangle} s_i s_j + \text{constant}, \qquad (11.6)$$

where we have put all the constant terms together. Comparing this expression with Equation (5.7), we see that the total energy of our collection of adatoms has exactly the same form as the Hamiltonian of the conserved-order-parameter Ising model with the spin–spin coupling J set equal to $\frac{1}{4}E_b$.

For the typical values of E_b found in metals, the critical temperature of this system is significantly higher than room temperature. To see this, recall that the critical temperature of the Ising model (see Equation (3.53)) is

$$kT_c = 2.269J = 0.567E_b. \qquad (11.7)$$

Typically binding energies between adatoms are on the order of about 0.3 eV, which is equivalent to about 5×10^{-20} joules. Substituting this figure into (11.7) we get a critical temperature of about $T_c = 2000\,\text{K}$, which is well in excess of most laboratory temperatures (room temperature is about $300\,\text{K}$), and indeed in excess of the melting point of most metals. Thus the adatoms on a metal surface will under most conditions be in the condensed phase of the COP Ising model, in which they tend to cluster together into "islands" rather than being dispersed like the atoms in a gas. We can make use of many of the results of Chapter 5 concerning the COP Ising model to tell us about the properties of the adatoms in this condensed phase. In this chapter however, we are more concerned, as most surface scientists are, with the out-of-equilibrium properties of the system, such as the way in which adatom islands form and grow. It is on these properties that we will concentrate for the remainder of the chapter.

11.1   Dynamics, algorithms and energy barriers

We showed in the last section that the total energy of a certain adatom configuration can be well approximated using the Hamiltonian of the conserved-order-parameter Ising model with an appropriately chosen spin–spin coupling. This however only tells us about the equilibrium behaviour of the system. As discussed in Chapter 9, the out-of-equilibrium behaviour of the system depends on the dynamical processes by which the adatoms move. Before we can create a simulation of our system, therefore, we need to work out what the appropriate dynamics should be. We break the discussion up into two parts: first we look at the dynamics of a single adatom on a crystal surface and design an algorithm to simulate this simplified problem, and then we consider how this algorithm needs to be modified when more than one adatom is present.

11.1.1 Dynamics of a single adatom

At typical laboratory temperatures (or even considerably higher) the observed motion of the adatoms on a metal surface is of infrequent hops of atoms from one site on the surface to another adjacent site. A single isolated atom on a (001) surface will stay in the same four-fold hollow site for comparatively long periods of time before making a quick hop to one of its four neighbour sites. We will assume that the only moves we need to take into account are hops of this kind between neighbouring sites.

When an adatom is in a four-fold hollow site, it is in contact with four surface atoms at once, which is the maximum number possible on a (001) surface. However, when it makes a jump from one site to another it has to pass through a succession of intermediate states where it is in contact with a smaller number of surface atoms, typically two in the (001) case. These intermediate states are therefore higher in energy than the preferred four-fold hollow sites, and constitute an energy barrier of the kind which we discussed in Section 9.1.1. It is because of these energy barriers that adatoms prefer to stay in the same place for relatively long periods of time.

The typical height B_{ij} of the energy barrier for hopping from a site i to a site j is significantly greater than the thermal energy kT available for crossing it. The **hopping rate** r_{ij} at which an isolated adatom hops from i to j is related to the barrier height by the **Arrhenius law**:

$$r_{ij} = \nu e^{-\beta B_{ij}}. \tag{11.8}$$

In other words, the hopping rate decreases exponentially with increasing height of the energy barrier. The quantity ν is called the **attempt frequency** and sets the overall time-scale for the movement of adatoms. The attempt frequency typically has a value on the order of 10^{11} Hz but can vary depending on, amongst other things, the local environment of the atom making the hop. Most of the variation in r_{ij}, however, comes from differences in the barrier heights; the variation with attempt frequency is quite small by comparison. For this reason, it is common to assume ν to be a constant, independent of the adatom environment, and we will make that assumption here as well.

There are a number of reasons why the Arrhenius law may not be perfectly obeyed in practice. An atom which crosses one barrier, for example, will have a lot of kinetic energy when it reaches the other side, which it may be able to use to cross another barrier immediately afterwards, an effect known as **transient mobility**. On surfaces with reasonably high energy barriers, however, this is a small effect and we will ignore it for the moment

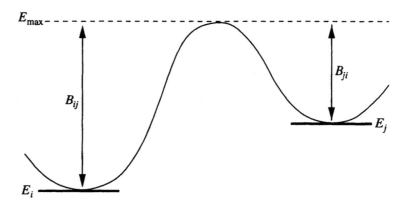

FIGURE 11.2   The energy at the highest point of the trajectory between sites i and j is $E_{max} = E_i + B_{ij} = E_j + B_{ji}$. Rearranging
we then get $B_{ij} - B_{ji} = E_j - E_i$. The condition of detailed balance,
Equation (11.10), follows as a result.

and assume that successive hops are uncorrelated and obey the Arrhenius
law.

Given these assumptions it is not hard to show that the hopping process
satisfies the normal condition of detailed balance, Equation (2.14). As Figure 11.2 shows, the energy barrier B_{ij} which an atom has to climb in order
to hop from site i to site j is related to the barrier B_{ji} for the reverse hop
by

$$B_{ij} - B_{ji} = E_j - E_i, \tag{11.9}$$

where E_i and E_j are the binding energies at the two sites. (For a single
adatom on a uniform surface these would be the same, but the proof works
without making this assumption. In the next section we consider systems of
many adatoms for which the binding energies vary, so it is worth our while
to prove the more general result.) Using Equation (11.8) we can then show
that the rates for hopping in the two directions satisfy detailed balance:

$$\frac{r_{ij}}{r_{ji}} = e^{-\beta(E_j - E_i)}. \tag{11.10}$$

The condition of ergodicity also applies to adatom hopping, since a given
atom can hop to any site on the lattice given long enough. Thus, each site
on the lattice should be occupied with the normal Boltzmann probability
$Z^{-1}e^{-\beta E_i}$ (see Equation (1.5)). This means that we can indeed apply the
results for the normal conserved-order-parameter Ising model to the equilibrium properties of our adatoms, as we claimed at the beginning of the
chapter.

So how do we use these results to create a Monte Carlo algorithm for simulating the motion of our adatoms? For the case of just a single isolated adatom on a (001) surface, one possible algorithm would be to choose a direction for our atom to hop in, and then make this move with an acceptance probability

$$A = e^{-\beta B_{ij}}, \tag{11.11}$$

where B_{ij} is the energy barrier for that move. Under this algorithm, each Monte Carlo step would be equivalent to an interval of real time

$$\Delta t = \frac{1}{4\nu}. \tag{11.12}$$

The factor of four comes from the four directions in which the adatom can hop.

We can do considerably better than this however. Notice that if all the energy barriers are high compared with kT then all moves will get rejected most of the time with this algorithm, which is a waste of CPU time. To get around this problem we can use a trick similar to the one we used for the Metropolis algorithm in Section 3.1 and multiply all our acceptance ratios by a constant chosen such that the largest acceptance ratio becomes equal to one, which is the highest value it is allowed to have. We can do this by setting

$$A = e^{-\beta(B_{ij} - B_{\min})}, \tag{11.13}$$

where $B_{\min}$ is the lowest energy barrier in the system. (Again, we are avoiding making the assumption that all the barriers are same, even though this will be the case in most single adatom systems.)

Making this change does not affect the dynamics of the system, but it does improve the efficiency of our simulation by a factor of $\exp(\beta B_{\min})$. It also changes the time-scale by the same factor. With this algorithm, each Monte Carlo step now corresponds to a time interval of

$$\Delta t = \frac{e^{\beta B_{\min}}}{4\nu}. \tag{11.14}$$

A better approach still to the problem is to use a continuous time Monte Carlo algorithm. As described in Section 2.4, this means that at each Monte Carlo step we choose one of the possible moves of our atom and perform it. The probability of choosing any particular move should be proportional to r_{ij} and a single Monte Carlo step now corresponds (see Equation (2.19)) to an amount of real time

$$\Delta t = \frac{1}{\sum_j r_{ij}}. \tag{11.15}$$

The sum here is over all sites j to which the adatom can hop and the r_{ij} are calculated from Equation (11.8). Notice that this time interval can vary from one Monte Carlo step to another.

With this algorithm, as with all continuous time algorithms, we never reject any moves at all, which makes it a very efficient use of our CPU time. It does have the disadvantage that we have to calculate the hopping rates of all possible moves at each time step, which makes the calculation more complicated. For the case of a single adatom, this is no great burden since there are only four possible moves at any one time, but in the next section, where we look at the case of many adatoms, it is more of a problem. In that case, we will need to use bookkeeping schemes of the type introduced in Section 5.2.1 to save us from having to recalculate all possible moves at every step.

11.1.2   Dynamics of many adatoms

To simulate a system in which there are many adatoms we can use much the same types of algorithms as in the previous section. One difference is that we must now reject any moves which involve hopping an atom to a site which is already occupied by another atom. The other main difference between the two cases is in the energy barriers between states. As we saw at the beginning of the chapter, the interactions between adatoms can affect the binding energies of atoms in particular sites and this affects the heights of the energy barriers; if an atom is bound more tightly to one site than to another, then presumably the energy barrier it has to cross to get away from that site is also higher. However, this effect does not account for all, or even most, of the interesting physics of adatom dynamics. The interactions between atoms also affect the energies of the intermediate states which atoms have to pass through in moving from one site to another, and this also changes the energy barriers, often quite dramatically.

It is a difficult task to calculate the energy barriers for adatom hopping exactly. In Monte Carlo simulations a variety of approaches have been used, ranging from the very simple and fast to extremely sophisticated atomic structure calculations. Most simulations fall into one of the following three classes.

Kawasaki-type energy barriers: The simplest choice is to set the energy E_{max} of the top of the barrier (see Figure 11.2) equal to the greater of the binding energies of the atom in the two sites i and j between which it is hopping, plus a constant B_0. In other words, the barrier B_{ij} for hopping from site i to site j is given by

$$B_{ij} = \begin{cases} B_0 + E_j - E_i & \text{if } E_j > E_i \\ B_0 & \text{otherwise.} \end{cases} \qquad (11.16)$$

Notice that this choice satisfies Equation (11.9), so that detailed balance will be obeyed once the system has come to equilibrium.

We can now feed these barrier heights into any of the algorithms proposed in the previous section to create a Monte Carlo algorithm for the many adatom system. The easiest way to go is to use the second algorithm we proposed. In this case, we would choose an atom at random and consider moving it in one of the four possible directions, also chosen at random. If the site we want to move it to is already occupied by an atom, then we cannot make the move. If it is empty, then we decide whether to make the move based on the height of the energy barrier for the move. The height of the lowest barrier is $B_{min} = B_0$, so the acceptance ratio, Equation (11.13), for the move becomes

$$A = \begin{cases} e^{-\beta(E_j - E_i)} & \text{if } E_j > E_i \\ 1 & \text{otherwise.} \end{cases} \tag{11.17}$$

This algorithm is very similar to the Kawasaki algorithm for the conserved-order-parameter Ising model (Section 5.1), and for this reason we refer to the energy barriers in Equation (11.16) as Kawasaki-type energy barriers.

The only subtlety with this algorithm is that a single Monte Carlo step of this algorithm now corresponds to an interval of real time

$$\Delta t = \frac{e^{\beta B_{min}}}{4n\nu}, \tag{11.18}$$

where n is the number of adatoms. (We need this factor of n, so that the attempt frequency of each atom for hopping in each direction is correctly ν per unit of real time.)

We can also use Kawasaki-type energy barriers with the continuous time algorithm proposed in Section 11.1.1. Combining Equation (11.16) with the Arrhenius law, Equation (11.8), we get

$$r_{ij} = \begin{cases} \nu e^{-\beta B_0} e^{-\beta(E_j - E_i)} & \text{if } E_j > E_i \\ \nu e^{-\beta B_0} & \text{otherwise.} \end{cases} \tag{11.19}$$

We can then employ this result in our continuous time algorithm. In practice this is the better way of simulating the system, although the continuous time algorithm is more complicated to implement. As discussed above, it requires a certain amount of careful programming to avoid having to recalculate lists of possible moves and their rates at every time step. In Section 11.2 we discuss implementation in more depth.

Both of these algorithms can be simplified if we approximate the total energies E_i and E_j of the system before and after the hop using Equation (11.1). (As we mentioned, this is usually a pretty good approximation.) In this case the energy change $E_j - E_i$ can be written as

$$E_j - E_i = (n_i - n_j)E_b, \tag{11.20}$$

where n_i is the number of occupied nearest-neighbour sites of site i, and n_j the corresponding number for site j. Then the energy barriers simplify to

$$B_{ij} = \begin{cases} B_0 + (n_i - n_j)E_b & \text{if } n_i > n_j \\ B_0 & \text{otherwise.} \end{cases} \qquad (11.21)$$

The bond-counting method: The trouble with Kawasaki-type energy barriers is that they are a rather poor representation of the physics of real adatoms on a crystalline surface. As discussed at the beginning of this section, the energy barriers in real systems vary from site to site depending on the configuration of adatoms in the nearest-neighbour sites. A slightly more sophisticated approach which approximates the effects of such neighbouring atoms is the **bond-counting method**.

Substituting Equation (11.20) into (11.9), we can write the difference in the energy barriers for hopping in either direction between sites i and j in terms of the number of occupied nearest neighbour sites n_i and n_j thus:

$$B_{ij} - B_{ji} = (n_i - n_j)E_b. \qquad (11.22)$$

One simple solution of this equation is

$$B_{ij} = B_0 + n_i E_b, \qquad (11.23)$$

where B_0 is again a constant. This defines the barrier heights in the bond-counting approach. Combining this result with the Arrhenius law, Equation (11.8), we obtain

$$r_{ij} = \nu e^{-\beta B_0} e^{-\beta n_i E_b}. \qquad (11.24)$$

As with the Kawasaki barriers, we can then use these rates to create a continuous time Monte Carlo algorithm for simulating the system.

Lookup tables: As we pointed out at the beginning of this section, the interactions between an atom and its neighbours when the atom is seated in a four-fold hollow site are not enough to explain the variation of barrier heights seen in real systems. If we want to perform a quantitatively accurate simulation of the dynamics of a collection of adatoms on a real crystal surface, neither the Kawasaki nor the bond-counting approach is adequate. The correct way to proceed is to make use of one of a variety of atomic structure methods to make an accurate estimate of the energy of a set of atoms in a given spatial configuration and then use this estimate to calculate the effective barrier height for the hopping of an atom between two given sites. These calculations are too complicated to be performed quickly during our Monte Carlo simulation, so the usual solution is to calculate beforehand all the barrier heights we are going to need for a particular simulation and store them in a "lookup table". As discussed at the beginning of the chapter, the

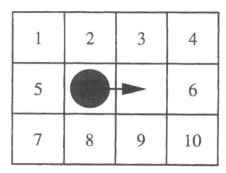

FIGURE 11.3   On a square lattice, there are ten nearest- and next-nearest-neighbour sites which need to be considered in a full lookup-table treatment of the energy barriers. This means that there are $2^{10} = 1024$ different possible arrangements of occupied and vacant neighbour sites.

energies of particular configurations of adatoms depend, to a good approximation, only on the arrangement of their nearest neighbours. If we want to be very accurate we can also include next-nearest neighbours. For an atom hopping to an adjacent site on a (001) surface there are ten sites which can be occupied by a nearest- or next-nearest-neighbour atom (Figure 11.3) and hence $2^{10} = 1024$ different possible arrangements of neighbouring atoms for which we have to calculate an energy barrier. Actually, this number can be reduced a bit by noting that some of the configurations are symmetry equivalent. Basically, however, the lookup table approach involves just calculating the barrier for each configuration of neighbours and storing it in the table. The calculation has to be repeated anew for each material we wish to study, since the heights of the energy barriers vary quite a bit from one material to another. The barriers calculated in this way can then be plugged into Equation (11.8) to give hopping rates which we can use, for example, in a continuous time Monte Carlo algorithm.

To illustrate the differences between the three choices of energy barriers—Kawasaki-type barriers, the bond-counting method and atomic structure calculation of lookup tables—we have calculated the energy barriers for movement of an adatom in the two situations depicted in Figure 11.4. In the first case the atom hops between two isolated sites with no occupied neighbour sites. In the second it hops one step along the edge of an island composed of other adatoms.

If we use Kawasaki-type barriers, the energy barrier crossed in each of these cases is the same—they are both equal to B_0 since the energies of initial and final states are the same in both cases

Using the bond-counting approach the energy barrier in the first situation is B_0 again, but in the second it is $B_0 + E_b$ since one of the nearest-neighbour sites of our adatom is occupied.

We have also calculated the energy barrier in each situation for the movement of a copper atom on a (001) copper surface using an atom-embedding

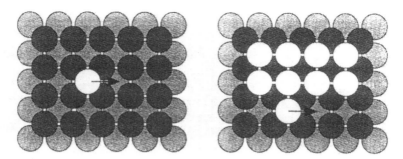

FIGURE 11.4   Adatoms moving as shown on a (001) surface have the same energy barriers and hence the same hopping rates under the Kawasaki-type approximation discussed in the text, but quite different hopping rates if we use the bond-counting method, or more sophisticated atomic structure methods.

method (Breeman *et al.* 1996). In the first situation we find an energy barrier of about 0.4 eV. In the second we find a barrier of about 0.2 eV. Note that the isolated atom has the higher barrier for hopping, which is the opposite of the prediction of the bond-counting approach.

To highlight the differences between the three cases further, let us calculate the ratio of the hopping rates for the two moves considered. For the Kawasaki-type barriers, the barriers in the two cases are the same, so the ratio of hopping rates is 1. For the bond-counting approach, Equation (11.8) tells us that the ratio of the rates will be $\exp(-\beta E_b)$. For copper, E_b is about 0.3 eV, which means that at room temperature the hopping of the isolated atom is more than 100 000 times faster than the hopping of the atom on the edge of the island.

Finally, plugging our atom-embedding results into Equation (11.8), we get hopping rates of $3 \times 10^{-4} \nu$ and $10^{-7} \nu$ for the two situations in Figure 11.4. Thus the atom on the edge of the island has the higher hopping rate (by contrast with the bond-counting approach, in which it has a lower rate) and hops about 3000 times faster than the isolated atom.

Thus we see that there is a factor of 3000 between the values of this hopping ratio in the atom-embedding calculations (which are the most accurate) and the Kawasaki-type approximation, and a factor of 3×10^8 when we go to the bond-counting method. Clearly such large factors have important implications for our simulation results. In Section 11.3 we give an example of the application of the methods described here to the problem of simulating molecular beam epitaxial growth and show that they do indeed give qualitatively different pictures of adatom dynamics. First however, we look at how these types of simulations are implemented.

11.2 Implementation

As discussed in the last section, there are a couple of different ways to formulate our Monte Carlo simulation of adatom hopping. The simpler way is to choose atoms at random and move them in a randomly chosen direction with an acceptance probability given by Equation (11.13), provided always that the site we wish to move them to is empty. The trouble with this approach is that most of the time the site we wish to move them to will not be empty. At typical laboratory temperatures, adatoms tend to phase separate and form islands, which means that our Monte Carlo algorithm wastes a lot of time picking atoms in the middle of islands and then not moving them because all the surrounding sites already have atoms in them. The solution, as we have already pointed out, is to use a continuous time algorithm. As it turns out, this is much easier to do for the Kawasaki and bond-counting methods discussed in the last section than it is if we wish to use energy barriers stored in a lookup table. This in fact is the main reason for the continued popularity of these simpler methods of calculating barriers despite their rather poor representation of the properties of real surfaces. Here we will discuss Kawasaki and bond-counting algorithms first. Lookup table algorithms are discussed in Section 11.2.2.

11.2.1 Kawasaki and bond-counting algorithms

Examining Equations (11.21) and (11.23), we see that in both the Kawasaki and bond-counting approaches the energy barriers take the general form

$$B_{ij} = B_0 + mE_b, \qquad (11.25)$$

where m is an integer which takes values in the range $0 \dots z$, with z being the lattice coordination number of the surface (four in the case of the (001) surface we have been considering). In fact, for both Kawasaki and bond-counting approaches, m can only equal z if the *spin* coordination number n_i at site i also equals z, in other words if all the nearest-neighbour sites of site i are occupied, in which case no moves are possible. So in practice we only need to consider values of m up to $z - 1$. (Note that in general the value of m will be different for any given move under the Kawasaki and bond-counting approximations.)

This suggests the following scheme for implementing our continuous time algorithm. For every possible move on the lattice, the barrier height, Equation (11.25), takes one of z possible values corresponding to $m = 0 \dots z - 1$. Suppose that at the start of our simulation we find all the possible moves and divide them into z lists, one for each value of m. Since the hopping rate, Equation (11.8), is entirely determined by the barrier height, all the moves in each list occur at the same rate

$$r_m = \nu e^{-\beta B_0} e^{-\beta m E_b}. \tag{11.26}$$

Our continuous time algorithm consists of three steps:

1. We choose a move at random from all the possibilities in proportion to the hopping rate. We can achieve this by first of all picking one of our lists at random with probability proportional to $N_m r_m$, where N_m is the number of moves in list m. Then we pick one of the moves from that list uniformly at random.

2. We perform the chosen move.

3. We add an amount Δt to the time, where Δt is the inverse of the sum of the hopping rates for all moves (see Equation (11.15)), which we can write as

$$\Delta t = \frac{1}{\sum_m N_m r_m}. \tag{11.27}$$

4. We update our lists of moves to take the move we have made into account.

The last step is the only tricky part of the algorithm. We don't want to recalculate all our lists of moves from scratch for the entire lattice after every move, since most sites on a large lattice will be far enough away not to be affected by the move. However, if our Monte Carlo step moves an adatom from site i to site j, then the hopping rates for all moves to or from nearest-neighbour sites of either i or j will be affected and the corresponding entries will have to be found in our lists and moved to different lists. In addition, some new moves will have become possible (ones into the site vacated by the atom which just moved) and others will no longer be allowed (ones into the site it now occupies). These moves will have to be added to or subtracted from the lists. As discussed in Section 5.2.1, all this bookkeeping can slow the algorithm down by as much as a factor of 40. On the other hand, no Monte Carlo step is wasted in a continuous time algorithm, and the resultant improvements in the efficiency of the simulation usually more than make up for the extra complexity of the algorithm. Depending on the temperature used for the simulation the continuous time algorithm can be a factor of 100 or more faster, on balance, than the simple algorithms described in Section 11.1.1.

11.2.2 Lookup table algorithms

If we want to perform our simulation using barrier heights stored in lookup tables, as described in Section 11.1.2, then the number of possible barrier heights becomes quite large, maybe as large as 1024 (see Figure 11.3). In order to use the algorithm of the last section in this case we would have to maintain 1024 different lists of possible moves. Unfortunately, the amount of time taken to perform one Monte Carlo step increases approximately linearly with the number of lists, and this makes the algorithm impractically slow. A better way of performing the simulation is to store the moves in a **binary tree**. (If you are unfamiliar with tree structures, you will find them discussed in detail in Section 13.2.4.) The idea is as follows.

We create a binary tree, each leaf of which contains one of the possible moves of an adatom. (The leaves of the tree are the nodes at the ends of the branches, the ones which have no child nodes.) We also calculate the hopping rate for each move and store it in the corresponding leaf. Let us denote by r_i the rate for the ith possible move, calculated using Equation (11.8). In the parent node of any two leaves i and j we store the sum $r_i + r_j$ of the hopping rates for the two children and in the grandparent the sum of those sums, and so forth up the tree until we reach the root node, which contains the sum $R = \sum_i r_i$ of all the rates. To choose a move at random in proportion to its hopping rate, we first generate a random real number in the range from 0 to R. If this number is less than the sum of rates R_1 stored in the first child of the root node then we proceed to that child. Otherwise we subtract R_1 from our random number and proceed to the second child. We repeat this process all the way down the tree until we reach a leaf. It is not hard to show that the probability of ending up in any particular leaf is then proportional to the hopping rate for the move stored in that leaf.

To complete the Monte Carlo step, we now perform the chosen move and then add an amount Δt to the time, which is given by the inverse of the sum of the rates for all possible moves, Equation (11.15). For this algorithm Δt is particularly easy to calculate, since the sum of the rates is just the number R stored in the root node of the tree:

$$\Delta t = \frac{1}{R}. \tag{11.28}$$

Finally, as with the lists in our Kawasaki and bond-counting algorithms, we have to update the tree to reflect the changes in the rates of possible moves and to add new moves or take away ones which are no longer allowed. Again, this is really the time-consuming part of the algorithm. For each move which has to be altered, added or removed we have to go all the way up the tree to the root node adjusting the sums of rates stored in all the nodes in the pedigree of the leaf which has been changed. As discussed in Section 13.2.4 however, such updates take a time which scales typically as the logarithm of the number of nodes in the tree. The only other time-consuming step in the algorithm, the selection of the move, also takes an amount of time which scales logarithmically with the size of the tree, and hence so does the time taken by the entire algorithm. This is a great improvement on the straightforward implementation of the continuous time algorithm in which both the move selection and the updating of the data structures take a time which goes linearly with the number of moves.

11.3   An example: molecular beam epitaxy

As an example of the methods discussed in this chapter, we look at the simulation of the growth of an adatom monolayer under **molecular beam epitaxy** (MBE). Molecular beam epitaxy is an experimental technique in which atoms, usually metal atoms, are evaporated from a heated source in vacuum, and diffuse through space to land on a crystal surface. Typical deposition rates in MBE experiments are low by comparison with the hopping rates of atoms on the surface, so we can simulate the process quite well using algorithms of the type described in this chapter if we simply add an extra adatom to the system at regular intervals of real time.

We have performed simulations for copper atoms deposited on the (001) surface of a copper crystal, using each of the three choices of barriers from Section 11.1.2 in turn. For the Kawasaki-type barriers and the bond-counting method we use the standard continuous time algorithm of Section 11.2.1. We also computed a full lookup table of barrier heights using atom-embedding methods and performed a simulation using the tree-structure algorithm of Section 11.2.2. To simulate the deposition process we added extra adatoms at regular intervals throughout the simulation at randomly chosen locations on the lattice. If the chosen location turns out already to be occupied by another adatom, the new atom diffuses around until it finds a vacant site on the surface. This ensures that the system has only a single layer of adatoms, as we have been assuming throughout this chapter, and is a reasonable approximation as long as the fraction of occupied four-fold hollow sites is quite small.

Starting the simulation with an empty surface, we typically see the first atoms deposited diffusing around freely until their number becomes sufficiently large that they start to meet one another and form stable pairs. As more atoms are deposited these pairs grow to form islands, and the islands themselves diffuse and join up to form larger islands.

In Figure 11.5 we show examples of the configurations of the adatoms in each of our three simulations after 20% of the sites on the surface are filled (left column) and after 50% are filled (right column). The top row shows the results of the simulation using Kawasaki-type energy barriers. In this simulation adatoms can diffuse quite easily along the edges of the islands, but there is a sizeable energy penalty to moving around a corner. For this reason, the islands tend to develop a "spiky" appearance, with pointed promontories growing out of them.

This effect is even more pronounced in the case of the bond-counting

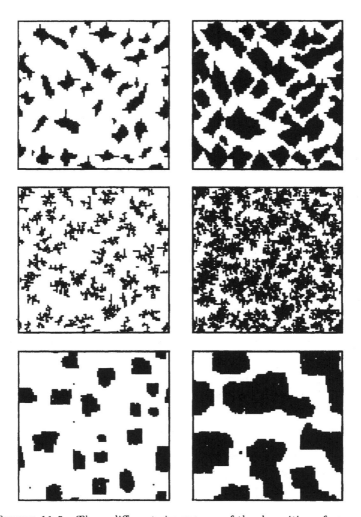

FIGURE 11.5   Three different simulations of the deposition of copper atoms on a copper (001) surface at room temperature (300 K) using molecular beam epitaxy. In the top row the energy barriers were calculated using the Kawasaki-type approach, in the middle row using the bond-counting approach and in the bottom row using a lookup table of values taken from atom-embedding calculations. In each case the atoms were deposited at a rate of $10^{-11}\nu$ per site, where ν is the attempt frequency. The frames on the left were taken at 20% filling and on the right at 50% filling.

simulation, which is shown in the second row. With the bond-counting approach atoms tend to stick to islands at the position where they first touch, since the presence of an occupied nearest-neighbour site creates a sizeable barrier to their making any further moves. (Recall that at the end of Section 11.1.2 we showed that the ratio of the hopping rates for isolated atoms and atoms on the edge of an island was more than 100 000 for the bond-counting approach.) This gives the islands a dendritic look similar to the fractal structures seen in diffusion limited aggregation studies (see, for instance, Mouritsen (1990)).

The bottom row of Figure 11.5 shows the results of the simulation using the lookup table of energy barriers calculated with atomic structure methods. In this case, as we pointed out at the end of Section 11.1.2, the barrier for hopping along the edge of an island is actually lower than that for diffusion of free adatoms, and for this reason the islands can easily change their shape and tend to adopt rounded shapes which minimize their energy. We also notice that the density of islands is lower in this simulation than in the others. This is probably caused by the ability of groups of two or three adatoms to diffuse with a velocity comparable to that of a single adatom, so that an island only becomes fixed when it contains four or more atoms.

Problems

11.1 The adatom hopping process discussed in this chapter is only one mechanism by which atoms diffuse on metal surfaces. Another common one is the **exchange mechanism** where an adatom pushes one of the atoms in the bulk of the metal up onto the surface and then takes its place in the bulk. On a (001) surface, the ejected atom usually ends up in a site which is a next-nearest-neighbour of the site occupied by the original adatom, so that the overall effect of the process is to move an adatom by a distance $\sqrt{2}$ times the lattice parameter. For some metals, such as copper, the attempt frequency for the exchange mechanism is actually higher than that for the hopping one, but it also has a higher energy barrier. Suppose that the attempt frequency of the exchange mechanism is a factor of ten greater and its barrier is $\Delta B = 0.2$ eV higher. At what temperature does exchange become the dominant transport mechanism?

11.2 Suppose that we perform an MBE simulation using the algorithm described in Section 11.2.2, which stores all the possible moves in a binary tree. If a given run on a 100×100 square lattice takes one hour of CPU time to reach a certain adatom density with a given deposition rate, about how long will it take to reach the same density on a 250×250 lattice with the same deposition rate?

12

The repton model

In the preceding chapters of this book we have given a number of examples illustrating the use of creatively designed Monte Carlo algorithms to answer questions about problems in statistical physics. In this chapter we will apply the same techniques to a problem of great importance in biophysics, the problem of polymer electrophoresis. We will study the problem, as we have the others in the book, by simulation of a simplified model which captures the essence of the processes involved. This model is called the "repton model" and, as we shall see, it is very similar in spirit to the models we are familiar with from statistical physics. This should come as no great surprise; the demarcation of science into the separate disciplines of physics, biology, chemistry and so forth is, after all, an artifact of our own creation, and has little to do with the science involved. Indeed, it is the wide applicability of the ideas of physics to problems which traditionally have been classified as biophysics or genetics or biochemistry, which has led an increasing number of physicists to desert their conventional fields of study for other equally fertile ground.

12.1 Electrophoresis

Electrophoresis is an experimental technique for separating mixtures of long-chain polymers according to their length. It has particular importance in the rapidly growing fields of molecular genetics and genetic engineering for refining DNA, which is just such a polymer. The model we will be discussing in this chapter is quite a good approximation to the behaviour of DNA. (It would not be so good, for instance, as a model of RNA—see Section 12.2.) The basic idea behind electrophoresis is to separate charged particles by pulling them through a stationary gel with an electric field. If we start with a mixture of different types of particles, each of which drifts with a different

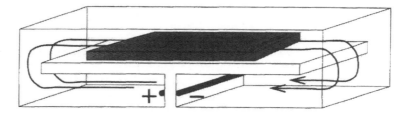

FIGURE 12.1 A schematic depiction of the typical gel box used in
DNA electrophoresis. The gel is placed on the table in the middle,
two electrodes are located at the bottom, and the box is filled with
buffer solution. When a voltage is applied between the electrodes, the
resulting electric field curves from the positive electrode through the
gel to the negative electrode, as denoted by the arrows. The DNA
samples are injected into the gel at the end closest to the negative
electrode, as indicated.

velocity under the influence of the electric field, then electrophoresis can be
used to separate these types from each other.

The experimental setup for DNA electrophoresis is simple and inexpen-
sive: all you need is a box full of gel and a way to produce an electric field.
The most common type of gel, and the one for which our model is best
suited, is the **agarose gel**, which is made by mixing one part agarose pow-
der (a culinary thickening agent made from seaweed) with about a hundred
parts of a buffer solution. The mixture is heated to around 50 °C which
makes the agarose dissolve in the buffer, and then it is poured into the gel
box. As the mixture slowly cools down again, the agarose condenses into
long strands, which eventually crosslink to form a layer of firm gel at the
bottom of the box. Microscopically, the resulting network looks like a three-
dimensional spider's web of thin threads. The DNA to be analysed is now
injected into small holes in the gel and more buffer is poured in to cover the
gel. The buffer solution imparts a roughly uniform charge per unit length
on the chains of monomers or **base pairs** which make up the DNA, so that
when an electric field is applied across the gel box the fragments of DNA
start to drift downstream through the gel along a "lane" away from the hole
where they were injected. A sketch of the experimental setup is given in
Figure 12.1.

As we will see, longer pieces of DNA experience more resistance than
shorter ones when moving through the gel and as a result drift with a lower
velocity. Usually, the DNA injected into the gel is a mixture of a number
of different types of DNA fragments of varying lengths. (The lengths are
normally measured as the number of base pairs (bp) or thousands of base
pairs (kb or "kilobase") in a fragment.) Thus, after we have waited a certain
amount of time, typically a few hours, the shortest fragments will have trav-

elled furthest down the lane, and the mixture of DNA will have separated into a number of bands, each containing DNA fragments with one specific length. The positions of these bands are a measure of the length of the fragments making them up, and the experiment also serves to separate the DNA by length; at the end of the experiment, we can cut one of the bands out of the gel, thus refining fragments of one particular length from the initial mixture of many lengths.

Figure 12.2 shows results from a typical DNA electrophoresis experiment, performed according to the procedure outlined above. In this experiment five different lanes were run in the same gel box, with the same mixture of DNA fragments being injected into successive lanes at regular intervals of two hours, starting with the rightmost lane. The electric field was turned off two hours after the last injection, so that, reading the figure from left to right, the lanes correspond to five individual electrophoresis experiments of length 2, 4, 6, 8 and 10 hours. If each fragment of DNA drifts with a constant velocity, a band in the four-hour lane will have drifted twice as far at the end of the experiment as a band of identical DNA in the two-hour lane. If we were to connect the middle points of all bands containing the same DNA we should get a straight line, and the slope of this line would tell us the drift velocity for that particular DNA fragment.

However, determining the drift velocity of a particular group of DNA fragments doesn't actually tell us what their length is—the experiment needs to be calibrated before we can calculate this. The customary calibration procedure involves injecting a standardized **DNA ladder** into an empty lane in the gel box. This ladder is a mixture of DNA fragments of known lengths, usually multiples of 100 or 1000 bp. During electrophoresis this ladder separates to form a set of bands which are then compared with the bands formed by the fragments in other lanes, to determine the lengths of those fragments. This procedure is advantageous because it provides an approximate estimate of fragment length which is independent of variations in the experimental conditions. If a band of unknown DNA fragments drifts about as fast as the standard 7 kb band in the DNA ladder, then it consists of fragments around 7 kb in length, regardless of changes in temperature, voltage or buffer concentration during the experiment or between one experiment and another. The DNA mixture that we used to produce Figure 12.2 was just such a DNA ladder.

12.2 The repton model

As a tool for analysing and refining samples of DNA, electrophoresis is of great practical importance in genetics and biophysics, and consequently there is a good deal of interest in gaining an understanding of the physical processes involved. In particular we would like to understand the relationship between

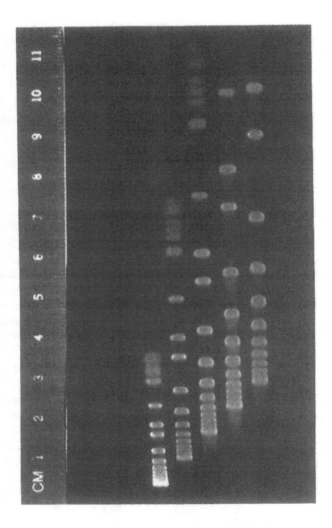

FIGURE 12.2   Electrophoresis of a mixture of DNA fragments. This
experiment was run for ten hours in an electric field of 4.4 V/cm. DNA
was injected in the rightmost lane at the start of the run and then in
successive lanes at intervals of two hours. For illustrative purposes,
we used DNA from a standard DNA "ladder" containing fragments
of length 1 kb, 2 kb and so on up to 12 kb. There were also some
fragments of length 1.6 kb added to help identify the bands, as well as
a few smaller fragments.

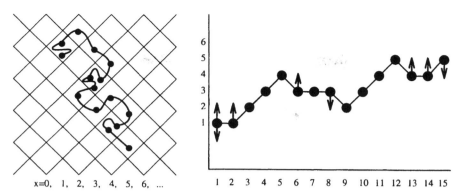

x=0, 1, 2, 3, 4, 5, 6, ...

FIGURE 12.3   A typical configuration of the repton model (left). If the reptons are numbered along the chain, starting with the end in the top-left corner, we can characterize this configuration by plotting the x-position as a function of the repton number. This is done in the right side of the figure. This is the projected model described in the text. Allowed moves are denoted by arrows.

fragment length and drift velocity, so as to put the calibration of the gel and the use of DNA ladders on a firmer scientific footing. In this chapter we discuss the mechanisms at work in electrophoresis and study a simple model of the process called the "repton model".

First it is important to have some idea of how DNA diffuses through a gel in the absence of an electric field. The fragments of DNA which we inject into the gel are usually quite long (between 1 and 10 kb) and only moderately flexible. One can think of the molecule as a piece of stiff rope sliding through the web-like gel. Sideways movement of the rope is blocked by the agarose strands, and its main mode of diffusion is motion along its own length. The end of the molecule makes random thermal movements around the gel, and the rest of the molecule follows behind, like a snake. This mechanism was first proposed by de Gennes (1971), who christened it **reptation**.

The **repton model** was introduced by Rubinstein (1987) as a simple model of polymer reptation. The model is illustrated on the left-hand side of Figure 12.3. The gel is represented as a square lattice; the plaquets on the lattice correspond to the pores in the gel. (The lattice is shown as two-dimensional in Figure 12.3, although the real system is three-dimensional. However, as discussed below, the properties of the model are independent of the number of dimensions of the lattice, so we may as well stick with two for clarity.) We represent the DNA or other polymer as a chain of N polymer segments or **reptons**, the black dots in the figure. (The reptons are not the same thing as base pairs. In fact, each repton corresponds to a string of about 150 base pairs, as discussed below in Section 12.2.2.) Adjacent reptons in

the chain can only occupy the same or nearest-neighbour pores on the lattice
and reptons move from pore to adjacent pore diagonally according to the
following rules:

(i) A repton in the interior of the chain can move to one of the adjacent
squares on the lattice, provided that one of its immediate neighbours
in the chain is already in the square to which it is moving, and the
other is in the square which it leaves. This rule reproduces de Gennes'
reptation motion; reptation is the only mechanism for diffusion in the
repton model. Note that if three or more neighbouring reptons in the
chain find themselves all in one pore, only the two with connections to
other reptons outside this pore are allowed to move.

(ii) The two reptons at the ends of the chain may move in any direction
to an adjacent square, provided that such a move does not take them
more than one lattice spacing away from their neighbouring repton on
the chain.

In real polymers there is the additional effect that the polymer has a finite
width, which limits how much of the polymer can fit into a certain space.
This self-avoidance property is assumed to have a small effect on the dynamics of real polymers and is not included in the repton model.

Since we are interested in the rate at which polymers drift through the
gel, we also need to define the time-scale on which moves take place. To do
this, we make the assumption that all the reptons are continually driven by
thermal fluctuations to attempt moves to the adjacent squares on the grid.
We assume that every such move is as likely as every other to be attempted
at any particular time, and we choose the time-scale such that each move
will be attempted once, on average, per unit time. (Not all of these moves
will actually take place; many of them will be rejected for violating one
of the rules above. However, the moves which are allowed each take place
with equal probability per unit time. There are no energies associated with
the different states, and so there are no Boltzmann weights making one
move more likely than another. The dynamics of the repton model is purely
entropic.)

The model described so far is, as we mentioned, just a model of polymer
diffusion. All the motions are random thermal ones and there is no applied
electric field. To make a model for electrophoresis, we first of all assign
to each repton the same negative electric charge, mimicking the charging
effect of the buffer solution in the experiment. Then we apply a uniform
electric field to the model along the x-axis (the horizontal direction in the
figure), breaking the spatial symmetry. As a result, instead of unit rates
for the allowed moves in the positive and negative x-direction, the rates
become $\exp(-E/2)$ in the positive direction and $\exp(E/2)$ in the negative
one, where E is a new parameter which is proportional to the applied field

(see Section 12.2.2). The resulting model describes the qualitative behaviour of DNA electrophoresis reasonably well. Many details of the real system are still missing however, such as mechanical properties of the polymer, the effects of counterions, the inhomogeneity of the gel and the gel concentration, and to be quantitatively accurate the model would have to include these features as well. However, we can still extract a lot of useful information from the simple model we have here.

12.2.1 The projected repton model

The drawing in Figure 12.3 is two-dimensional, but real electrophoresis experiments are three-dimensional. Does this matter? As we mentioned briefly above, it does not, for the following reason. We are interested in calculating the drift velocity of the polymer for the case of a uniform electric field in the x-direction, which means we are interested only in the x-coordinates of the polymer; the values of the other coordinates are irrelevant. Thus, we might as well project the polymer onto the x-axis and ignore the other coordinates. We do precisely this on the right-hand side of Figure 12.3, where we have numbered the reptons along the length of the chain and then plotted the x-coordinate as a function of repton number. This one-dimensional projection of the model proves very convenient for simulation purposes, and we will be working mainly with this form of the model from here on.

Comparing the states of the projected model with those of the original higher-dimensional version, we can work out how the dynamics looks after projection. First, reptons in the projected chain must either be at the same height as, or one place higher or lower than their neighbours, and each possible move consists of moving one repton either up or down one place. However, just satisfying these constraints is not enough. Consider, for instance, the repton numbered 7 in Figure 12.3. This repton is not allowed to move either up or down, even though the final configuration of the projected chain following such a move would be a perfectly legitimate one. This fact can easily be verified by going back to the higher-dimensional representation and examining repton 7. In fact there is essentially only one type of move allowed in the projected model, in which a repton which has one of its neighbours on the level above or below it and one on the same level is allowed to move up or down respectively. The reader may like to verify that such moves are indeed the equivalent of our update rules for the higher-dimensional version of the model. For the particular configuration shown in Figure 12.3, the allowed moves for the projected model are shown with arrows.

This projection of the repton model onto one dimension works just as well from three dimensions as from two, or indeed from any higher number of dimensions. The final rules for the dynamics of the projected model are just the same in each case, so we conclude that the properties of the model are independent of the dimensions of the lattice on which it is built.

12.2.2 Values of the parameters in the model

Before we can perform a simulation of the repton model, we need to get an idea of the sort of values that the free parameters N and E should take. The parameter N measures the length of the polymer, but it is not the number of monomers or base pairs in the molecule. Instead it represents the number of **persistence lengths** from one end of the chain to the other. The persistence length is a measure of the stiffness of the polymer, and essentially it is the typical scale on which the polymer bends.[1] DNA is quite stiff, and has a persistence length of around 150 bp, or about 400 Å. It also becomes stiffer as the strength of the buffer solution used in the electrophoresis is increased, and in real experiments can be as much as 800 Å.

But N also plays another role in the repton model. As we said above, the diagonal lattice of squares in Figure 12.3 represents the pores in the agarose gel and the lattice spacing represents the typical pore size, the distance from one agarose strand to the next. And in this model, the maximum distance from one repton to the next is one lattice spacing, so N is also effectively the length of the polymer in multiples of the pore size. The pore sizes in agarose gels have not been determined very accurately yet, but for a 1% agarose gel they are estimated to be in the range of 1000 to 3000 Å.

We are fortunate that the persistence length of DNA and the typical pore size in agarose gels have reasonably similar values. If this were not the case, there would be an inconsistency between these two different roles which the parameter N plays, and little hope that the repton model would make reasonable predictions about real DNA electrophoresis. For the electrophoresis of RNA, another polymer of great biological interest, the persistence length is only one or two bp, or about 3 to 5 Å, and again the agarose pore size is

---

[1] To put this on a quantitative footing, we can define a vector $\hat{n}_i$ which is the unit direction vector from the ith monomer or base pair to the $(i + 1)^{\text{th}}$. Then we define the direction–direction correlation function G_{ij} thus:

$$G_{ij} = \langle \hat{n}_i \cdot \hat{n}_j \rangle,$$

where $\langle \ldots \rangle$ represents an average over the states the molecule passes through over a long period of time. When the base pairs i and j are very close together, their corresponding direction vectors will be pointing in almost the same direction, so G_{ij} will be close to unity. When they are far apart there will be no correlation between the directions, and the product of the two vectors is as likely to be negative as positive, so that after averaging over many states of the system, the correlation function will be zero. As we approach this long length-scale limit, the correlation function will decay exponentially:

$$G_{ij} \sim \exp(-|i - j|/\xi).$$

The length-scale ξ for this decay is the persistence length (measured here in multiples of the spacing between monomers).

1000 to 3000 Å; these two vastly different length-scales cannot be represented well by the one parameter in the repton model, and as a result the model does not make good predictions about the results of RNA experiments.

The dimensionless parameter E is defined so as to be proportional to the electric field applied to the system. The quantity $\exp(E)$ is the ratio between the probability for a repton to move one step down the slope of the electric potential and the probability for moving one step up it, and should be equal to the ratio of the Boltzmann factors for making those two moves. This means that

$$E = \frac{\sqrt{2}aqE_f}{kT},$$ (12.1)

where a is the lattice parameter (i.e., the pore size), q is the charge per repton (i.e., per persistence length) and E_f is the applied electric field. The numerator here represents the energy needed to move a repton a distance $\sqrt{2}a$ (the x-distance between two nearest neighbours of the same lattice site) against the electric force qE_f acting on it.

The charge q is equal to one electron charge e per base pair, or about $150 \times 1.6 \times 10^{-19}$ C per repton.[2] If we take a typical room-temperature value of 300 K for T and a value of about 2000 Å for the pore size a, we find that

$$E \simeq 10^{-3}E_f$$ (12.2)

when E_f is measured in $\mathrm{V\,m^{-1}}$. The experiment in Figure 12.2 for example, was performed with an electric field of 4.4 $\mathrm{V\,cm^{-1}}$ or 440 $\mathrm{V\,m^{-1}}$, which corresponds to a value of about $E = 0.4$.

12.3 Monte Carlo simulation of the repton model

There is no known analytic solution for the properties of the repton model, except for rather short chains,[3] and so we turn instead to Monte Carlo simulation. In the rest of this chapter we discuss the design of an algorithm to simulate the repton model efficiently.

---

[2] Each base pair consists of two nucleotides. These are single acids, meaning that each one is capable of donating one H^+ ion to the buffer, leaving the acid with a net negative charge of e. We might imagine that this would mean that each base pair would have a total negative charge of $2e$, but in fact when the base pairs form, one of the H^+ ions is used up in forming a hydrogen bond between the two nucleotides, leaving only one free for donation to the buffer.

[3] For chains up to $N = 20$ a numerically exact solution has been given by Newman and Barkema (1997). $N = 20$ corresponds to about $20 \times 150 = 3000$ base pairs in DNA, which is a non-trivial length. DNA fragments of a few kilobase are of experimental interest. However, the bulk of experiments focus on longer fragments, and for these a Monte Carlo method is appropriate.

To begin with, let us consider the case of the repton model in zero electric field E. In this case, we attempt each possible move of a repton once in each unit of simulated time. We will work with the projected model, in which case these moves consist of shifting the reptons up or down in Figure 12.3, for a total of $2N$ possible moves (some of which, as we have said, will be rejected, because they don't satisfy the rules set out in Section 12.2.1). We should attempt each of the possible moves with equal probability, and we should attempt them at a total rate of $2N$ attempted moves per unit of simulated time.

The simplest Monte Carlo simulation of the repton model then goes like this. In each Monte Carlo step we randomly select one of the N reptons and one of the two directions for it to move in (up or down). If the move is allowed according to the rules given in Section 12.2, it is accepted. Otherwise, it is rejected. The probability for each move to be selected in one Monte Carlo step equals $1/(2N)$, and after $2N$ moves each possibility will have been tried once on average, corresponding to one unit of time. Instead of making $2N$ Monte Carlo moves and then adjusting the time t on our clock by one, it is more elegant to add $\Delta t = 1/(2N)$ to t after each step.

To simulate electrophoresis we have to include the effects of an electric field. We can achieve this if we keep the probability for *attempting* each upward or downward move of a repton the same, but now make the upward moves a factor of $\exp(E/2)$ more likely to be *accepted* and the downward moves $\exp(-E/2)$ less likely. As with the Metropolis algorithm of Section 3.1, we can get the ratio of these acceptance probabilities right, and at the same time maximize the overall rate at which moves are accepted, if we always accept upward moves, so long as they do not violate the rules of the dynamics, but only accept downward moves with probability $\exp(-E)$. Of course, moves which violate the rules will still be rejected outright.

This algorithm makes the ratio between the upward and downward moves right, but it doesn't get the rates exactly correct. As it stands the algorithm accepts allowed upward moves exactly as often as in the zero-field case, whereas we'd like them to be accepted a factor of $\exp(E/2)$ more often. To achieve this, we simply add

$$\Delta t = \frac{\exp(-E/2)}{2N} \tag{12.3}$$

to the time t at each step. This also makes the downward moves $\exp(-E/2)$ less likely (per unit time) than in the zero-field case (instead of $\exp(-E)$ less likely) which is exactly what we want.

12.3.1   Improving the algorithm

This algorithm is not a bad one. In a typical run about 20% of the attempted moves are accepted, and simulated times on the order of a few million can be reached with a moderate investment of CPU time. However, as we now demonstrate, there is considerable room for improvement in the algorithm which will allow us to reach longer simulated times with the same expenditure of CPU time. One way of speeding up the calculation is to make use of "multispin coding", a low-level programming technique which is described in more detail in Chapter 15. In Section 15.4 we discuss in detail how to create a multispin-coded Monte Carlo algorithm for the repton model. In this chapter, however, we look at more conventional approaches.

First, we note that it is inefficient to spend valuable CPU time in selecting a move, only to reject it for one reason or another. We want to avoid rejecting moves wherever possible. In the algorithm outlined above, upward moves are only rejected if they fail to satisfy the rules of Section 12.2, but downward moves are rejected with probability $1 - \exp(-E)$ even when they satisfy the rules. Rather than rejecting all these downward moves and wasting CPU, can we modify our algorithm so that it just proposes fewer downward moves?

Suppose that, instead of choosing upward and downward moves with equal probability of a half on each move, we chose the downward moves with probability α, and the upward ones with probability $1 - \alpha$. (The repton that we propose to move is still selected completely at random.) In that case, the probability of choosing a *particular* downward move becomes α/N, and for a particular upward move it becomes $(1 - \alpha)/N$. The ratio of the probabilities for upward and downward moves will be $(1 - \alpha)/\alpha$. We want to make this ratio equal to $\exp(E)$. Setting the two expressions equal and solving for α we get

$$\alpha = \frac{1}{1 + \exp E}. \tag{12.4}$$

If we chose upward and downward moves with these probabilities, then the probability of selecting any particular upward move on a given Monte Carlo step is $1/[N + N\exp(-E)]$. As before, we want this to be equivalent to $\exp(E/2)$ per unit time on average. If M is the number of Monte Carlo steps which correspond to one unit of time, then that means we want to choose M such that

$$\frac{M}{N[1 + \exp(-E)]} = \exp(E/2), \tag{12.5}$$

or

$$M = 2N \cosh(E/2). \tag{12.6}$$

We can achieve this by increasing t by

$$\Delta t = \frac{1}{2N \cosh(E/2)} \tag{12.7}$$

after each Monte Carlo step.

How much exactly does this speed our algorithm up by? Well, let us assume first of all that the time taken to perform each Monte Carlo step is the same for this new algorithm as it was for our original one. This seems a reasonable assumption; the only extra work we have to do in the new algorithm that we didn't have to do in the old one is the calculation of the quantities α and Δt from the formulae above. However, we only have to calculate each of these once for the whole simulation, so they don't add any effort to the individual Monte Carlo steps.

The new algorithm is more efficient than the old one because some of the steps that were previously rejected are now accepted. How much of a difference does this make? This is measured by the quantity Δt in Equation (12.7) above, which is the amount of time which is simulated in one Monte Carlo step. If the steps in both algorithms take the same amount of CPU time, then Δt is also a measure of the amount of simulated time you can get through with a given amount of CPU time. The corresponding quantity in the original algorithm was $\exp(-E/2)/(2N)$ (see Equation (12.3)) so the factor by which we have speeded up our algorithm is

$$\frac{2N \exp(E/2)}{2N \cosh(E/2)} = \frac{2}{1 + \exp(-E)}. \tag{12.8}$$

When $E = 0$ this is 1 so there is no speed-up, as we would expect since when $E = 0$ the basic algorithm doesn't reject any downward moves except those that violate the rules of the dynamics. However, as E becomes bigger, our new algorithm becomes better and better, and in the limit of large E the new algorithm is a factor of two faster than the old one. In other words, for large fields it can reach simulated times nearly twice as long in the same amount of CPU time.

12.3.2   Further improvements

The algorithm described in the previous section still generates a lot of moves that are later rejected because they contravene the rules of the repton model dynamics. If we could arrange for these moves simply not to get chosen, we would again save ourselves some CPU time (though we would have to be careful to increase the time t by the appropriate amount to account for the steps we are not now performing). We can do this by creating a list of all the allowed moves by going through the reptons in the chain one by one and checking what moves, if any, each one can make. Better still, we can create two lists, one for the upward moves and one for the downward ones. Since all upward moves are equally likely, and all downward moves are equally likely, we then just have to pick one list or the other with the correct probability, and then choose any move from that list at random and perform it, and that's our entire algorithm.

The only questions we need to answer are, what is the probability for choosing each of the lists, and how big is the increment Δt that we have to add to the time at each step of the simulation? The probability of choosing one list or the other is obviously related to the probability α in the previous version of the algorithm, but it's not quite equal to α, because if it were, the probability of choosing any one particular allowed move would go up as the number of possibilities on the corresponding list went down, which is not how we want things to work at all.

Suppose the number of allowed upward and downward moves are n_u and n_d respectively. At each step of the simulation we either decide, with probability γ, to take a move from the downward list, or we decide, with probability $1 - \gamma$, to take one from the upward list. The probability of selecting any particular downward move at a given time is then γ/n_d, and the probability of selecting a particular upward one is $(1 - \gamma)/n_u$. We want the ratio of these probabilities to be $\exp(-E)$, which when you work it out means that

$$\gamma = \frac{n_u}{n_d + n_u \exp(E)}. \tag{12.9}$$

Note that this value of γ is not a constant; it has to be adjusted each time the number of upward or downward moves changes.

To calculate Δt, we note that the rate γ/n_d at which a particular downward move is made should be $\exp(-E/2)$ per unit time, which means that we have to make $n_u \exp(E/2) + n_d \exp(-E/2)$ Monte Carlo steps to simulate one unit of time. Or alternatively, we increment the time-scale by

$$\Delta t = \frac{1}{n_u \exp(E/2) + n_d \exp(-E/2)} \tag{12.10}$$

at each step. (Since we have already fixed the probability ratio between up and down moves at $\exp(E)$ by our choice of γ, we are ensured that the rate for upward moves will now be $\exp(E/2)$, as we want it to be.)

There is one more point we have to consider with this algorithm. The process of going through the entire chain and making a list of all the possible moves could be quite a lengthy one, and there is little point in constructing this elaborate algorithm to save CPU time making moves if we simply squander all the time saved constructing lists instead. In fact, if we had to reconstruct the lists after every move, this would not be an efficient algorithm at all. However, just as with the similar algorithms used in Chapters 5 and 10, it turns out that we do not have to reconstruct the lists completely at every Monte Carlo step. When we make one move, most of the repton chain stays exactly as it was and the possible moves are just the same as they were. The possible moves of only three reptons are affected at each step (or two if it's one of the reptons at the end of the chain that moves). So the efficient thing to do is just update the entries for those reptons and keep everything else unchanged. If we move repton number i, we only have to check the possibilities for reptons $i-1$, i and $i+1$. Nonetheless, keeping the lists up to date does use up some CPU time, and in practice we find that, depending on the exact implementation of the algorithm, each Monte Carlo step takes around 50% longer than before. To decide whether this sacrifice is worthwhile, we need to look again at the time increment Δt which, as we pointed out, is a measure of how much simulated time we get out of each Monte Carlo step. Unfortunately, we cannot evaluate Δt directly from the expression (12.10) in this case, since we don't know the values of the numbers n_u and n_d. However, in the next section, we make use of a mapping of the projected repton model onto a particle model to calculate the average values of these quantities, and thus demonstrate that Δt is actually at least a factor of 4.5 bigger than in the previous version of the algorithm. So allowing for the extra CPU time needed to keep our lists up to date, this new algorithm is about $4.5/1.5 = 3$ times more efficient than the next best algorithm we have considered, and up to six times more efficient than the simple one which we started out with.

12.3.3   Representing configurations of the repton model

There are several ways of representing configurations of the chain in the repton model. The simplest way is to use the x-coordinates x_i of all the reptons, where $i = 1 \ldots N$. Alternatively, we can keep the x-coordinate of just one of the reptons, say number 1, plus the position s_i of each subsequent repton relative to its neighbour: $s_i \equiv x_{i+1} - x_i$ $(i = 1 \ldots N - 1)$. Given this information, the configuration of the repton chain can easily be reconstructed. The constraint on the repton model that the x-coordinates of neighbouring reptons on the chain differ by at most 1 means that the quantities s_i can take just three possible values, -1, 0 or $+1$, corresponding to segments going down, remaining flat or going up in Figure 12.3. This makes the representation in terms of the s_i considerably more economical than the one in terms of the x_i.

The model can also be represented as a collection of particles of two different types, which we denote by A and B, moving around on a one-dimensional lattice. We identify $s_i = +1$ with the presence of a particle of type A at site i, $s_i = -1$ with a particle of type B, and $s_i = 0$ with the absence of a particle (a vacancy) at site i. The mapping from the reptons to these particles is illustrated in Figure 12.4 for one particular configuration of the polymer. The rules for the repton model are easily expressed in the language of these particles:

(i) In the interior of the chain the particles are conserved: they can move from site to site on the lattice but they are not created or annihilated.

(ii) A particle can hop to a neighbouring site if that site is not already

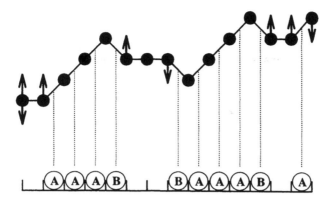

FIGURE 12.4   Illustration of the link between a configuration of the
projected repton model and a configuration in the model with two
species of particles.

occupied by another particle.[4]

(iii) A particle on one of the two end sites of the chain can leave the chain—
it simply vanishes from the lattice.

(iv) If an end site of the chain is empty, a particle of either type can appear
there.

In the absence of an electric field, the rates for all particle moves are
equal. On average, each particle on the chain tries to hop both to the left
and to the right once per unit of simulated time, type A particles try to
enter the chain at both the left and right ends once per unit of time, and so
do type B particles. If an electric field is introduced, type A particles will
hop to the left with a rate of $\exp(E/2)$ moves per unit of time and to the
right at a rate of $\exp(-E/2)$. Conversely, type B particles will have rates of
$\exp(-E/2)$ and $\exp(E/2)$ moves per unit time for hopping to the left and
right respectively. Thus, an electric field draws A-particles to the left, and
B-particles to the right.

This new picture of the repton model is useful, since some things which
are not obvious in our other pictures are obvious with this one (and *vice
versa*). For example, once we understand that the electric field will push
type A particles one way and type B ones the other way, it is intuitively
reasonable if we start off with a mixture of the two types of particles, that
after a while one or more traffic jams are going to build up on the chain, since

---

[4]The particles are a bit like fermions in this respect: no two can coexist at the same
site. Notice however that if they were true fermions, particles of type A and type B would
be able to coexist, because they are distinguishable. This is not the case in this model,
where each site can only ever contain one particle.

the two particle types are trying to go in opposite directions but cannot pass one another. Furthermore, these traffic jams will be unstable to fluctuations. If one side of the jam becomes larger than the other—if there are more particles of type A for example than there are of type B—then the larger half will "push" the smaller half back, and, ultimately, off the end of the chain. When this happens, we can end up with a configuration of the chain dominated by one type of particle or the other. What does this situation correspond to if we project it back onto the original repton model? It turns out that it corresponds to a polymer that has got itself lined up along the electric field, so that the reptation motion along the line of the molecule is in exactly the direction the field is trying to push it. And in fact, this is a situation that frequently occurs in electrophoresis experiments.

We can use the particle form of the repton model to calculate the ratio of the number $2N$ of attempted moves to the number of moves that are actually possible according to the rules governing the dynamics. When there is no applied electric field, there is no physical process to distinguish particles of the two types, and the equilibrium densities of A- and B-particles will be identical. They will also be independent of position along the chain, since the rates at which they enter and leave the chain are in equilibrium. Let us define θ_A and θ_B to be the equilibrium densities of the two types of particles, with $\theta_A = \theta_B$ when $E = 0$. Both types of particles can enter at either end of the chain if the corresponding end site is empty. Summed over both ends, the rate at which particles enter the chain is thus equal to the probability $1 - \theta_A - \theta_B$ of an end site being unoccupied, times 2 for the number of ends, and times another 2 for the two different types of particles. The rate at which particles leave the chain summed over both ends equals $\theta_A + \theta_B$ for the probability that there will be a particle in one end site, times 2 for the number of ends. In equilibrium these rates are equal, which implies that $\theta_A + \theta_B = \frac{2}{3}$. In zero electric field $\theta_A = \theta_B$, so we should find that the densities of A-particles, of B-particles and of vacancies are all equal to $\frac{1}{3}$.

The probability that a particle can move from a site i to a site $i + 1$ is given by the probability that there is a particle at site i times the probability that there is a vacancy at site $i + 1$. In zero field this is just $\frac{2}{3} \times \frac{1}{3} = \frac{2}{9}$. Thus the ratio of the total number of attempted moves and the total number of allowed moves will be $\frac{9}{2} = 4.5$ on average. We used this property in our estimation of the efficiency of our algorithm in the previous section.

12.4   Results of Monte Carlo simulations

So, what can we do with the algorithm we have developed? First, let's consider the case of zero electric field.

12.4.1 Simulations in zero electric field

Following scaling arguments given by de Gennes (1971), it is believed that the diffusion constant D for a reptating polymer in zero electric field varies with the length N of the polymer as $D \sim N^{-2}$, or alternatively $DN^2 = \text{constant}$. We would like to see this behaviour reproduced in our simulation of the repton model. Prähofer and Spohn (1996) have argued that the constant in the above expression should be $\frac{1}{3}$ for the model, and if this is the case we should be able to see this in our simulations too. The exact calculations of Newman and Barkema (1997) give the diffusion constant for chains of up to 20 reptons, but even for $N = 20$ the value of DN^2 is still almost twice this conjectured value of $\frac{1}{3}$. Can we do any better with our Monte Carlo method?

The diffusion constant is defined by

$$D = \frac{\langle r^2 \rangle}{2dt},\qquad(12.11)$$

where $\langle r^2 \rangle$ is the mean square displacement after time t of a repton chain which started off at the origin, and d is the dimensionality of the system, which is usually 3. Since we are working with the projected model, our simulations actually only give us the x-coordinates of the reptons. However, in zero field there is nothing to distinguish one direction from another, so

$$\langle r^2 \rangle = \langle x^2 + y^2 + z^2 \rangle = 3\langle x^2 \rangle,\qquad(12.12)$$

which we can calculate and then plug into Equation (12.11). The results of simulations to determine the zero-field diffusion constant in this way are shown in Figure 12.5 for values of N up to 250. In this figure $DN^2 - \frac{1}{3}$ is plotted as a function of N on logarithmic scales. On the same axes we have also plotted the exact results for $N = 3 \ldots 20$. From this figure it is clear that the diffusion constant for large polymers approaches $D = \frac{1}{3}N^{-2}$, as suggested above. The slope of the line through the data is $-\frac{2}{3}$, which suggests that for smaller N there is a correction to de Gennes' scaling form which goes like $N^{-2/3}$. This is actually rather surprising, since similar corrections in other particle models usually go like N^{-1}. The line in Figure 12.5 is given by

$$DN^2 = \frac{1}{3}\left(1 + 5N^{-2/3}\right).\qquad(12.13)$$

12.4.2 Simulations in non-zero electric field

Now what happens when we introduce the electric field? The chain will start to drift down the potential gradient of the electric field with some average drift velocity v. (There will be fluctuations in the velocity because the chain

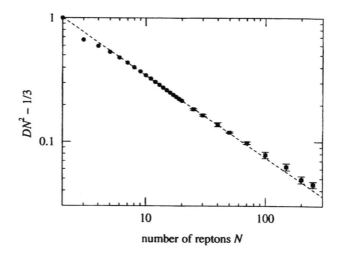

number of reptons *N*

FIGURE 12.5   This figure shows how the diffusion constant scales with
the length N of the repton chain. The circles indicate the known exact
results for chain lengths up to $N = 20$ and the squares are the Monte
Carlo results. The straight line is given by $DN^2 = \frac{1}{3}\left(1 + 5N^{-2/3}\right)$.
After Newman and Barkema (1997).

will still have some random diffusive behaviour as in the zero-field case,
but if we average over a long enough time we can get an accurate measure
of v.) The electric field exerts a force on each repton proportional to the
parameter E, and the total force on the polymer is thus proportional to NE.
The Nernst–Einstein relation tells us that, if we exert a small force F on a
particle with diffusion constant D, its drift velocity will be $v = FD$. In our
model, the diffusion constant is given by $D \sim N^{-2}$ to leading order, so for a
small electric field, the drift velocity should go like $v = NED \sim E/N$. In this
regime then, the drift velocity depends on the length of the polymer, which
is exactly what we want, since it makes the separation of DNA fragments
of different lengths possible. It is known from experiments however that
for strong fields the drift velocity becomes independent of the length of the
fragment, making such fields useless for separating DNA. Note also that, if
the direction of the electric field is reversed, the polymers will drift with an
equal velocity in the opposite direction. For this reason it has been argued
that the velocity should be expressed as a sum of odd powers of E, so that
when E changes sign, so will v. Combining this idea with the results above,
the form

$$v = c_1 E/N + c_3 E^3 \tag{12.14}$$

has been suggested for the drift velocity, where c_1 and c_3 are constants.

This expression contains only odd powers of E, has the right form when E is small, and becomes independent of N for large E. For many years it was believed that this expression was essentially correct.

In simulations of the repton model using the algorithms described in this chapter (and also in simulations of other related models), it is indeed observed that in a sufficiently strong electric field, the drift velocity is independent of N. However, the velocity appears to scale as $v \sim E^2$ instead of the proposed E^3. Recent more careful examination of the experimental results has shown that in DNA electrophoresis the velocity for strong electric fields actually does go like $v \sim E^2$. The combination of these two observations, plus some supporting scaling arguments (Barkema *et al.* 1994) indicate that the received wisdom on this matter is, after all, wrong.[5]

So what is the appropriate formula linking the drift velocity to N and E? Well, there are a number of plausible possibilities, but the one that seems to fit the Monte Carlo data best is

$$v^2 = c_2^2 \frac{E^2}{N^2} + c_4^2 E^4, \tag{12.15}$$

where c_2 and c_4 are two new constants. An elegant graphical demonstration of the accuracy of this formula is shown in Figure 12.6. In this figure we have performed a "data collapse" (see Section 8.3.2) as follows. By rearranging, we can write Equation (12.15) in the form

$$\frac{vN^2}{\beta} = \left[\left(\frac{NE}{\alpha} \right)^2 + \left(\frac{NE}{\alpha} \right)^4 \right]^{1/2}, \tag{12.16}$$

where $\alpha = c_2/c_4$ and $\beta = c_2^2/c_4$. This equation expresses the product vN^2 as a function of just one quantity NE. If our formula is correct then, we should be able to take the Monte Carlo data and plot the values of vN^2 against NE and they should all fall on the line given by this equation (or "collapse" onto it, as the jargon goes). This is precisely what we have done in Figure 12.6, where the circles represent our data and the solid line is Equation (12.16). The best fit of the line to the simulation data is given by $\alpha = \frac{5}{6}$, $\beta = \frac{5}{18}$ or equivalently $c_2 = \frac{1}{3}$, $c_4 = \frac{2}{5}$. On the left-hand side of the figure where the electric field is small, vN^2 increases linearly with NE. This corresponds to the expected behaviour $v = \frac{1}{3}E/N$ for small fields. On the right-hand side, vN^2 goes quadratically with NE: we have $v = \frac{2}{5}E^2$ for large fields, independent of the value of N, in accordance with the experimental

---

[5] How does this fit with the argument about reversing the electric field and the odd powers of E? Well, we simply write the large E drift velocity in the form $v \sim E|E|$, which goes like E^2 for positive E and like $-E^2$ for negative E. If we do this, v is no longer analytic, but there is no requirement that it should be. This is the root of the error in the old argument.

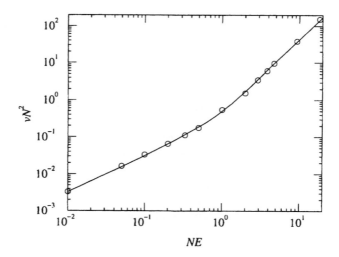

FIGURE 12.6   Monte Carlo results from simulations on the repton model. The circles give the measured values of vN^2 as a function of NE. The solid line is the scaling form, Equation (12.16).

observations. The cross-over between the two regimes occurs when the two terms on the right-hand side of Equation (12.16) are approximately equal, or in other words around $NE = \frac{5}{6}$. Thus for longer and longer chains, we have to work at smaller and smaller fields in order to be sure of falling in the regime in which v is N-dependent and hence separation by length occurs.

So what does this tell us about real experiments? Well, having obtained a convincing collapse for the data from our simulations, we should perhaps try to apply the same idea to the experimental data. A difficulty that arises is that experiments have an additional parameter, the concentration p of the gel. However, it turns out that we can still get a collapse of the data if we use the same collapse formula that we used for the simulation data modified to include variation in p too. In Figure 12.7 we show experimental values of the product $p^{5/2}vN^2$ as a function of pNE for experimental DNA electrophoresis data published by Heller *et al.* (1994). As in the repton model, the left-hand side of the figure shows linear behaviour, while the right-hand side is quadratic. The solid line running through the points is inspired by the collapse function that we obtained from the Monte Carlo simulations and is given by

$$\frac{p^{5/2}vN^2}{\beta} = \left[\left(\frac{pNE}{\alpha} \right)^2 + \left(\frac{pNE}{\alpha} \right)^4 \right]^{1/2} \qquad (12.17)$$

(These powers of p were just found by trial and error; unlike the powers of

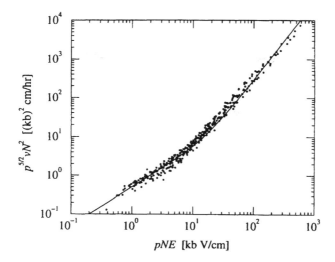

FIGURE 12.7   Experimental data for DNA gel electrophoresis collapse onto a single line if $p^{5/2}vN^2$ is plotted as a function of pNE, where p is the gel concentration (in % agarose). The data are taken from Heller *et al.* (1994).

N and E, our Monte Carlo simulation does not give us any hints as to what the correct forms are for the variation with gel concentration.)

The fact that we can obtain a good collapse of the experimental data using essentially the same method which worked for the repton model suggests that the important processes which drive the dynamics of DNA electrophoresis are the same ones which went into the repton model, namely entanglement of the polymer in the gel and reptation as the dominant mechanism of movement.

Problems

12.1   We obtained a collapse of the data from our repton model simulations using the scaling form given in Equation (12.16). What would be the appropriate scaling form for a system which obeyed Equation (12.14)?

12.2   De Gennes (1971) proposed a model of reptation slightly different from the one discussed in this chapter. In his model the polymer chain is once again represented as a line of particles or reptons on a lattice. Unlike the repton model however, successive reptons in the chain are prohibited from occupying the same square on the lattice; they can only occupy adjacent squares. The possible moves of the chain are that (i) either of the two reptons at the ends of the chain can move to any other square, as long as the resulting configuration is an allowed one, and (ii) any two adjacent links in the chain which are anti-parallel can move together to another position, provided the resulting configuration is lawful. (a) Verify that the chain does indeed move by reptation. (b) The most straightforward Monte Carlo algorithm for this model is to choose at each step one of the two end links of the chain or a pair of links in the middle, and move them at random to a new allowed position. What increment of time should be attributed to one such elementary move for the two-dimensional version of the model? (c) In the two-dimensional case what is the average probability that any proposed move will be accepted?

Part III

Implementation

13

Lattices and data structures

In the preceding chapters of this book we have looked at quite a number of problems and models in statistical physics, and described a variety of different Monte Carlo methods for their study. For many of these Monte Carlo methods we have also given some details about how they can be most efficiently implemented on a typical modern computer. Many details of implementation however are common to a lot of different algorithms, and in the remaining four chapters of this book we examine some of these general implementation issues. For instance, pretty much all of the problems discussed in this book are defined on a lattice, and so we need to know how to represent such a lattice on our computer. Also, most of our algorithms require us to store a certain amount of data regarding, for example, spin clusters or other constructs, and for this we need to know about the efficient implementation of data structures. In this chapter we describe the basic techniques for handling these issues in Monte Carlo programs. If you are an experienced computer programmer, some of this material may already be familiar to you, in which case you will probably want to skip over those parts.

In the following two chapters we look at two more specialized topics, which are very useful for certain types of problems: in Chapter 14 we describe methods for implementing Monte Carlo simulations on parallel computers, and in Chapter 15 we discuss "multispin coding" which is a way of exploiting a CPU's low-level bit manipulation functions to speed up critical parts of a program.

Finally, in Chapter 16, we discuss a problem of great importance to all Monte Carlo simulations, the generation of random numbers.

13.1   Representing lattices on a computer

Most models in statistical physics are defined on lattices. These lattices can
have a variety of different geometries. The most common lattices are square
and cubic lattices, but many others are important for particular models,
including especially the triangular lattice in two dimensions and the fcc and
bcc lattices in three dimensions. In the following sections we describe how
these and a number of other more complex lattices can be represented in a
computer program, and also illustrate some general principles which can be
used to construct representations of more complex lattices still.

13.1.1   Square and cubic lattices

Square and cubic lattices are not only the commonest types of lattices, they
are also the easiest to represent on a computer. Since all high-level computer
languages allow us to define multi-dimensional arrays, we already have a
built-in mechanism for representing square, cubic or even hypercubic lattices
with no work at all. The lattice of spins in an Ising model on an $L \times$
L square lattice for example could be represented in FORTRAN using a
two-dimensional array of integers declared thus: INTEGER S(L,L). In C the
declaration int s[L][L] would achieve the same result. Then we can simply
refer to a spin s_{ij} with lattice coordinates i and j as S(I,J) or s[i][j].

For many calculations we would really like to work on an infinite lattice.
This is not possible on a computer of course, since we don't have an infinite
amount of memory or CPU time, so we are constrained to working on the
largest practical lattice of finite size. However, as discussed in Section 3.1.1,
we can make an approximation to the behaviour of an infinite lattice by
applying periodic boundary conditions to our finite system. In other words,
the system is considered to "wrap around" from one side to another, so that
spins on one edge of the lattice are neighbours of the corresponding spins
on the opposite edge. On an infinite square lattice the nearest neighbours
of a site with coordinates (i, j) would be $(i \pm 1, j)$ and $(i, j \pm 1)$. On a finite one with dimension L and periodic boundary conditions these become
$((i \pm 1) \bmod L, j)$ and $(i, (j \pm 1) \bmod L)$, where mod L indicates the modulo operation,[1] i.e., remainder after division by L. (For languages such as
FORTRAN in which array indices start at 1, you need to add 1 to these
formulae.) This ensures that there are no special sites on the borders of the
lattice; all sites are equivalent under periodic boundary conditions and ap-

---

[1] One must be little careful about using the modulo operation. The effect of the modulo
function on negative numbers varies from one computer to another. To be sure that you are
getting the right answer it is therefore prudent to make sure that you feed the function a
positive first argument. For this reason the neighbours of a site are usually actually written
as $((i + 1) \bmod L, j)$, $((i + L - 1) \bmod L, j)$, $(i, (j + 1) \bmod L)$, and $(i, (j + L - 1) \bmod L)$
on a square lattice. A similar form should be used for cubic or higher-dimensional lattices.

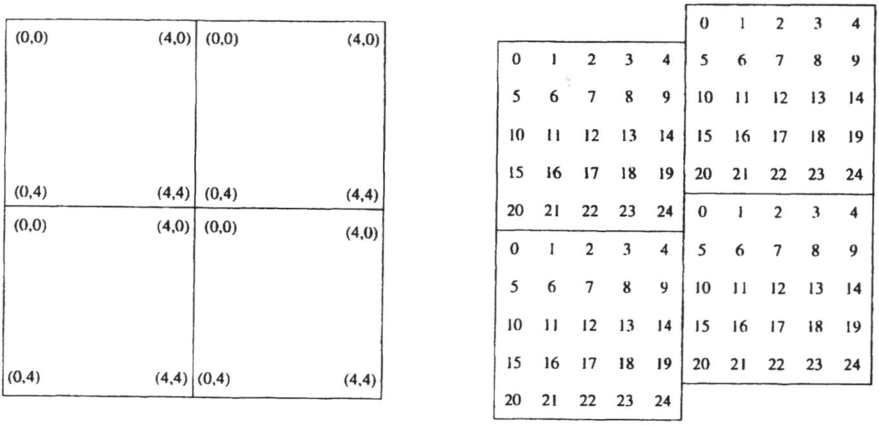

(0,0)	(4,0)	(0,0)	(4,0)
(0,4)	(4,4)	(0,4)	(4,4)
(0,0)	(4,0)	(0,0)	(4,0)
(0,4)	(4,4)	(0,4)	(4,4)

					0	1	2	3	4
0	1	2	3	4	5	6	7	8	9
5	6	7	8	9	10	11	12	13	14
10	11	12	13	14	15	16	17	18	19
15	16	17	18	19	20	21	22	23	24
20	21	22	23	24	0	1	2	3	4
0	1	2	3	4	5	6	7	8	9
5	6	7	8	9	10	11	12	13	14
10	11	12	13	14	15	16	17	18	19
15	16	17	18	19	20	21	22	23	24
20	21	22	23	24					

FIGURE 13.1   Illustration of periodic boundary conditions (left) and helical boundary conditions (right) for a 5 × 5 square lattice.

proximate reasonably well to sites drawn from the interior of a large system (see Figure 13.1).

For many simulations on square and cubic lattices a representation of this type is completely adequate. However, for some simulations where speed is critical it can be inefficient. The problem is that the memory of a computer is only one-dimensional. Each memory location is labelled with a single number. To represent an $L \times L$ array in memory therefore, the computer lays out the elements of the array one after another in memory and the value of a spin with lattice coordinates i and j is found at the variable $iL + j$ from the beginning of the array.[2] This means that every time you retrieve the value of a particular spin, the computer has to perform a multiplication to work out where that value is. And multiplications are slow. Often this is only the tiniest contribution to the total time taken by the simulation. Sometimes, however, as in the Metropolis algorithm for the Ising model (Section 3.1), it can be quite a significant portion of the total time because the rest of the algorithm is very simple and can be programmed with great efficiency. In cases like this, it often pays to use **helical boundary conditions**.

Helical boundary conditions are a variation on the idea of periodic boundary conditions in which each lattice site is indexed by just a single coordinate i. For the case of a two-dimensional $L \times L$ lattice for instance, i would run from 0 to $L^2 - 1$ in C, or from 1 to L^2 in FORTRAN. (Henceforth, we will use the C-style convention of indices starting at zero, since the formulae work out a little simpler in this case. The FORTRAN version can easily be

---

[2]FORTRAN uses so-called "column major" ordering of arrays (as opposed to the "row major" ordering used in C) and stores the element at $i + jL$ instead.

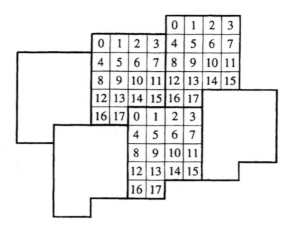

FIGURE 13.2 With helical boundary conditions, the number of sites does not have to be a multiple of the width of the lattice. Here, a lattice with 18 sites and width 4 is drawn, where the neighbours of site i are $i \pm 1$ and $i \pm 4$, modulo 18.

derived from the C one.) On the square lattice i increases along one row of the lattice and then along the next and so on until the entire lattice is filled. On a cubic lattice each plane is filled in this fashion and then the next plane, and so on down the z-axis. The idea can also be generalized to higher dimensions still. In Figure 13.1 we contrast helical boundary conditions with ordinary periodic ones for the case of the square lattice. As the figure shows, in both systems each site is equivalent to all the others, and each site approximates to a site drawn from the interior of a large system.

The advantage of using helical boundary conditions is that no multiplication is necessary to find the value of a spin or other variable at a particular site. The spins on the lattice are now represented by a one-dimensional array, regardless of the actual dimensionality of the lattice, and the value of spin i is given simply by s[i] or S(I). Since there is no multiplication involved (and in addition, only one index i to deal with) most modern computers can retrieve this value from memory about twice as fast as in the case of periodic boundary conditions. The wrapping around of the lattice is handled in a way very similar to the periodic boundary condition case: the four neighbours of the spin at site i are $(i \pm 1) \bmod L^2$ and $(i \pm L) \bmod L^2$. In practice, it is often advantageous to work with a lattice where the total number of sites is a power of 2. In this case, the modulo operation can be performed very simply by setting the highest order bits to zero, which we can do with a bitwise AND operation, which is usually much faster than the modulo function.

Although they may seem a little asymmetric at first, helical boundary

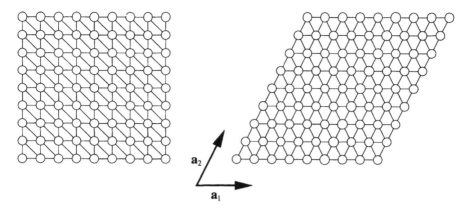

FIGURE 13.3   If we add diagonal bonds running in one direction to an ordinary square lattice (left) we get a lattice with the same topology as the triangular lattice (right). In this way we can represent a triangular lattice using a square array.

conditions are just as good at approximating an infinite system as periodic boundary conditions and there is no reason to avoid using them if there is a significant speed gain to be won. Our only word of caution is to be careful when using them with anti-ferromagnetic models, in which the combination of helical boundary conditions and the need to have a well-defined ground state can sometimes require that we use a rectangular lattice of proportions $L \times (L + 1)$, rather than $L \times L$. The simulation will still work fine; you just have to be careful to choose your lattice correctly.

In fact, if we use helical boundary conditions, there is in general no reason to stick to lattices which are perfectly rectangular. Helical boundary conditions work fine for lattices with any number of sites, regardless of the length of the individual rows. For instance, we can construct a lattice with 18 sites but with rows of length four and still have it fit together fine, as is shown in Figure 13.2. By using this trick we can if we wish make use of the speed gain derived from making the number of sites a power of two, while still retaining some control over the dimensions of the lattice.

13.1.2   Triangular, honeycomb and Kagomé lattices

After the square and cubic lattices, probably the next most common lattice used in Monte Carlo simulations is the triangular lattice. At first it might appear that the triangular lattice is going to be much harder to represent in our computer programs than the lattices of the previous section. However, it turns out that there is a simple trick which makes it very nearly as easy as the square lattice. As we show in Figure 13.3, if you add next-nearest-neighbour diagonal bonds along one diagonal of each square on a square lattice, you

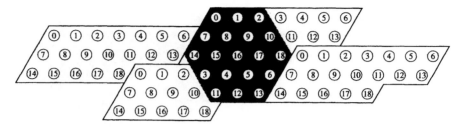

FIGURE 13.4 By using an elongated system of appropriately chosen dimensions and number of sites we can arrange for the repeated cell in the system to be a regular hexagon (shaded area). The hexagon has the same symmetries as the infinite lattice, which makes this a better choice of geometry for many applications.

produce a lattice which has the same topology as the regular triangular lattice; a simple shear will take the lattice into a triangular one. Since it is only the topology of the lattice that we are interested in, we can thus represent a triangular lattice in exactly the same way as we did the square lattice, using a square array. The only thing which changes is which sites are nearest neighbours of which others. If we use a two-dimensional array to store the lattice, the neighbours of site (i, j) are $(i \pm 1, j)$ and $(i, j \pm 1)$ as before, but there are also two new neighbours $(i + 1, j - 1)$ and $(i - 1, j + 1)$. If we use periodic boundary conditions then all these coordinates need to be calculated modulo L, where L is the dimension of the lattice. If we store our lattice in a one-dimensional array and use helical boundary conditions, which is usually more efficient (see Section 13.1.1), then the neighbours of spin i are $i \pm 1$, $i \pm L$ and $i \pm (L - 1)$, all modulo N, where N is the total number of sites on the lattice (which is usually L^2, though it need not be).

One small issue with this representation of the triangular lattice is that the overall shape of the lattice in Figure 13.3 is that of a rhombus. This is a little unsatisfactory because the rhombus does not have the same symmetries as the lattice itself. An infinite triangular lattice has a six-fold rotational symmetry, whereas the rhombus only has a two-fold one. For large systems this is not usually a problem, but for smaller ones it can give rise to unwanted finite-size effects, particularly if we are calculating directional quantities for which the symmetries are important, such as correlation functions. A useful trick for getting around this problem is illustrated in Figure 13.4. As the figure shows, if instead of an $L \times L$ system we use one which is longer on one side than the other, and we stagger the boundary conditions, we can arrange for the shape of the repeated cell in the lattice to be a hexagon, which has the same six-fold symmetry as the infinite lattice. The figure shows how to do this for a system with helical boundary conditions, although the same

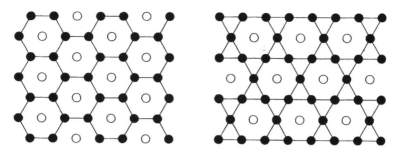

FIGURE 13.5   The solid circles and lines depict the honeycomb (left) and Kagomé (right) lattices. As the empty circles indicate, both these lattices are subsets of the sites in a triangular lattice.

trick works for periodic ones.[3]

Two other common two-dimensional lattices with six-fold rotational symmetry are the honeycomb and Kagomé lattices, which are illustrated in Figure 13.5. The **honeycomb lattice** is a lattice made up entirely of hexagonal plaquets; the **Kagomé lattice** is a regular array composed of both hexagons and triangles. The simplest way to represent these lattices is as a subset of a triangular lattice. As Figure 13.5 shows, both the honeycomb and Kagomé lattices map exactly onto a triangular lattice with the same nearest-neighbour distance, except that some of the sites are missed out. (The missed out sites are represented by open circles in the figure.) Thus we can construct a representation of these lattices by defining an array to hold a triangular lattice, as above, but simply not using some of the entries in the array. In the case of a honeycomb lattice stored in a two-dimensional array with periodic boundary conditions, the missed out sites are those sites (i, j) for which $i-j$ is a multiple of three, or, if you like $(i-j) \bmod 3 = 0$. You can derive a similar rule for systems with helical boundary conditions. In either case, you should be careful about the way the boundary conditions wrap around the lattice. Both periodic and helical boundary conditions work fine for the honeycomb lattice with either rhombic or hexagonal lattice shapes, but there are some restrictions on the dimensions of the lattice. The lattice shown in Figure 13.4 wouldn't work, for instance, because the number of sites is not a multiple of three. However, one can construct a similar lattice with 18 sites and it will work fine.[4]

---

[3]With periodic boundary conditions it is in fact not possible to contrive for the system to be a regular hexagon and we must content ourselves with a lattice some of whose edges differ slightly in length from one another. This however has no effect on simulation results as long as the set of vectors to periodic images possesses all the symmetries of the hexagonal lattice, which it does.

[4]In this case the periodic cell would be an irregular hexagon rather than a regular one, even with helical boundary conditions, but this is not a problem for large systems.

For a Kagomé lattice with periodic boundary conditions, the missed out sites are those for which i and j are both odd (or both even, if you prefer). Helical boundary conditions will *not* work for the Kagomé lattice, and so we are restricted to using periodic ones. (Readers might like to take another look at Figure 13.5 to convince themselves of this.) Below we describe an alternative representation of the Kagomé lattice which does not suffer from this drawback.

These representations of the honeycomb and Kagomé lattices are simple and relatively straightforward to implement. They are, however, wasteful. They require us to define arrays to hold our lattices which are bigger than the number of sites on the lattice, the extra elements of the array simply not getting used. In the case of the honeycomb lattice for example, this increases the space needed to store all the spins or other variables on the lattice by 50%. For smaller simulations this is unlikely to bother us, but for large lattice sizes we might run into problems with the memory capacity of our computer. As it turns out there are other ways of representing these lattices which are more economical with memory and no slower than the simple methods given above. What's more, studying these representations will give us a good idea of how to represent more complex lattices still, such as the diamond lattice of Section 13.1.3.

The fact that we can represent a triangular lattice in the simple fashion described at the beginning of this section is a special consequence of a more general property of lattices. The triangular lattice is one of the five **Bravais lattices** in two dimensions. Bravais lattices (in two dimensions) are ones for which the coordinates of every site can be written in the form $\mathbf{r} = i\mathbf{a}_1 + j\mathbf{a}_2$, where i and j are integers. The vectors $\mathbf{a}_1$ and $\mathbf{a}_2$ are linearly independent and their magnitudes are the **lattice parameters** a_1, a_2 of the lattice. The simplest two-dimensional Bravais lattice is the square lattice, for which $\mathbf{a}_1$ and $\mathbf{a}_2$ are the perpendicular unit axis vectors $\hat{\mathbf{x}}$ and $\hat{\mathbf{y}}$. In the triangular case $\mathbf{a}_1$ and $\mathbf{a}_2$ are again unit vectors, but this time set at a 60° angle to one another, as shown in Figure 13.3. The other two-dimensional Bravais lattices are the rectangular, centred rectangular and oblique lattices, none of which you are ever likely to use in a Monte Carlo simulation. However, we can make use of the Bravais lattice idea to construct more complicated periodic lattices such as the honeycomb and Kagomé lattices, which *are* useful for Monte Carlo work.

A **periodic lattice** is any infinite lattice which maps exactly onto itself when shifted along a correctly chosen translation vector. Periodic lattices account for almost all the lattices used in Monte Carlo simulations. Rare exceptions are the random lattices used in some quantum simulations, and quasiperiodic lattices, which are of interest in the study of quasicrystals. We discuss the representation of lattices like these in Section 13.1.4. Periodic lattices can be represented in terms of Bravais lattices (of which they are

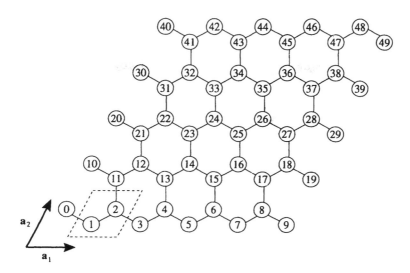

FIGURE 13.6   Representation of a honeycomb lattice. The unit cell consists of two sites, and is repeated along the directions a_1 and a_2, $60°$ apart. The lattice contains $2L^2$ sites, with $L = 5$ in this case.

a superset) as a **unit cell** which is "decorated" with a finite number of sites in some spatial arrangement. The unit cell is repeated over and over throughout space on the sites of an appropriately chosen Bravais lattice. This construction is illustrated in Figure 13.6 for the the honeycomb lattice. In this case, the unit cell (indicated by the dotted line) consists of two atoms, repeated over and over at the vertices of a triangular lattice. We also show how this construction can be used to create a logical numbering scheme for the sites of the lattice which does not waste any memory space. This particular scheme would be appropriate for a system with helical boundary conditions, although we can easily construct a similar one for periodic boundary conditions. Under this scheme, the nearest neighbours of site i on a honeycomb lattice are as follows:

- If i is even, its neighbours are $i \pm 1$ and $i - 2L + 1$.

- If i is odd, its neighbours are $i \pm 1$ and $i + 2L - 1$,

where L now represents the linear dimension of the lattice in unit cells.

For the Kagomé lattice, the unit cell consists of three sites arranged in an equilateral triangle. The underlying Bravais lattice is again a triangular one. In Figure 13.7 we have used this construction to create a numbering scheme for the Kagomé lattice, appropriate for use with helical boundary conditions. In this case the nearest neighbours of site i are as follows:

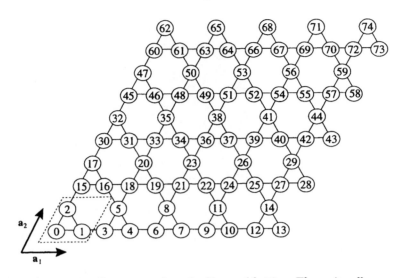

FIGURE 13.7   Representation of a Kagomé lattice. The unit cell consists of three sites in an equilateral triangle, and is repeated along the directions a_1 and a_2, 60° apart. The lattice contains $3L^2$ sites with $L = 5$ in this case.

- If $i \bmod 3 = 0$, its neighbours are $i + 1$, $i \pm 2$, and $i - 3L + 2$.

- If $i \bmod 3 = 1$, its neighbours are $i \pm 1$, $i + 2$, and $i - 3L + 4$.

- If $i \bmod 3 = 2$, its neighbours are $i - 1$, $i - 2$, $i + 3L - 2$, and $i + 3L - 4$,

with L again being the dimension of the lattice in unit cells.

Enforcing the helical boundary conditions is straightforward. Just as before, we take the index i of the site modulo N, where N is the total number of sites on the lattice. The only constraint is that N must be a multiple of the number of sites in the unit cell. If we want we could also use the more symmetric scheme depicted in Figure 13.4 so that the overall shape of the lattice is that of a hexagon.

13.1.3   Fcc, bcc and diamond lattices

Representing three-dimensional lattices is in principle no more complicated than the two-dimensional examples we have seen above. In practice, however, it can be quite confusing because thinking in three dimensions is much harder than thinking in two. The most common three-dimensional lattice used in simulations is the cubic lattice, which we discussed in Section 13.1.1. After this the most common lattices are the face-centred cubic (fcc) which is the usual lattice structure for the majority of crystalline metals, body-centred cubic (bcc) which is the lattice structure for iron and the alkali metals, and the diamond lattice which, as its name indicates, is the lattice adopted by the carbon atoms in diamond, amongst other examples. If you are not familiar with these lattices, a good description can be found in, for example, Ashcroft and Mermin (1976).

The representation of these lattices can be tackled in two different ways, exactly akin to those used for the honeycomb and triangular lattices in Section 13.1.2. The simpler though more squanderous way is to notice that the sites in all of these lattices are subsets of the sites in a cubic lattice. So we can represent them as three-dimensional cubic arrays in which some of the elements are not used. The appropriate rules are as follows:

- A point (h, k, l) on the cubic lattice[5] is a member of the fcc lattice if $h + k + l$ is even.

- A point (h, k, l) is a member of the bcc lattice if h, k and l are either all even or all odd.

- A point (h, k, l) is a member of the diamond lattice if h, k and l are either all even or all odd *and* $(h + k + l) \bmod 4 = 0$ for the even case or $(h + k + l) \bmod 4 = 3$ for the odd case.

While these representations are relatively easy to use, they are wasteful, in the same way that the equivalent representations of the honeycomb and Kagomé lattices were wasteful. In the case of the fcc lattice, a half of the lattice goes unused, meaning that a half of the memory allocated to store lattice variables is wasted. In the bcc case, $\frac{3}{4}$ of the space is unused, and in the diamond lattice a full $\frac{7}{8}$ goes to waste. This could become a real problem for larger lattices, so it pays to find a more economical representation.

Let us look at the case of the fcc lattice first. Although we tend to think of the fcc lattice as being a decorated cubic lattice (and indeed it is possible to construct a representation this way, using a four-site unit cell), it is in fact a Bravais lattice in its own right. The position r of any site on the lattice can be written in the form $r = i a_1 + j a_2 + k a_3$, where the three basis vectors are $a_1 = (1, 1, 0)$, $a_2 = (0, 1, 1)$ and $a_3 = (1, 0, 1)$. (Notice that this form automatically satisfies the condition given above.) The nearest neighbours of a site are not only those along the vectors $\pm a_1$, $\pm a_2$ and $\pm a_3$, but also those along the vectors $\pm(a_1 - a_2)$, $\pm(a_2 - a_3)$ and $\pm(a_3 - a_1)$, and thus the lattice is twelve-fold coordinated. We can then construct a lattice of $L \times L \times L$ sites and apply periodic boundary conditions by taking the three coordinates i, j and k modulo L. This gives a lattice in the form of

---

[5] For some reason, it has become conventional to use the letters h, k and l for these indices. We're not quite sure why this is, but we'll stick with the convention anyway.

a rhombohedron. Alternatively, and more efficiently, we can label each site with an index $n = i + jL + kL^2$ and apply helical boundary conditions. The twelve nearest neighbours of site n are then $n \pm 1$, $n \pm L$, $n \pm L^2$, $n \pm (L-1)$, $n \pm (L^2 - 1)$ and $n \pm (L^2 - L)$.

The bcc lattice is also a Bravais lattice, with basis vectors $\mathbf{a}_1 = (1, 1, -1)$, $\mathbf{a}_2 = (-1, 1, 1)$ and $\mathbf{a}_3 = (1, -1, 1)$. We can again construct a rhombohedral $L \times L \times L$ system with periodic boundary conditions along each principle direction. Each site has eight nearest neighbours along the vectors $\pm\mathbf{a}_1$, $\pm\mathbf{a}_2$, $\pm\mathbf{a}_3$ and $\pm(\mathbf{a}_1 + \mathbf{a}_2 + \mathbf{a}_3)$. Alternatively, we can label each site with the index $n = i + jL + kL^2$ and apply helical boundary conditions. The nearest neighbours of site n are then $n \pm 1$, $n \pm L$, $n \pm L^2$, and $n \pm (L^2 + L + 1)$.

The diamond lattice is more complicated because it is not a Bravais lattice. In fact the diamond lattice has a unit cell of two atoms at positions $(0, 0, 0)$ and $(1, 1, 1)$, repeated over and over at the vertices of a bcc lattice of double the usual size. We can represent the sites on this bcc lattice in the form $\mathbf{r} = i\mathbf{a}_1 + j\mathbf{a}_2 + k\mathbf{a}_3$ with $\mathbf{a}_1 = (2, 2, -2)$, $\mathbf{a}_2 = (-2, 2, 2)$ and $\mathbf{a}_3 = (2, -2, 2)$ and then derive the sites of the diamond lattice by adding $(0, 0, 0)$ or $(1, 1, 1)$ to these points. A practical way of doing this which lends itself to helical boundary conditions is to label each site with an index $n = 2(i + jL + kL^2) + m$, where m is either zero or one depending on which site in the unit cell we are labelling. Using this scheme, the neighbours of site n are as follows:

- If n is even, its neighbours are $n + 1$, $n + 2L - 1$, $n + 2L^2 - 1$ and $n - 2L - 2L^2 - 1$.

- If n is odd, its neighbours are $n - 1$, $n - 2L + 1$, $n - 2L^2 + 1$ and $n + 2L + 2L^2 + 1$.

13.1.4   General lattices

All periodic lattices can be represented as a unit cell superimposed on a Bravais lattice, as we have done with the lattices of Sections 13.1.2 and 13.1.3. Occasionally however, it may be necessary to construct representations of non-periodic lattices, such as quasiperiodic or random lattices. In this case the best approach, if we want our Monte Carlo simulation to run quickly, is to create and store lists of the neighbours of each site. Suppose for example that we have a random lattice in which each site is connected to a small number of other nearby ones. If site number 1 is a neighbour of sites 2, 3, 10 and 11, say, then we would store these four numbers in an array which is associated with site 1. Site 2 would have a similar array which would have 1 as one of its entries, along with whatever other sites are neighbours of site 2. And so on for the entire lattice. This set of lists defines the topology of the lattice, and allows us to perform quick evaluations of the Hamiltonian or any other observable defined on the lattice: we simply run through the sites on the lattice adding terms for the interactions of each site with each of its neighbours. If we have three-point or higher interactions in the Hamiltonian we don't need to store any further lists; we can calculate all orders of interactions from just the topological information contained in our lists of nearest neighbours.

13.2 Data structures

Virtually all Monte Carlo simulations require us to store quite large amounts of data in the memory of our computer. For lattice models, as discussed in Section 13.1.1, we need to store the state of the spins or other variables defined on the lattice. For random lattices we need to store lists of the nearest neighbours of each site. For cluster algorithms we need to keep a record of the spins in each cluster. For the conserved-order-parameter algorithms of Chapter 5 we need to keep lists of which spins are up and which down. And so on. Each case requires us to maintain a **data structure** to store the relevant information. Data structures come in a variety of types, different types being appropriate in different situations. The commonest types of data structures are arrays, linked lists, trees and buffers. In following sections of this chapter we describe each of these in turn and explain how to implement them in a computer program. First however, we discuss briefly the different variable types available in various computer languages. It is from these variables that our data structures will be built.

13.2.1 Variables

All high-level computer languages provide a variety of different variable types which can be used to represent the variables in our physical models. The most common are **integer variables** and real or more correctly **floating-point variables**. We assume that the reader is already familiar with these types of variables and their implementation in the language of his or her choice. Some languages, notably FORTRAN, also provide **complex variables**. For languages in which there is no complex variable type we can usually create one using a structure or class (see below). Other variable types which occur in some languages include Boolean (true/false) variables, character (single letter) variables and strings (rows of letters such as words or sentences). All of these may find occasional use in Monte Carlo calculations.

Many computer languages provide a selection of different precision ranges for variables. Integers, for instance, may be stored in 16, 32 or 64 bits and may be signed or unsigned. Floating-point numbers typically come in **single precision** (32-bit) or **double precision** (64-bit) incarnations, while newer computers often provide 80-bit or even longer representations. At the time of the writing of this book, almost all new computers come with special-purpose hardware for performing fast calculations with double precision floating-point numbers, which makes these calculations just as fast as their single precision counterparts. Unless memory space is an important consideration therefore, there is usually no reason not to use double precision variables in your simulations.

Two common extensions of the basic idea of a program variable are **pointers** and **references**. Pointers, which occur in C and Pascal and derivative languages, are variables which store not the *value* of a quantity, but the *location* of that value in the memory of the computer. Pointers are useful, amongst other things, for efficient implementation of the linked list and tree data structures which are discussed in following sections. References are in some ways similar to pointers. A reference is a variable for which we can specify where in the memory of the computer its value is stored. (Often we are only allowed to do this indirectly by specifying that the memory location should be the same as that of some other variable we already have.) References allow us to do pretty much the same things as pointers do, but are sometimes more convenient to work with. References occur most notably in C++, but are also found in a more limited form in some older languages, such as FORTRAN and Pascal.

Two other useful features found in many modern computer languages are **structures**[6] and **classes**. A structure is a collection of variables which we refer to by one name. For example, in a program which used complex variables, we could define a structure consisting of two floating-point numbers to represent the real and imaginary parts of a complex number. This is merely a convenience: there is nothing that can be done with structures which cannot also be done with ordinary variables. However, one should not underestimate the importance of such conveniences. By making programs simpler and easier to read, structures can greatly reduce the number of bugs introduced into a program when it is written, and help us to spot more easily those that we do introduce.

Classes are found in object-oriented languages and are an extension of the idea of structures. Like a structure, a class contains a number of variables which can all be referred to by one name. but a class can also contain functions which perform operations on those variables. This again is just a convenience, but a very useful one. A properly written class of complex variables, for instance, would allow us to perform straightforward arithmetic on our complex numbers without ever worrying about the real and imaginary parts; we could simply add or multiply our variables together and every-

---

[6]The word "structure" here refers to a specific construction in a high-level computer language and should not be confused with our more general use of the phrase "data structure" throughout this chapter to refer to ways of storing the data in our Monte Carlo simulations.

thing would be taken care of automatically. Again, this can greatly improve the readability of our programs and help us to avoid introducing too many bugs.[7]

In the rest of this chapter we will look at a variety of data structures for storing large numbers of variables in efficient ways. The variables in these data structures can be any of the types described above. The techniques are the same whether you are using integers or real numbers or complex numbers or any other type of data.

13.2.2 Arrays

The most common data structure is the **array**, which is a collection of variables stored in a block of memory on the computer and indexed by one or more integer labels. Arrays are so important that all high-level computer languages provide them as a feature of the language. They also provide the perfect way of representing the variables on a periodic lattice, as described in the first part of this chapter. We will assume that our readers are already familiar with the idea of arrays and how they are used.

13.2.3 Linked lists

A common problem which crops up in many computer programs is that of storing lists of values of one kind or another. The examples given most often are of database applications which require the storage of ever-changing lists of, for instance, names and addresses. In our Monte Carlo codes we are more likely to want to store lists of numbers, such as the positions and values of the non-zero elements in a sparse matrix or vector. We could store such a list in an ordinary array, but this has problems. First, the length of a list may vary, but arrays have a fixed size, which means that we have to decide beforehand what the largest likely number of items is going to be. And if the list ever grows longer than this estimate our program will probably malfunction. More importantly, we often need to be able to add or remove values from any point in a list. If the list is stored in an array, then adding or removing a value somewhere in the middle can only be done by moving all higher elements in the list up or down one place to make a hole or close a hole up. For a large list such movements can take a long time.

One solution to these problems is to use a **linked list**. A linked list is a data structure which allows us to store lists of arbitrary length and to quickly insert or delete elements from them at any point in the list. These features come at a price however. The main disadvantage of a linked list is

---

[7]Typically classes also provide a number of other features, such as scope limitation and inheritance, which are useful for large software development projects but may be of less use for small Monte Carlo programs.

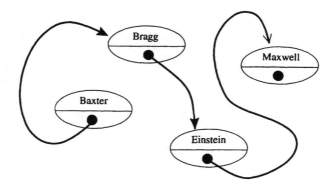

FIGURE 13.8 A linked list used, in this case, for the alphabetical storage of names.

that it is not possible to read any particular element in the list without first going through each of the preceding elements in turn. This makes linked lists good for algorithms which only require sequential access to items in a list (such as sparse matrix multiplications), but poor for algorithms where "random access" is required. (For cases in which random access is required, the tree structures described in Section 13.2.4 are often a good solution.)

A linked list is a set of data items, stored in any order in the memory of the computer, with each of which is associated a pointer which points to the memory location of the next item in the list. As long as we know the location of the first item, we can then easily run through all of them by starting at the first, reading its value, and then following its pointer to the second item and reading that, and so on. We also need some way of indicating when an item is the last one in a list. Typically this might be done by setting the pointer for that item to a special value such as 0 or -1.

Figure 13.8 is a schematic illustration of a linked list of names. In this example, the names are stored in alphabetical order. If we want to add a new name to the list, it should be inserted in the correct place in the alphabet. This is actually very simple to do, as Figure 13.9 shows. To insert a new element after an existing element e_n, we create a variable containing the new element, set the pointer of element e_n to point to it, and set its pointer to point to e_{n+1}. There is no need to shift any data around the memory of the computer, making the list much quicker to change than an equivalent array. Sometimes it may also be necessary to insert a new element *before* an existing one e_n. This is a little more complicated. What we would like to do is create a variable containing the new element and set the pointer of element e_{n-1} to point to it. Unfortunately, we can't easily do this because we can't go backwards from element e_n to get to element e_{n-1}; the pointers only point one way along the list. Instead we use the trick illustrated in Figure 13.10.

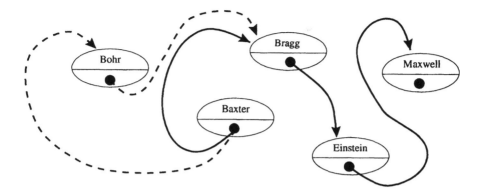

FIGURE 13.9 To add a new element ("Bohr") to a linked list after an existing element ("Baxter"), we change the pointer of the existing element to point to the new element, and set the pointer of the new element to point to the next element in the list ("Bragg").

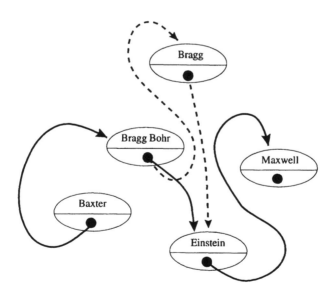

FIGURE 13.10 To add a new element ("Bohr") to a linked list before an existing element ("Bragg"), we create a new element and copy the existing element and its pointer into it. Then we put our new value ("Bohr") where the old one ("Bragg") was and set its pointer to point to the newly created element.

We create a new variable and copy both the value and the pointer from element e_n into it. Then we set the pointer of element e_n to point to this new element, and we store our new value in e_n. This only involves looking ahead along the list and not backwards, and achieves the desired result of inserting the new value into the list ahead of the nth element.

We use a similar trick to remove an element from the list. If we wish to remove element e_n, then what we would really like to do is set the pointer of element e_{n-1} to point to e_{n+1} and just throw e_n away. This again would involve moving backwards in the list, which we cannot easily do. So instead we set the value of e_n equal to the value of e_{n+1}, set the pointer of e_n to point to e_{n+2}, and throw away element e_{n+1}. This achieves the desired result while only looking ahead along the list.

Sometimes, there will be cases where we really do need to be able to move backwards as well as forwards through a list. In this case we can use a **doubly linked list** in which each element includes pointers to both the next and previous elements of the list. Making changes to such a list takes longer than in the case of the singly linked list, since two pointers have to be maintained for each element. For this reason we should use a singly linked list wherever possible, reserving the doubly linked one for those cases where the need to move backwards is unavoidable.

These methods, as we have described them, rely on two crucial features of the computer language used: it should have pointers and it should allow dynamic memory allocation. At the time of the writing of this book, this restricts us to using C or C-like languages. In other languages a more primitive implementation of the linked list is possible, using an array and storing array indices instead of pointers to elements. Such structures have most of the advantages of linked lists but their size is limited by the size of the array used to store them, and so cannot expand arbitrarily, as a true linked list can.

13.2.4   Trees

Both arrays and linked lists are good for certain types of task, but both have their problems. As we saw in Section 13.2.3, a linked list allows us to add or remove elements efficiently at any point in the list, and the list can grow or shrink arbitrarily as the number of elements we wish to store changes. Arrays do neither of these things, having a structure and size which cannot be changed after they are first declared. On the other hand, arrays are good for some other things which linked lists are not, such as allowing us to read a specified value quickly without having to run through all the preceding elements of the array.

The obvious question to ask then is whether there is some other data structure which combines the good features of both arrays and linked lists.

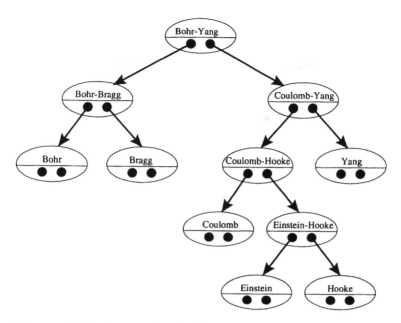

FIGURE 13.11   An example of a binary tree structure. In this case
the tree stores an alphabetically ordered list of names.

Sadly, there is no perfect solution to this problem. However, in many cases
the data structure we call a **tree** will do what we want. A tree is more
complicated to implement than either an array or a linked list, and in addition has some technical problems, particularly the tendency to become
unbalanced (see below), which can make it inefficient under certain circumstances. However, there are many cases in which an algorithm can be made
enormously faster by the use of a tree structure.

A tree is similar to a linked list in being made up of elements stored
in memory and associated pointers which point from one value to another.
The difference is that in a tree, each element or **node** can be accompanied
by more than one pointer. The most common type of tree is the **binary
tree** in which each node has two pointers to other nodes. This gives rise to
a branching structure like that illustrated in Figure 13.11. In this section
we will illustrate the use of trees with the example of a binary tree, but the
methods described can easily be generalized to other types of trees.

Typically we use genealogical terms to describe the relationships between
nodes in a tree: the **parent** of a node is the node which points to it, and
the **child** of a node is one which it points to. Sometimes we also use the
language of real green trees: the node at the very beginning of the tree is
often called the **root node** and the nodes at the very end, which don't point

to anything, are called **leaves**.

Let us look again at the simple example of the storage of a collection of names. Typically we want to be able to add and remove names from the tree and to check to see if a particular name exists in the tree. Neither an array nor a linked list can do all of these things quickly, but a tree can.

The best way to organize a tree to store data of this kind is to put all of the data items in the leaves of the tree, i.e., in the nodes at the very end of the tree's branches. We don't store data in the nodes along the branches. Instead we store information in these nodes which allows us to find the right leaf quickly when we need to. In the case of the names, one way to proceed would be to organize the tree alphabetically, and store at each node the alphabetic range covered by the names which are descendents of that node. This scheme is illustrated in Figure 13.11.

To find a name in the tree, we start at the root node and perform the following steps:

1. We check whether the alphabetic ranges covered by either of the children of the current node include the name we are looking for. If neither of them do, then our name is not in the tree. If one of them does, we proceed to that child.

2. We check whether the current node is a leaf, i.e., the end of a branch. If it is, then it contains the name we are looking for. If it isn't, we go back to step 1.

Typically, the number of steps needed to complete this search goes as the logarithm (base 2 in the case of a binary tree) of the number of names or other values stored in the tree. The best method for searching an array containing the same ordered list of names is to use a binary search, which also takes a time which scales logarithmically with the length of the array. So in this respect the tree is as efficient as the array. (And it is much more efficient than the linked list. It would take a time which scales linearly with the number of names to search the equivalent linked list.)

To add a new name to the tree, we again start at the root node and perform the following steps:

1. We check the alphabetic ranges covered by each of the children of the current node. If our name falls in one of them, we proceed to that child. If it doesn't, we extend the alphabetic range of one of the children to cover the new name and then proceed to that child. (We must be careful to do this in such a way that the ranges covered by the two children of a node never overlap.)

2. We check whether the current node is a leaf, i.e., the end of a branch. If it is not, we go back to step 1. If it is, then we add two new children

to the current node. To one of them we assign the old name that used to be stored in the current node, and to the other we assign the new name.

This procedure also takes an amount of time which scales as the logarithm of the number of names or other values stored in the tree. This is much better than the time taken to add a new name to an alphabetically ordered array, which goes linearly with the number of names. However, it is not as good as the case of the linked list, for which the time taken to add a new name is independent of the length of the list. For this reason you should use linked lists where possible, reserving the binary tree for cases in which it is important to be able to search for a particular value quickly.

Finally, to remove a name from the tree, we start once more at the root node and perform the following steps:

1. We find the leaf which contains the name we want to remove, using the search method described above. As we move through the tree, we record the location of each node we pass through. These locations are needed in the next two steps.

2. From the leaf containing our name, we retrace our tracks to the grand-parent node of that leaf and alter it so that it points to the *brother* of the node we are removing.

3. We check whether the alphabetic range recorded by the grandparent node still correctly describes its children. If it does, we are finished. If not, we correct it, and then move up to the great-grandparent, and so on, repeating this step until we reach a node for which no changes need to be made, or we reach the root node.

This process, like the addition of a name, takes a time which increases logarithmically with the number of names or other values stored in the tree.

In fact, it is not always true that all these operations take an amount of time going logarithmically with the size of the tree. This is the typical result, but it is possible for trees to get into a pathological state in which it takes much longer to perform searches or updates. If the number of leaves which you can get to by choosing either path from a node is roughly the same, then the logarithmic rule applies. Such trees are said to be **balanced**. If we happen to add names to the tree in the wrong order however, it is quite possible that the tree will become unbalanced and the efficiency of our algorithms will fall dramatically. In the worst case it can take an amount of time which scales linearly with the size of the tree to both search and update it, which is worse than either an array or a linked list. There are a variety of strategies for preventing this situation. Often we have some flexibility in the way in which we add values to the tree. In the example given above, for instance, we will sometimes add a new name which falls between the alphabetic ranges covered by the two children of a node. In this case we are free to choose the branch which we add the new name to. If we always add the name to the branch which leads to fewer leaves this will help to keep the tree balanced. There are times, however, when strategies like this are insufficient to maintain balance. In these cases we can employ one of a number of algorithms which have been developed for rebalancing trees. The working of rebalancing algorithms is rather technical in nature however, and we won't go into it here. A good discussion is given by Sedgewick (1988).

An example of the use of a tree data structure in a Monte Carlo algorithm can be found in Section 11.2.2, where we describe an efficient algorithm for simulating the diffusion of adatoms on a crystal surface.

13.2.5 Buffers

A **buffer** is a data structure used to store the values of variables temporarily, and retrieve them later. Buffers come in a number of different varieties. Here we describe the two which occur most commonly in Monte Carlo programs, the **last in/first out** (LIFO) buffer or **stack**, and the **first in/first out** (FIFO) buffer or **queue**. Both of these can be created easily using arrays. The descriptions here assume arrays whose indices start at zero as in C, rather than at one as in FORTRAN. However, it is very straightforward to adapt the same ideas for use in FORTRAN programs.

A last in/first out buffer operates exactly as its name suggests. We "push" the values of as many variables as we like into the buffer, and at any point we can "pop" a value out again. The values come out in the opposite order to the way we put them in, so that the last one which went in is the first one out. To create a LIFO buffer in a computer program we make a one-dimensional array of the appropriate variable type with enough elements to store the maximum number of values which the buffer will have to contain—let us call this number L—and we also create an integer variable, m say, to store the number of values in the buffer. Initially $m = 0$. When we push a value onto the stack, we place the new value in the mth element of the array and then increase the value of m by one. When we want to pop a value out again we decrease m by one and read the value in the mth element of the array. At any time the number of values in the buffer is equal to the value of m. If m ever reaches L, then the buffer is full and no more values can be added. If $m = 0$ then the buffer is empty and no more values can be removed.

A first in/first out buffer is one in which the values stored pop out in the *same* order as they were put in, so that the first one in is also the first one out, just as the name says. One simple way to implement a FIFO buffer in a computer program would be to create an array of L elements and an integer variable m to store the number of values in the buffer, just as in the LIFO case, with $m = 0$ initially. Just as before, when we add a value to the buffer we place the new value in the mth element of the array and then increase m by one. Now however, when we want to pop a value out of the buffer, we read the *first* element of the array (which would have index zero in C), decrease m by one, and then move all the other elements down one place, so that a new one becomes the first element, ready to be popped out in its turn.

Simple though it is, this is not a good way to implement our buffer. It will work, but it will be slow. Every time we pop a value out of the buffer we have to reassign the values of m elements in the array. This takes an amount of time proportional to m and if m is very large it could be a very slow process. Luckily, there is a much better way to make a FIFO buffer, using **circular buffering**. Circular buffering also makes use of a single array of size L to store the contents of the buffer, and in addition it uses two integers m and n which point to the first and last values in the buffer. Initially m and n are both set to zero. When we want to add a value to the buffer we place the new value in the mth element of the array and increase m by one, modulo L. When we want to remove a value we read the nth element of the array and then increase n by one modulo L. The number of values in the buffer at any time is equal to $m - n$ (possibly plus L because of the modulo operations).

This way of implementing the FIFO buffer is much more efficient than the first one we proposed. Each push or pop operation requires us to read or write the value of only one variable, and can be completed in an amount of time which remains constant as the number of values in the buffer grows. There are a couple of things we need to be careful about however. First, just as in the case of the stack, a FIFO buffer will overflow if we try to put more than L values into it. To guard against this we should compare the values of m and n after each value is added to the buffer. If $m = n$ after a value has been added, then the buffer is full and adding any further values will probably cause our program to malfunction. Second, the buffer can underflow if we try to remove a value from it when there are none there. To guard against this we should compare the values of m and n after removing each value. If $m = n$ after removing a value then there are no more values left in the buffer.

As an example of the use of buffers in a Monte Carlo algorithm, consider the Wolff cluster algorithm for the Ising model, which was discussed in Section 4.2. We summarized a step of the algorithm as follows in Section 4.2.1:

1. Choose a seed spin at random from the lattice.

2. Look in turn at each of the neighbours of that spin. If they are pointing in the same direction as the seed spin, add them to the cluster with

probability $P_{\text{add}} = 1 - e^{-2\beta J}$.

3. For each spin that was added in step 2, examine each of *its* neighbours to find the ones which are pointing in the same direction, and add each of them to the cluster with the same probability P_{add}. This step is repeated as many times as necessary until there are no spins left in the cluster whose neighbours have not been considered for inclusion in the cluster.

4. Flip the cluster.

The simplest way for a computer program to carry out this procedure is to store the spins of the current cluster in a buffer as follows. Starting off with an empty buffer, we first choose one of the spins on the lattice at random to be our seed. We look at each of the neighbours of this spin, to see if it points in the same direction as the seed itself. If it does, we generate a random real number r between zero and one, and if $r < P_{\text{add}}$, we add that spin to the cluster *and* we add its coordinates to the buffer. When we have exhausted the neighbours of the seed spin we are looking at, we pop a spin out of the buffer, and we start checking *its* neighbours one by one. (If some of its neighbours have already been added to the cluster then we should miss those out, otherwise the algorithm will never stop.) Any new spins added to the cluster are again also added to the buffer, and we go on popping spins out of the buffer and looking at their neighbours in this way until the buffer is empty. This algorithm guarantees that we consider the neighbours of every spin added to the cluster, as we should.

This method will work equally well with either a FIFO or a LIFO buffer and, statistically speaking at least, we should get the same answers whichever we use. Furthermore neither buffer type gives an algorithm significantly more efficient than the other. However, it is worth noting that there are differences between the two in the way in which the clusters grow. Figure 13.12 shows how the same cluster grows using each type of buffer. With the FIFO buffer, the cluster grows in a spiral fashion, remaining, on average, roughly isotropic throughout its growth. With the LIFO buffer, the cluster first grows along a line in one direction, and then backs up along its tracks and begins to grow sideways. Because of these differences in cluster growth patterns, these two versions of the Wolff algorithm can provide useful checks against potential problems in the program, such as bugs or deficiencies in the random number generator used. (The Wolff algorithm is known to be unusually sensitive to imperfections in random number generators (Ferrenberg *et al.* 1992).) In addition, there some variations on the basic Wolff algorithm for which one or other buffer type may be more efficient. For example, if we wish to place constraints on the maximum size of a cluster, as is done in the **limited cluster flip algorithm** (Barkema and Marko 1993), then an implementation making use of a FIFO buffer is usually faster.

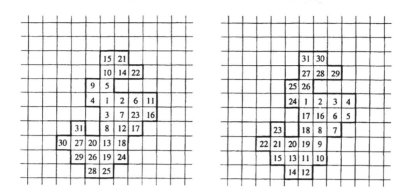

FIGURE 13.12 The numbers indicate the order in which sites are added to a cluster in the Wolff algorithm. If a FIFO buffer is used (left) the sites are added in a spiral around the initial seed spin. If a LIFO one is use (right) the cluster tends to grow first in one direction, and only later spreads out.

Problems

13.1 If we pack a two-dimensional space with identical circular disks, what fraction of the space can we fill if the disks are packed with their centres lying on (a) a square lattice and (b) a triangular lattice?

13.2 As described in Section 13.1.3, the set of points with integer coordinates (h, k, l) where h, k and l are either all even or all odd lie on a bcc lattice. If we stretch this lattice in the z-direction by a factor of $\sqrt{2}$, we get another periodic lattice. What lattice is this?

13.3 A common problem that comes up in many Monte Carlo algorithms is choosing a single value at random from a list. If the values are stored in consecutive elements of an array this is easy, but it is harder if they are stored in a binary tree of the type described in Section 13.2.4. Suppose we have a number n of values stored in the leaves of such a tree. Devise an algorithm which selects one of them uniformly at random, given a random integer $0 \leq r < n$. (Hint: you will need to add to the data stored in the tree.)

13.4 The time taken to find a particular element stored in a balanced binary tree varies as the number of levels of the tree which we have to search through, which is $\log_2 N$, where N is the number of elements stored. If a binary tree is unbalanced in such a way that one child of each node has roughly twice as many descendants as the other, the average search time still scales as $\log_2 N$ but with a larger multiplicative prefactor. How much larger is this prefactor?

14

Monte Carlo simulations on parallel computers

Some of the simulations we would like to perform require so much computing power that it is not possible to finish them in a reasonable length of time on any available computer. Under these circumstances we have two choices: either we can wait a few years for a more powerful computer to become available, or we can perform the calculation on several computers. Since waiting a few years for the answer to a problem is not always a practical course of action, the latter approach is becoming increasingly popular and has led to the development of hardware and software for **distributed** and **parallel computing**. In this chapter we explore the ways in which these techniques can be applied to Monte Carlo simulation.

The term "distributed computing" refers to the linking up of many individual computers to perform a task. A parallel computer simplifies our job by physically bringing many CPUs together in one box, often with additional electronics to allow them to communicate with one another at high speed. From the point of view of someone developing Monte Carlo programs, the problems presented by parallel computers and distributed computers are essentially the same and we will not distinguish between them in this chapter.

Parallel computers come in two principal varieties: those with **distributed memory** and those with **shared memory**. In a parallel computer with distributed memory, each CPU or processor has its own memory which cannot be read or written in by any of the other processors. This means that if our algorithm requires that a processor know the contents of another processor's memory, the appropriate data must be sent in the form of a "message" from the one processor to the other. One day, such messages may be handled automatically by compilers or other standard software, but as the technology stands at the moment they have to be put in by hand by the programmer, which adds an extra level of complexity to our programs. In addition, the transfer of messages between processors is rarely as fast as the transfer of data from memory to processor, and so can slow the program down. These problems can be overcome by using a parallel computer with shared memory. In such computers all the processors share the same memory and can all read from and write to it simultaneously. For the purposes of Monte Carlo simulation at least, shared memory computers are undoubtedly superior to distributed memory ones. Unfortunately, they are also much harder to build. As a result, almost all the parallel computers commercially available at the time of the writing of this book are distributed memory machines. This may change in the future, but for the moment at least, if we are going to make use of parallel computers, we will have to learn the techniques of programming using distributed memory architectures. It is on these techniques that we focus in this chapter. All distributed computers necessarily also use the distributed memory model. So the techniques of this chapter are applicable to distributed calculations as well as parallel ones.

The holy grail of parallel programming is to devise an algorithm which displays **linear speed-up** in the number of processors used. In other words, we want the time taken to complete a certain calculation to halve every time we double the number of processors used to run the program. Often people quote the **efficiency** of the algorithm, which is the ratio of the fraction by which the calculation speeds up when we increase the number of processors used, to the fractional increase in the number of processors. If the efficiency is 100%, we have found our holy grail. The main reason why we don't usually achieve an efficiency of 100% is because, as discussed above, the different processors need to send one another messages as part of the operation of the algorithm and this places a **communication overhead** on the program. In particular, if the average amount of data which a processor needs to send and receive goes up with increasing number of processors (as it usually does) then we can never achieve a linear speed-up.

There are three primary classes of techniques for creating a parallel algorithm to perform a task: **trivial parallelization, functional decomposition** and **domain decomposition**. Trivially parallel problems are ones which can be split up into separate tasks which are completely unrelated and can be performed by separate programs running on isolated processors without the need for inter-processor communication. For more complicated problems we have two ways to go. Functional decomposition refers to the breaking up of a task into a variety of *different* jobs: each processor in a program which carries out such a task is performing a different type of work, and the contributions of all the processors put together perform the complete task. We could imagine for example delegating separate processors in a Monte Carlo calculation to evaluation of a Hamiltonian or other observable, construction of clusters, updating lattice variables, or any number of other elementary chores. This however is not a very common approach to the kinds of calculations this book is concerned with. Monte Carlo simulations almost always use domain decomposition, which means breaking up the calculation into separate simulations of different regions of the system under study. Since Monte Carlo simulations are usually concerned with systems defined on a lattice this approach is often quite straightforward to implement. The processors in such a calculation all perform essentially the same task on different parts of the lattice, and indeed are often running exactly the same program.

14.1   Trivially parallel algorithms

The aim of most Monte Carlo calculations is to generate a large number of statistically independent configurations of the system of interest, in order to make an estimate of some observable of interest. In many cases there is no reason why these independent configurations need all be generated by a single simulation. We could just as well run our program N times and generate s samples each time as run it once and generate Ns samples. For problems of this type, a very simple form of parallel algorithm suggests itself: we can run our program simultaneously on N processors, thereby dividing the time taken by the calculation—in the ideal case—by N. Such algorithms are called **trivially parallel** algorithms. They are also sometimes called **embarrassingly parallel** algorithms, presumably by people who consider themselves to be above using such simple tricks. We however, as physicists, should never be embarrassed to use such an approach. Trivially parallel algorithms are easy to write and if they get you the answer you want quickly, then you should use them. We might even go so far as to say that trivially parallel algorithms are the ideal application for parallel computers.

There are however many cases in which the trivially parallel approach will not work. In the case of an equilibrium simulation, performing many simulations of a system instead of one long simulation means that we have to equilibrate our system many times, once for each processor. This means that N times as much CPU time is wasted by the parallel program as by the serial one on equilibration. If the equilibration time is only a small fraction of the total time taken by the simulation then this is not a problem. However, if it is a large fraction of the total time then using the parallel computer may not help us much.

Another problem with trivially parallel simulations is that they cannot simulate any larger a system than a single, serial simulation. Suppose we want to study a system so large that we cannot get any results for it using a serial approach, either because the correlation or equilibration time of the system is too long, or simply because the system will not fit into the memory of a single computer. In this case it is clear that a trivially parallel simulation will do no better, since none of the individual simulations running on separate processors will give any results either.

Occasionally, we may also run into the problem that the quantity we want to measure depends on the very long time behaviour of an observable in our simulation, such as an autocorrelation function (see Section 3.3.1). In this case again, performing many short simulations will not give us the same answer as performing one long one.

For Monte Carlo simulations, the solution to all of these problems is to develop an algorithm using domain decomposition. We address this topic in the next few sections.

14.2 More sophisticated parallel algorithms

In the following sections we study parallel Monte Carlo algorithms which work by domain decomposition, splitting up the lattice or other space occupied by a model and giving each processor a separate portion to work on. Unfortunately, programs to perform such simulations are considerably more complicated to write than the trivially parallel ones discussed in the last section, and there are no universal techniques which will work for all problems. Different models require different approaches, and a certain amount of ingenuity is needed to find an algorithm which works in any given case. Here we give two examples of domain decomposed parallel algorithms for simulating the Ising model in equilibrium, one based on the Metropolis algorithm (Section 3.1) and the other based on the Swendsen–Wang cluster algorithm (Section 4.4.1). We hope that these examples will give you a hint about the types of approaches that work with parallel computers.

14.2.1 The Ising model with the Metropolis algorithm

As we saw in Chapter 3, the fastest way of simulating the Ising model away from its critical point is to use the single-spin-flip Metropolis algorithm (Section 3.1). It actually turns out to be quite simple to get this algorithm to work on a parallel computer using domain decomposition.

Suppose we partition our lattice into a number of domains, as illustrated in Figure 14.1.[1] We can then store one of the domains in the memory of each of our processors, and have that processor perform Metropolis-type spin flips on only the spins in its own domain. The only problem with this approach comes when we want to flip the sites on the edge of a domain. The formula for deciding whether to flip a spin or not involves knowing the values of all

---

[1]The word "domain" is used in a different sense here from its use in Chapter 4 to describe groups of similarly oriented spins. Here it just refers to the division of the lattice between processors.

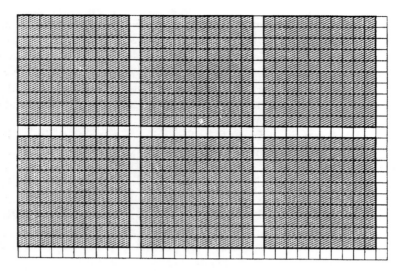

FIGURE 14.1   Different processors can simulate different regions of a lattice, as long as these domains are separated by strips of sites in which the spins are fixed. Here, six processors are used for the simulation of a model on a 33×22 lattice. Each processor takes care of a 10×10 domain (the gray areas), and these areas are separated by strips of fixed spins (the white areas).

the surrounding spins (see Equation (3.10)), and for the spins on the edge of a domain the values of one or more of the surrounding spins are stored in the memory of another processor and are not available. One solution to this problem might be for every processor to keep a record of the values of all the spins just outside the border of the domain which it covers. This would require each processor to inform its neighbour whenever it changed the value of a spin on its border, so that the neighbour can keep its records up-to-date. This approach has its problems however, the main one being that it requires all the processors to run in perfect synchrony. Since this rarely occurs in real computers, one would have to ensure synchronization by having processors wait after each Monte Carlo step until the slowest amongst them had caught up with the rest. This is wasteful of CPU time and besides, there is a much better way of doing it.

The efficient way to perform the simulation is simply not to flip the spins on the border of a domain. We flip all other spins as usual, but not the spins on the borders (see Figure 14.1 again). This doesn't affect detailed balance, since the forward and backward probabilities for flipping a border spin are both zero, so that Equation (2.12) is still obeyed. It does however destroy the ergodicity of our algorithm, since there are some states (the ones in which the border spins have been flipped) which can never be reached. To get around this problem we move the domain borders periodically, so that all spins get a chance to be in the interior of a domain. When we move a border, some spins pass out of the domain covered by one processor and into that covered by another, which means that we have to send the current values of those spins from the old processor to the new. This of course requires us to send some inter-processor messages, which, as we pointed out at the beginning of the chapter, is a comparatively slow process. On the other hand, we don't have to send such messages very often. The ideal time interval between successive moves of the domain boundaries turns out in fact to be the correlation time of a single simulation the size of one of our domains. Under any circumstances this time is at least one Monte Carlo step per site, which means we still get to do quite a large chunk of simulation in between one reorganization of the domains and another.

14.2.2 The Ising model with a cluster algorithm

As we saw in Chapter 4 the Metropolis algorithm becomes a very inefficient way of simulating the Ising model close to the critical point. In such cases cluster algorithms like the Wolff and Swendsen–Wang algorithms of Sections 4.2 and 4.4.1 work much better. In this section we take the example of the Swendsen–Wang algorithm and look at how it can be made to work on a parallel computer, assuming again that, for one reason or another, the system simulated needs to be spread over many processors by domain decomposition, rather than employing the trivially parallel approach of Section 14.1.

Recall that a step of the Swendsen–Wang algorithm consists of making "links" between similarly oriented nearest-neighbour spins on a lattice with probability $P_{\text{add}} = 1 - e^{-2\beta J}$ so that the entire lattice is divided up into clusters of sites joined by these links. Then we go through each cluster in turn, randomly setting all its spins to point either up or down with probability $\frac{1}{2}$. This algorithm presents more of a challenge to the programmer of a parallel computer than does the Metropolis algorithm of the previous section. We can break our system up into domains as illustrated in Figure 14.1 just as before. Then each processor can make the required links between nearest neighbour spins in its own domain. It can also find the groups of spins which are joined into clusters by these links, but only within its own domain. If a cluster extends across the domains covered by two or more processors, then it will appear as several separate sub-clusters in the memories of different processors. To get around this problem we need a "master" processor which oversees the process and joins up sub-clusters to form the correct Swendsen–Wang clusters. The same master process can also then decide, with probability $\frac{1}{2}$, whether the spins in each cluster should be flipped, and communicate this decision to the other processors so that they can carry out the actual spin flips.

This approach can speed up the Swendsen–Wang algorithm by about a factor of three (Barkema and MacFarland 1994), but it is difficult to do any better than this, since the speed of the program is limited by the time the master processor takes to do its part of the job. Since this part of the algorithm is performed by just one processor, increasing the number of processors will make no difference to the time taken, and the efficiency of the algorithm is guaranteed to tend to zero as the number of processors becomes large.

Above the critical temperature, where the average cluster size is relatively small, we can use a different approach. Since the clusters are small, many of them fall entirely within a single domain. It turns out that it works quite well to flip these clusters with probability $\frac{1}{2}$ as before, but to leave any clusters which cross the domain boundaries untouched. As with our parallel version of the Metropolis algorithm in Section 14.2.1, this satisfies the condition of detailed balance for the same reasons the original Swendsen–Wang algorithm did, but does not satisfy the condition of ergodicity. And, as with the Metropolis case, we can get around this by periodically shifting the boundaries between the domains. Again only the shifting of the boundaries requires communication between processors, while the cluster-flips themselves can be performed by the individual processors independently. This approach leads to a sizeable increase in the speed of the simulation, and one which scales approximately linearly with the number of processors employed.

A similar approach works well for creating a parallel version of the Wolff algorithm of Section 4.2. As long as we are working at temperatures high enough that the average cluster size is well below the size of the domain covered by a single processor, we can grow clusters separately in each domain and flip only those which do not cross the domain boundaries. By moving the domain boundaries periodically we can ensure that the algorithm is ergodic as well as respecting detailed balance.

Unfortunately, no simple parallel version of the Wolff algorithm exists which works well for lower temperatures at which the typical cluster size is on the order of or larger than the size of a domain. In this regime, the Swendsen–Wang algorithm is a better choice.

Problems

14.1 (a) The lattice in Figure 14.1 is divided into six regions of equal size by fixing 126 spins. Is it possible to divide it into six such domains but fix fewer spins? (b) What is the biggest lattice that can be divided into eight domains of equal size with 1000 fixed spins?

14.2 In Figure 14.1 a 33×22 lattice is divided into six square regions.

The Monte Carlo algorithm of Section 14.2.1 for the Ising model will not be ergodic if the boundaries are not moved, since the boundary spins are never changed. Can we make the algorithm ergodic by alternating between two different positions of the boundaries which are shifted with respect to one another?

14.3 In the Ising model algorithm described in Section 14.2.1, we shift the boundaries between regions of the lattice about once every τ Monte Carlo steps per site, where τ is the correlation time of a simulation performed on a lattice the size of the region covered by a single processor. Calculate how the fraction of time spent on communication scales with the number of processors for fixed system size, in terms of the dimension d of the system and the dynamic exponent z. Now perform the same calculation for the other algorithm suggested in Section 14.2.1, in which the values of the spins on the borders of each region are transmitted to neighbouring processors every time they are changed.

15

Multispin coding

Multispin coding is a programming technique which allows us to speed up some Monte Carlo simulations (or other programs) by a significant factor by making use of the low-level bit manipulation functions provided by the CPU of our computer. These functions include logical operations like AND ($\wedge$), OR ($\vee$) and exclusive-OR ($\oplus$), as well as things like shifts, additions and subtractions.

The basic idea behind multispin coding is to store the states of several variables, such as spins in a spin model for example, in the bits of a single word on our computer. (A "word" is the number of bits of information the CPU of the computer can process at once, and is usually, at least at the time of the writing of this book, either 32 or 64 bits.) Then we write our Monte Carlo algorithm in terms of operations which act on an entire word at a time. The result is that we can perform operations on several variables at once, rather than performing them sequentially, which improves the speed of our calculation, thereby giving us better results for a given investment in CPU time.

There are a number of different ways of actually putting this idea into practice. The most common ones are **synchronous update** algorithms in which the bits in a word represent several spins or other variables on the lattice of a single system so that one multispin-coded Monte Carlo step updates many sites at once, and **asynchronous update** algorithms in which the bits within a word represent spins or other variables on separate lattices so that one Monte Carlo step simultaneously updates spins in many different systems. Asynchronous algorithms are usually simpler to program than synchronous ones and it is primarily on these algorithms that we will concentrate in this chapter. Synchronous algorithms are discussed briefly in Section 15.5.

Asynchronous algorithms effectively perform a number of independent simulations simultaneously and for this reason are suitable for the same types of problems as the "trivially parallel" Monte Carlo simulations discussed in Section 14.1, i.e., ones for which performing many separate simulations will give us results as good as performing one long simulation. As with our trivially parallel simulations, if the equilibration time of our system is a large fraction of the total simulation time, or if we want results for particularly large systems or long times, then this type of multispin coding will probably not work. Moreover, don't even think about using multispin coding if your Monte Carlo algorithm is a complex one. If the number of lines in the core part of the algorithm is larger than about a hundred, then multispin coding is not the way to go. Multispin-coded programs are hard to write, hard to debug, and hard to alter if you decide to change the way the algorithm works. Getting such a program to work for a complicated Monte Carlo algorithm would be difficult and painful, and there's a good chance that you would never get it to work at all.

However, if you have a simple Monte Carlo algorithm which you wish to implement—the Metropolis algorithm is a good example—and if your problem is simply that you can't run your simulation for long enough to get good statistics for the quantity you want to measure, then multispin coding could be the answer to your difficulties.

15.1   The Ising model

As a first example of multispin coding we demonstrate in this section how one would go about writing a multispin-coded Monte Carlo program for the Ising model using the single-spin-flip Metropolis algorithm of Section 3.1. First of all we look at the problem in one dimension, where everything works out very simply, and then in two dimensions, which is a more realistic problem and also a harder one.

15.1.1   The one-dimensional Ising model

In this section we develop a multispin-coded Monte Carlo algorithm for the one-dimensional Ising model. No one really wants to simulate the one-dimensional Ising model—the problem has already been studied very thoroughly using analytic techniques—but the model illustrates in a nice simple context all the important ideas of multispin coding. If you can follow the developments of this section, then you will have a good understanding of how an asynchronous multispin algorithm works.

Multispin coding requires that we store the lattice variables of the system being simulated—spins in this case—as bits in the words of our computer's memory. For the Ising model this is particularly simple because the spins have only two states, up and down, which we can represent as single bits,

0 or 1. Let us decide that up-pointing spins will be represented by 1s and down-pointing spins by 0s. (For many other models the representation of the system in terms of bits is not as simple as this, but it is usually possible to find a reasonably convenient representation which works in any particular case.)

One step of the Metropolis Monte Carlo algorithm for the Ising model consists of choosing a spin from the lattice and calculating the change ΔE in the energy of the system if we were to flip that spin. If $\Delta E \leq 0$ we definitely flip the spin over. If $\Delta E > 0$ we flip it with probability $\exp(-\beta\Delta E)$. Another way of phrasing this in the case of the one-dimensional Ising model is that we always flip the chosen spin if either one or both of its neighbouring spins are pointing in the opposite direction to it. If both of the neighbouring spins are pointing in the same direction as the chosen spin then we flip it over with probability $\exp(-4\beta J)$, where J is the spin–spin coupling constant.

Suppose that we denote the bit representing the spin at site i by σ_i. (We use σ rather than s here in order to distinguish variables which take the values 1 and 0, rather than $+1$ and -1.) If we take the bitwise exclusive-OR of two such spins at different sites $\sigma_i \oplus \sigma_j$ we get 1 if and only if the two spins are not aligned. Thus the expression $(\sigma_i \oplus \sigma_{i-1}) \vee (\sigma_i \oplus \sigma_{i+1})$ is 1 if either one or both of the neighbours of spin i are pointing in the opposite direction to it. If in addition we can generate a random bit r which is 1 with probability $\exp(-4\beta J)$, then the expression $(\sigma_i \oplus \sigma_{i-1}) \vee (\sigma_i \oplus \sigma_{i+1}) \vee r$ is 1 if we should flip spin i and 0 otherwise. In other words, if we choose a spin i from the lattice and set its value to σ_i', where

$$\sigma_i' = \sigma_i \oplus [(\sigma_i \oplus \sigma_{i-1}) \vee (\sigma_i \oplus \sigma_{i+1}) \vee r] \tag{15.1}$$

then we have performed one step of our Monte Carlo algorithm. We call this expression the **master expression** for our multispin-coded Monte Carlo algorithm. This particular master expression is quite a simple one. As we will see in later sections of this chapter, deriving a master expression for a more complicated model is often quite hard work.

The point of all this is that we can now put many bits, typically 32 or 64 of them, representing many different spins, into one word on our computer, and use Equation (15.1) to perform our Monte Carlo step on all of them simultaneously. Notice that we can just as well apply (15.1) to an entire word full of bits as to a single bit at a time; the expression applies the same operation to each bit in the word independently and the bits to do not interfere with one another.

In practice, the simplest way to perform the simulation is to simulate many systems at once asynchronously, as described above. Let us assume that the computer we are using employs 32-bit words. Then we would simulate 32 different systems simultaneously by storing bit i of each of the 32 systems as one of the bits in the ith word of a lattice made up of such words, and applying Equation (15.1) directly to these words. Each step of such a Monte Carlo algorithm takes only slightly longer than an equivalent step of the normal Metropolis algorithm, but generates 32 times as many measurements of the magnetization or energy, or whatever quantity it is that we are interested in. In our own experiments with the one-dimensional Ising model we have measured an effective increase in the speed of the simulation by a factor of 28 on a 32-bit computer. Using a 64-bit computer we have measured an increase of a factor of 56.

One point which we haven't dealt with is that in order to use the algorithm described, we need to be able to generate a random word of 32 or 64 independent bits in which each bit is 1 with a given probability $\exp(-4\beta J)$. We discuss how such words can be generated in Section 16.3, after we have described how normal random numbers are generated.

15.1.2   The two-dimensional Ising model

In this section we tackle a more realistic problem, the construction of a multispin version of the Metropolis algorithm for the two-dimensional Ising model on a square lattice. As well as being more realistic, it is also more difficult.

The Metropolis algorithm for the Ising model on a square lattice can be described as follows. We choose a spin from the lattice and if two or more of its nearest-neighbour spins are pointing in the opposite direction to it, we definitely flip it over. If only one neighbouring spin is pointing in the opposite direction, we flip our spin with probability $\exp(-4\beta J)$, where J is the coupling between spins. If no nearest-neighbour spins at all are pointing in the opposite direction to the spin we chose, then we flip it with probability $\exp(-8\beta J)$. The master expression for a spin with lattice coordinates i and j (the equivalent of Equation (15.1) in the one-dimensional case) is thus

$$\sigma'_{ij} = \sigma_{ij} \oplus [R_{\geq 2} \vee (R_1 \wedge r_1) \vee (R_0 \wedge r_0)]. \tag{15.2}$$

Here R_0 and R_1 are logical expressions which are 1 if zero or one of the spin's nearest neighbours are anti-aligned with it respectively, and $R_{\geq 2}$ is 1 if two or more are anti-aligned. The variables r_0 and r_1 are random bits which are 1 with probability $\exp(-8\beta J)$ and $\exp(-4\beta J)$ respectively.

This equation will work fine, but it is slightly unsatisfactory, mainly because the expression for R_1 turns out to be quite complicated. An alternative and equivalent expression is

$$\sigma'_{ij} = \sigma_{ij} \oplus [R_{\geq 2} \vee (R_{\geq 1} \wedge r'_1) \vee (R_{\geq 0} \wedge r'_0)]. \tag{15.3}$$

Here $R_{\geq 0}$ is 1 if zero or more neighbours are anti-aligned and $R_{\geq 1}$ is 1 if one or more are anti-aligned. In order to make this expression work we must redefine our random bit variables. In fact, r_0' is just the same as r_0; it is a random bit which is 1 with probability $\exp(-8\beta J)$. However, r_1' is not the same as r_1. Notice that if $R_{\geq 1} = 1$, then necessarily $R_{\geq 0} = 1$ also, since if there are one or more anti-aligned spins, then there must logically also be zero or more anti-aligned spins. In this case the probability of the expression inside the square brackets being 1 is the probability that either one or both of r_0' and r_1' is 1, and we would like this probability to be equal to $\exp(-4\beta J)$. If we write the individual probabilities of r_0' and r_1' being 1 as p_0 and p_1, then the probability of either of them being 1 is[1]

$$p = p_0 + p_1 - p_0 p_1. \tag{15.4}$$

Setting this equal to $\exp(-4\beta J)$, putting $p_0 = \exp(-8\beta J)$ and rearranging, we find that r_1' should be 1 with probability

$$p_1 = \frac{e^{-4\beta J} - e^{-8\beta J}}{1 - e^{-8\beta J}}. \tag{15.5}$$

Equation (15.3) is easier to work with than Equation (15.2) because $R_{\geq 0}$ and $R_{\geq 1}$ are relatively simple to evaluate. In fact, since we always have zero or more anti-aligned neighbours of any spin we pick we can immediately see that $R_{\geq 0} = 1$, so we can write the master expression as

$$\sigma_{ij}' = \sigma_{ij} \oplus [R_{\geq 2} \vee (R_{\geq 1} \wedge r_1') \vee r_0']. \tag{15.6}$$

$R_{\geq 1}$ is a little more complicated. We first define four quantities thus:

$$a_1 = \sigma_{ij} \oplus \sigma_{i+1,j}, \qquad a_2 = \sigma_{ij} \oplus \sigma_{i-1,j},$$
$$a_3 = \sigma_{ij} \oplus \sigma_{i,j+1}, \qquad a_4 = \sigma_{ij} \oplus \sigma_{i,j-1}, \tag{15.7}$$

which are 1 if and only if the corresponding pair of spins are pointing in opposite directions. In terms of these variables we can then write $R_{\geq 1}$ as

$$R_{\geq 1} = a_1 \vee a_2 \vee a_3 \vee a_4. \tag{15.8}$$

$R_{\geq 2}$ can also be expressed in terms of the a variables as

$$R_{\geq 2} = (a_1 \wedge a_2) \vee (a_1 \wedge a_3) \vee (a_1 \wedge a_4)$$
$$\vee (a_2 \wedge a_3) \vee (a_2 \wedge a_4) \vee (a_3 \wedge a_4). \tag{15.9}$$

With a little juggling, this can be simplified to[2]

$$R_{\geq 2} = [(a_1 \vee a_2) \wedge (a_3 \vee a_4)] \vee [(a_1 \wedge a_2) \vee (a_3 \wedge a_4)]. \tag{15.10}$$

---

[1] To prove this we just note that the probability that neither of them is 1 is given by $1 - p = (1 - p_0)(1 - p_1)$.

[2] In order to demonstrate this, recall that the logical AND and OR operators satisfy the same rules of distribution and association as multiplication and addition in ordinary algebra.

Combining all of these expressions and feeding them into Equation (15.6) we now have our complete algorithm. As in the one-dimensional case, the simplest way to proceed is to simulate many different systems separately with the corresponding spins of all systems stored in the bits of one word on the computer. An example program to perform a simulation using this algorithm is given in Appendix B.

Since the logical expressions we need to evaluate at each Monte Carlo step are relatively complex, the gain in efficiency we get from the multispin coding is not as great in the two-dimensional case as it was in the simpler one-dimensional one. Running on a computer with a 32-bit CPU we have measured an effective increase in the speed of the simulation by a factor of about 21. On a 64-bit computer we measure a factor of 38.

15.2 Implementing multispin-coded algorithms

Once we have derived a master expression for our multispin Monte Carlo algorithm, such as Equation (15.1) or Equation (15.6), implementing it in a computer program is a very straightforward matter of translating the expression into the computer language of our choice. The only catch is that not all computer languages provide the necessary bit manipulation operators for performing this translation. In fact, at the time of writing, the only mainstream scientific computer languages which provide the necessary operators are C and its extensions, such as C++ and Objective C. In particular, neither FORTRAN nor Pascal provide suitable bit manipulation operators. If you wish to program in either of these languages, you may be able to find subroutine libraries which add the necessary operators to the language, or you may be able to write such libraries yourself using another language such as C or assembly language. You may even be able to write the necessary parts of the multispin algorithm itself in assembly language. As things stand at the moment, however, the best solution is to write the program in C or one of its derivatives. In Table 15.1 we list the bitwise operators commonly used in multispin-coded algorithms, along with their representations in C. Note that in addition to the basic AND, OR, XOR and NOT operators, we have also included left and right shifts in the table. We will use these operators in one of the algorithms described later in this chapter.

15.3 Truth tables and Karnaugh maps

In the Ising model examples of Sections 15.1.1 and 15.1.2 we derived the logical expressions which went into the master expressions intuitively. These examples were simple enough that we could see quite easily what form the

operation	symbol	in C
AND	$\wedge$	&
OR	$\vee$	|
XOR	$\oplus$	^
NOT	$-$	~
shift left	$\ll$	<<
shift right	$\gg$	>>

TABLE 15.1   Bit manipulation operators used in this chapter, and their representation in the computer language C.

master expression should take and how to construct it out of the available variables such as spins or random bits. For more complicated models however, it can often be quite difficult to just write down the equations in this way, and so a number of analytic tools have been developed to help us. The most important of these are truth tables and Karnaugh maps.

A **truth table** is a list of the values of one or more Boolean or single-bit variables which are functions of other such variables. If we wish to find a compact logical expression for a certain quantity, such as the quantity $R_{\geq 2}$ defined in Equation (15.2), then a good first step is to write down in a truth table the values we would like it to take for all possible values of the independent variables. In Table 15.2 we give an example of such a truth table for a fictitious quantity A which is a function of four other quantities $B_1 \ldots B_4$. As you will see, the left part of the table lists every possible combination of values of the B variables, and the right part lists the values, 0 or 1, which we would like A to take for each combination of inputs. If there are values of the Bs for which we don't care about the value of A then we mark these with an X. Often this occurs because we know that for one reason or another that combination of Bs will never arise in our algorithm.

Once we have a truth table for the quantity we want, we can immediately write down a logical expression which will calculate that quantity for us. To do this we use the **disjunctive normal form** of the expression. Let us illustrate what this means with the example of the quantity A above. There are nine 1s in the rightmost column of Table 15.2. For each of these we can write an expression involving the independent variables $B_1 \ldots B_4$ which is 1 only for the values in that row, and 0 for every other row. For example, the first 1 in the truth table occurs in row two. The expression $\overline{B}_1 \wedge \overline{B}_2 \wedge \overline{B}_3 \wedge B_4$ is 1 only for the values of the Bs in this row and not for any other. The corresponding expression for the fourth row is $\overline{B}_1 \wedge \overline{B}_2 \wedge B_3 \wedge B_4$, and so on. Combining nine such expressions, one for each of the nine 1s, we derive the

B_1	B_2	B_3	B_4	A
0	0	0	0	X
0	0	0	1	1
0	0	1	0	0
0	0	1	1	1
0	1	0	0	1
0	1	0	1	1
0	1	1	0	0
0	1	1	1	1
1	0	0	0	1
1	0	0	1	1
1	0	1	0	0
1	0	1	1	X
1	1	0	0	1
1	1	0	1	1
1	1	1	0	0
1	1	1	1	0

TABLE 15.2   An example of a truth table.

expression

$$\begin{aligned}
A = \ & [\overline{B}_1 \wedge \overline{B}_2 \wedge \overline{B}_3 \wedge B_4] \vee [\overline{B}_1 \wedge \overline{B}_2 \wedge B_3 \wedge B_4] \\
& \vee [\overline{B}_1 \wedge B_2 \wedge \overline{B}_3 \wedge \overline{B}_4] \vee [\overline{B}_1 \wedge B_2 \wedge \overline{B}_3 \wedge B_4] \\
& \vee [\overline{B}_1 \wedge B_2 \wedge B_3 \wedge B_4] \vee [B_1 \wedge \overline{B}_2 \wedge \overline{B}_3 \wedge \overline{B}_4] \\
& \vee [B_1 \wedge \overline{B}_2 \wedge \overline{B}_3 \wedge B_4] \vee [B_1 \wedge B_2 \wedge \overline{B}_3 \wedge \overline{B}_4] \\
& \vee [B_1 \wedge B_2 \wedge \overline{B}_3 \wedge B_4].
\end{aligned} \tag{15.11}$$

The trouble with this approach, as you will no doubt agree, is that Equation (15.11) is rather complicated. This makes it slow to evaluate on our computer and prone to the introduction of errors. To get around these problems we would like to simplify the equation to give a more compact expression for A which we can evaluate more quickly. One way to do this is just to make use of the rules of Boolean algebra to find a good simplification. This however is a somewhat hit-or-miss operation, and it too is prone to error. A quicker and more reliable tool for simplifying logical expressions is the **Karnaugh map**.

Figure 15.1 depicts the Karnaugh map corresponding to Table 15.2. Each of the sixteen squares in this map corresponds to one of the sixteen possible sets of values of the four independent variables $B_1 \ldots B_4$ in our truth table; the four black bars along the edges of the map denote the rows and columns

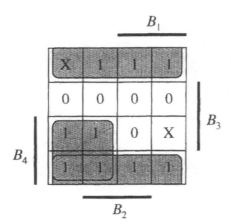

FIGURE 15.1   The Karnaugh map corresponding to Table 15.2. Using this map, we find that
$A = \overline{B}_3 \vee (B_4 \wedge \overline{B}_1)$.

in which the corresponding independent variable is equal to 1. The reader might like to verify that this arrangement does indeed give exactly one square for each possible set of values. It also has the special property that when we cross any line between two of the squares in the map only one of the B variables changes value. This is also true if we walk off one side of the map and come back onto it at the opposite side: Karnaugh maps have periodic boundary conditions. This property of only changing one bit at a time is the key to using the map to find a simple expression for A. It means that all rectangular blocks on the map which have sides whose lengths are powers of two correspond to simple equations in terms of the four B variables. This includes blocks which wrap around the boundary conditions. Some examples are: all 2×4 blocks on the map correspond to either $B_i = 1$ or $\overline{B}_i = 1$ for some i; all 1×4 and 2×2 blocks correspond to expressions of the form $B_i \wedge B_j = 1$, or similar expressions with $\overline{B}_i$ or $\overline{B}_j$; and all 1×2 blocks correspond to $B_i \wedge B_j \wedge B_k = 1$ or similar. Here is how we make use of these observations to find a simple expression for A.

First, we fill in all the squares in our Karnaugh map with the corresponding values of A taken from the truth table. Just as in the truth table, we place an X in any squares for which we don't care about the value of A. Next, we attempt to group all the 1s on the map into as few blocks as possible, where each block is a rectangle with sides of length a power of two, as discussed above. We also try to make the blocks as large as possible, since larger blocks correspond to simpler expressions. In choosing our blocks we can if we wish cover the same squares with more than one block. We can also include in a block any of the squares with an X in, if that will make things easier, although we can also leave them out if we want to. The only rule is that all the 1s should end up in a block, and all the 0s should be outside the blocks.

Finally, we write down the logical expressions corresponding to each of our blocks, and take the logical OR of all of them. In Figure 15.1 we show how our quantity A can be represented using just two blocks on the Karnaugh map, one 2×2 block, which corresponds to the expression $B_4 \wedge \overline{B}_1$, and one 2×4 block (which extends off the top of the map and comes back on at the bottom), corresponding to the expression $\overline{B}_3$. Our final expression for A is therefore

$$A = \overline{B}_3 \vee (B_4 \wedge \overline{B}_1). \tag{15.12}$$

As you can demonstrate for yourself by referring to the original truth table, this expression is indeed correct, and it is clearly much more satisfactory than Equation (15.11).

Karnaugh maps can be used for problems with other numbers of independent variables as well. For a function of three variables we would use a 2×4 map. For a function of five variables we need to use a three-dimensional map with two layers each comprised of a 4×4 map of the kind we have used here. A function of six variables requires a $4 \times 4 \times 4$ map and larger numbers of variables than this require maps with four or more dimensions. High-dimensional maps are quite hard for humans to visualize, which makes finding blocks by hand a difficult task when the number of variables is large. Computers, however, have no such problems, and algorithms have been developed to allow a computer to find the best choice of blocks on a high-dimensional Karnaugh map. An example is the Quine–McCluskey algorithm, which is described by Booth (1971).

15.4 A multispin-coded algorithm for the repton model

In Chapter 12 we looked at the repton model of gel electrophoresis as an example of an application of the Monte Carlo method to a problem in biophysics. This model is a good candidate for a multispin-coded algorithm of the type we have been considering in this chapter; the states at the sites of the lattice are of only a small number of types (three in fact) and the rules for changing the state at a particular site can be expressed as quite a simple set of logical conditions. In the rest of this chapter we show in detail how to create a multispin algorithm for this model. As we will see, the resulting algorithm is quite a bit more complicated than the algorithms for the Ising model which we looked at in Sections 15.1.1 and 15.1.2.

To create a multispin-coded algorithm for the repton model we go through exactly the same steps as we did for the Ising model. We first choose a representation of the states of the system in terms of bits stored in words, and then we try to devise an algorithm which uses only bitwise operators like AND and OR acting on these words to perform a Monte Carlo step.

u	a_1	b_1	a'_1	b'_1
0	0	0	X	X
0	0	1	1	1
0	1	0	1	0
0	1	1	1	0
1	0	0	X	X
1	0	1	0	1
1	1	0	1	1
1	1	1	0	1

TABLE 15.3   Truth table for moves of the leftmost repton in the chain.

As suggested in Section 12.3.3, we can represent the state of a repton chain by storing the coordinate x_1 of the first repton on the chain, plus the displacement $s_i = x_{i+1} - x_i$ of each subsequent repton relative to its neighbour. The dynamics of the repton model limits the variables s_i to just three values: $s_i = -1$, 0 or $+1$. To represent these three values in our multispin algorithm we need two bits, which we will call a_i and b_i. As it turns out, a convenient representation is to use $(a_i, b_i) = (0,1)$, $(1,1)$ and $(1,0)$ for $s_i = -1$, 0 and 1 respectively.[3]

A Monte Carlo step for the repton model consists of choosing a repton at random from the chain and a random direction in which to move it. If the chosen repton is allowed by the dynamics of the model to move in the chosen direction then we make the move, otherwise we reject it. For our multispin algorithm we need to examine three different cases: moves of the reptons on the left and right ends of the chain, and moves of reptons in the interior of the chain. Let us start with the reptons on the ends of the chain, since these are simpler than the interior ones.

Suppose then that the repton we choose to move turns out to be repton number 1, the one on the leftmost end of the chain. To decide in which direction we are going to move it, we generate a random bit u. We will say (quite arbitrarily) that $u = 1$ represents an upward move (one in the direction of increasing x), and $u = 0$ represents a downward move. Whether this move is allowed or not depends only on u and on the relative position of the first and second reptons in the chain, which is stored in the bit variables a_1 and b_1. If the move is allowed, then we need to perform it, and update

---

[3]There are a number of other possible choices, of course. The particular choice we have made here turns out to give relatively simple logical expressions for the final algorithm. There is however no way to know this beforehand. The process of developing a multi-spin algorithm is often one of trial and error; you just have to experiment with different representations of your system until you find one which gives simple Karnaugh maps.

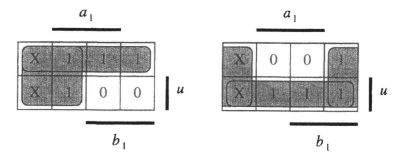

FIGURE 15.2 Karnaugh maps for moves of the leftmost repton in the chain. Using these maps we find: $a_1' = \bar{u} \vee \bar{b}_1$ (left map), and $b_1' = u \vee \bar{a}_1$ (right map).

a_1 and b_1 accordingly. If it is not, a_1 and b_1 will keep their current values. In Table 15.3 we list all the possible combinations of values for u, a_1 and b_1, and the correct new values a_1' and b_1' of the latter two which should result. In Figure 15.2 we show these values as two Karnaugh maps, with our choice of blocks shaded in. The master expressions for a_1' and b_1' which we derive from these two maps are

$$a_1' = \bar{u} \vee \bar{b}_1, \tag{15.13}$$
$$b_1' = u \vee \bar{a}_1. \tag{15.14}$$

For the rightmost repton on the chain the calculation follows similar lines. In this case acceptance of the move depends on the direction u in which we propose to move the repton and the values of the bits a_{N-1} and b_{N-1} which represent the relative position of the last two reptons. The truth table and Karnaugh maps are very similar to the ones for the leftmost repton and the end result is that

$$a_{N-1}' = u \vee \bar{b}_{N-1}, \tag{15.15}$$
$$b_{N-1}' = \bar{u} \vee \bar{a}_{N-1}. \tag{15.16}$$

Readers may wish to check this result themselves. It doesn't take long.

The rules for updating the rest of the reptons in the chain—all the ones in the middle—are a little more complex, but deriving them involves exactly the same steps. Suppose we choose a repton somewhere in the middle of the chain as the one we are going to move. Let us call this repton i, where $1 < i < N$. As before, we also generate a random bit u to represent the direction in which we propose to move the repton. To decide whether or not the proposed move is an allowed one we need to take into account the value of u and the relative position of the two reptons to either side of repton i, which are represented by the bits a_{i-1}, b_{i-1}, a_i and b_i. In Table 15.4 we

u	a_{i-1}	b_{i-1}	a_i	b_i	a'_{i-1}	b'_{i-1}	a'_i	b'_i
0	0	0	0	0	X	X	X	X
0	0	0	0	1	X	X	X	X
0	0	0	1	0	X	X	X	X
0	0	0	1	1	X	X	X	X
0	0	1	0	0	X	X	X	X
0	0	1	0	1	0	1	0	1
0	0	1	1	0	0	1	1	0
0	0	1	1	1	0	1	1	1
0	1	0	0	0	X	X	X	X
0	1	0	0	1	1	0	0	1
0	1	0	1	0	1	0	1	0
0	1	0	1	1	1	1	1	0
0	1	1	0	0	X	X	X	X
0	1	1	0	1	0	1	1	1
0	1	1	1	0	1	1	1	0
0	1	1	1	1	1	1	1	1
1	0	0	0	0	X	X	X	X
1	0	0	0	1	X	X	X	X
1	0	0	1	0	X	X	X	X
1	0	0	1	1	X	X	X	X
1	0	1	0	0	X	X	X	X
1	0	1	0	1	0	1	0	1
1	0	1	1	0	0	1	1	0
1	0	1	1	1	1	1	0	1
1	1	0	0	0	X	X	X	X
1	1	0	0	1	1	0	0	1
1	1	0	1	0	1	0	1	0
1	1	0	1	1	1	0	1	1
1	1	1	0	0	X	X	X	X
1	1	1	0	1	1	1	0	1
1	1	1	1	0	1	0	1	1
1	1	1	1	1	1	1	1	1

TABLE 15.4   Truth table for moves of a repton i in the middle of the chain.

show what the correct values a'_{i-1}, b'_{i-1}, a'_i and b'_i of these four variables should be after the move has been made or rejected, and in Figure 15.3 we show the corresponding Karnaugh maps. The resulting expressions are:

$$a'_{i-1} = (a_i \wedge a_{i-1}) \vee \overline{b}_{i-1} \vee (a_{i-1} \wedge u) \vee (a_i \wedge b_i \wedge u) \qquad (15.17)$$

$$b'_{i-1} = (b_i \wedge b_{i-1}) \vee \overline{a}_{i-1} \vee (b_{i-1} \wedge \overline{u}) \vee (a_i \wedge b_i \wedge \overline{u}) \qquad (15.18)$$

$$a'_i = (a_i \wedge a_{i-1}) \vee \overline{b}_i \vee (a_i \wedge \overline{u}) \vee (a_{i-1} \wedge b_{i-1} \wedge u) \qquad (15.19)$$

$$b'_i = (b_i \wedge b_{i-1}) \vee \overline{a}_i \vee (b_{i-1} \wedge u) \vee (a_{i-1} \wedge b_{i-1} \wedge u). \qquad (15.20)$$

Using Equations (15.13–15.20), we can create an asynchronous update multispin Monte Carlo algorithm for the repton model as follows. For each site in a chain of N reptons we create two words of n bits, where n will normally be either 32 or 64, depending on the architecture of the computer used. One of these two words holds the a_i bits for the corresponding position on the chain for each of n different systems. The other holds the b_i bits. We also need to create an array of n integers to hold the positions x_1 of the leftmost repton in each system. It doesn't matter greatly what values we give our variables initially, but a sensible choice would be to set $x_1 = 0$ for each system and set all of the bits a_i and b_i to 1, which represents a perfectly flat repton chain.

One step of our algorithm then goes like this. First we pick a repton i at random from the N possibilities, and we generate a word of n random bits, which tells us whether the corresponding reptons are to be moved up or down. Now if $i = 1$ or $i = N$, then we are either at the left end or the right end of the chain, and we should apply Equations (15.13) and (15.14), or Equations (15.15) and (15.16). Otherwise, we are in the middle of the chain and we should apply Equations (15.17–15.20). The only additional step is that every time we move the repton at the leftmost end of the chain we must also update our record x_1 of the absolute position of this repton. This is the only part of the simulation which we cannot perform in a multispin-coded fashion. However, we only have to do it once every N Monte Carlo steps on average, so if N is reasonably large the time wasted in this way is a small fraction of the total time taken by the simulation.

If we want to perform a simulation of the repton model in zero electric field, then we should generate our random bits u with equal probability to be 0 or 1. If we want a non-zero electric field E then, as described in Section 12.3.1, we should generate upward moves (1s) with probability $\alpha = [1 + \exp E]^{-1}$, and downward moves (0s) with probability $1 - \alpha$. The generation of random words with the bits biased in this fashion is discussed in Section 16.3.

Our multispin-coded Monte Carlo algorithm for the repton model allows us to simulate n different systems simultaneously, without expending much more CPU power than the straightforward algorithms described in Chap-

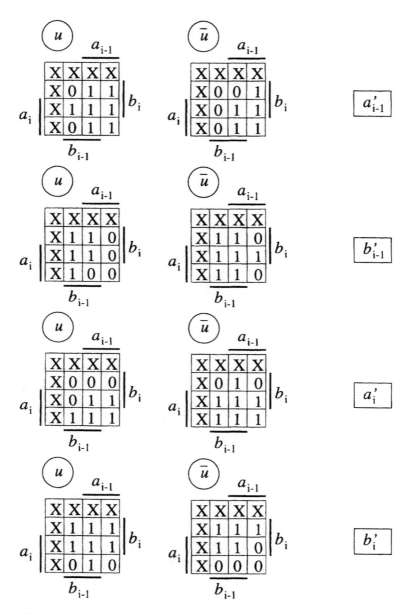

FIGURE 15.3 The four Karnaugh maps for moves of reptons in the middle of the chain. Note that each map is a three-dimensional $2 \times 4 \times 4$ one. We have not shaded the blocks of 1s in this case, since this is rather hard to do clearly in three dimensions.

ter 12. Certainly there is some overhead arising from the complexity of expressions like Equations (15.17–15.20). But in practice we find that the algorithm performs very well. For large chain lengths ($N > 100$ or so), we have compared our multispin algorithm with the standard algorithm of Section 12.3 and find that the multispin algorithm outperforms the standard one by a factor of about 27 on a 32-bit computer, and by a factor of about 48 on a 64-bit one.

15.5 Synchronous update algorithms

Almost all of this chapter has been devoted to asynchronous update multi-spin algorithms—ones which use the bits of a word to represent parts of separate systems. This approach is particularly simple because these separate systems do not interact with one another, so our multispin algorithm operates on each bit in a word separately. Occasionally however, simulating many systems in parallel is not as good as simulating one large system. In these cases, we may wish to use a synchronous multispin algorithm in which the different bits within a word represent parts of the same system, so that many parts of the one system get updated at each Monte Carlo step. In this section we describe briefly one technique which can be used to create such algorithms.

The simplest way of creating a synchronous algorithm is to start with an asynchronous one which simulates n different systems in parallel, and then "glue" these n systems together in some way to make a system n times larger. Consider, for example, the case of the one-dimensional Ising model. Normally we apply periodic boundary conditions to this model so that, in a system of size L, site L is a neighbour of site 1. If, however, we have n such systems and we can arrange for site L in system k to be the neighbour of site 1 in system $k + 1$ (modulo n) then we have achieved our goal. This situation is illustrated in Figure 15.4.

An elegant way to do this in practice is to make a copy of the word containing the Lth spin of each system and place it at the other end of the system, as an extra spin 0 (see the figure). In addition, in order that the n systems are joined together correctly, we need to rotate this word round by one bit[4] using the shift operator $\gg$. We also need to make a copy of the word which contains the first spin of each system, rotate that one bit in the opposite direction and add it as an extra spin $L + 1$ to the other end of the system. Note that we have to update these copies every time we perform a Monte Carlo step on spins 1 or L.

---

[4]Depending on the computer language used, shift operators may or may not perform a real rotation. Some such operators just discard the last bit in a word and introduce a zero in its place at the other end. This is the case in C for example. In this case, you

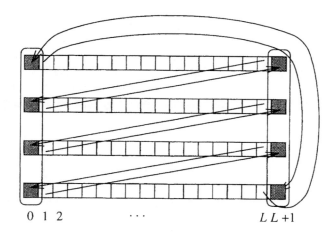

0 1 2   · · ·   *L L* +1

FIGURE 15.4   A way of "gluing" four independent simulations together to make one simulation of four times the size.

Problems

15.1 Consider the "asymmetric exclusion process", which is defined as follows. We have m particles located on a one-dimensional lattice of N sites
with periodic boundary conditions. No two particles can occupy the same
site. Particles attempt to hop to the site immediately to their left with average rate p per unit time, and to the site immediately to their right with
average rate $q \leq p$. Hops are only allowed if the target site is unoccupied.
An ordinary Monte Carlo algorithm to simulate this model might work as
follows. We select at random a pair of adjacent sites i and $j = i + 1$ (modulo N) and attempt to exchange their values. If site i is empty and site j
is occupied, we accept the move with probability 1. If i is occupied and j
is empty we accept it with probability q/p. Otherwise we reject the move.
(a) Suppose we represent the state of the lattice by variables s_i which are 1
for occupied sites and 0 for unoccupied ones. Write down the truth tables
for the variables s_i and s_j giving their values after their contents have been
exchanged. Use these to derive the master expression for the case $p = q$.
(b) If r is a random variable which is 1 with probability q/p and 0 otherwise, write down the master expression for the simulation in the general case
where $q \leq p$.

15.2 Consider de Gennes' model for polymer diffusion in two dimensions as
described in Problem 12.2. Suppose we represent the ith link in the chain
by two bits a_i and b_i with $(a_i, b_i) = (0,0)$, $(1,1)$, $(0,1)$ and $(1,0)$ signifying

---

must move the last bit in the word explicitly yourself.

up, down, left and right respectively. (a) Write down the truth table for the variable h that equals 1 if and only if links i and $j = i + 1$ are anti-parallel. Hence obtain a logical expression for h. There is another simpler way of expressing h which involves the XOR operation. Can you find it? (b) Using the previous result, write down a master expression for moves in the interior of the chain. (Hint: you will need to use some random bits for this.)

16

Random numbers

The fundamental constant which links all Monte Carlo methods together, the common factor which makes them all Monte Carlo methods, is that they contain an element of randomness. Things happen in a probabilistic fashion in a Monte Carlo calculation, and it is only the average over many unpredictable events that gives us an (approximate) answer to the question we are interested in. To generate such unpredictable events, all Monte Carlo methods require a source of random numbers. There are many ways of generating random numbers. In this chapter we examine some of the most common of them.

16.1 Generating uniformly distributed random numbers

The basic problem in random number generation is to generate a sequence of random real numbers r uniformly distributed in the range $0 \leq r < 1$. If we can do this, then techniques exist for transforming these numbers into random numbers with any other distribution we might be interested in. For example, a random real number R uniformly distributed in some other range $R_{\min} \leq R < R_{\max}$ can be derived from r thus:

$$R = R_{\min} + (R_{\max} - R_{\min})\,r. \tag{16.1}$$

A random integer i in the range $i_{\min} \leq i < i_{\max}$ can be generated by operating on this expression with the "floor" function:[1]

$$i = \lfloor i_{\min} + (i_{\max} - i_{\min})\,r \rfloor. \tag{16.2}$$

---

[1] The floor function $\lfloor x \rfloor$ is defined as the largest integer not greater than x. Basically, it's just what you get when you chop the digits after the decimal point off a number. One must be careful when dealing with negative numbers however. The floor function

Another situation which arises frequently in Monte Carlo methods is when we want to perform some action (such as the acceptance of a move) with a certain probability p. In that case, we generate a random real number r as above, and then we perform the action if $r < p$, and not otherwise.

Sometimes we also need to generate random numbers from some more complicated distribution, such as a Gaussian distribution. There are a number of methods which allow us to do this given a source of random reals in the interval zero to one, as above. We will deal with these methods in detail in Section 16.2. For the moment however, we address the issue of generating uniformly distributed random real numbers.

Almost all the random numbers used in Monte Carlo calculations are in fact only "pseudo-random" numbers, which means that they are generated by a computer program using a deterministic algorithm. The algorithm is chosen so that the sequence of numbers generated covers the desired interval (in this case the interval between zero and one) uniformly, and the numbers will certainly have no pattern discernible to the casual observer. As we will see in the next few sections, these pseudo-random number generators can be quite sophisticated and for the most part are adequate for our simulations. However, the fact still remains that the numbers are generated by a deterministic process, and therefore their sequence is predictable if one knows the algorithm used. Thus, there are, in a sense, correlations between the members of the sequence, and these correlations could possibly have an adverse effect on our Monte Carlo calculations. In a 1992 paper, Ferrenberg, Landau and Wong employed a handful of different, widely available pseudo-random number generators as the source of random numbers for Monte Carlo simulations of the two-dimensional Ising model using the Wolff cluster algorithm described in Chapter 4. Since the properties of the 2D Ising model have been solved exactly, they were able to compare their results with the known correct answers, and found that, while some of their random number generators produced answers in good agreement with the exact results, others produced answers as much as a hundred standard deviations away. Clearly such an error cannot be ascribed merely to chance, and the authors concluded that subtle correlations in the sequence of pseudo-random numbers generated were responsible for the errors. The majority of the techniques we describe in this section will be aimed precisely at reducing correlations of this kind in pseudo-random number generators. However, before we tackle that subject, let us address briefly the possibility of generating *true* random numbers.

---

always rounds downwards, which means that negative real numbers get rounded away from zero. In C the floor function is implemented using the integer cast (int), or the function floor(). In FORTRAN it is implemented using the INT function.

16.1.1   True random numbers

For a few special purposes, pseudo-random numbers are not sufficient; any
random number generator whose sequence is even in theory predictable is
problematic. A classic example is cryptography. Military and other organizations require sources of true random numbers for use as "one-time pads"
for encoded communications. To meet this need, companies exist which
market CD–ROMs or DVDs which contain nothing more than hundreds of
millions of random numbers, one after another. These numbers are usually
generated from measurements of the products of radioactive decay.[2] Although the overall rate of decay of a radioactive isotope can be measured
very accurately, the precise moment at which any unstable nucleus will decay
is unknown, and indeed unknowable, depending as it does on the probabilistic nature of quantum mechanics. All we know is that the probability per
unit time of such a decay is a constant. This fact can be used to generate
truly random numbers from the timing of the decay of individual atoms in
a sample, and these numbers can be stored by a computer and eventually
recorded onto CD–ROM. More recently, it has become possible to purchase
a circuit board which you can plug into your computer which contains a
(low-grade) radioactive sample and a detector, along with some associated
processing machinery, which can generate for you a sequence of as many true
random numbers as you like.

Important though these techniques may be for cryptographic purposes
however, none of the Monte Carlo methods we have discussed in this book
requires random numbers of this quality, and the expense of buying CD–
ROMs full of them, or specially built random number generator boards, is
not justified. In fact, even in cryptographic circles such random number
sources are now rare as a result of new encryption schemes whose security is
based on the problem of factoring the products of very large prime numbers.
Indeed, in a rather delightful twist, Press *et al.* (1988) have pointed out that
these same cryptographic schemes can be turned on their head and used
as a source of very high-quality pseudo-random numbers. In their excellent
book *Numerical Recipes*, they describe a method whereby the common code
known as the Data Encryption Standard (DES) can be employed as a random
number source. Unfortunately, the DES is a rather complex algorithm and
a large amount of CPU time would be needed to generate each random
number, which makes the idea impractical for our applications.

---

[2] An alternative technique was, and still is, used by the now ancient British computer
ERNIE, which picks the numbers of winning premium bonds (essentially a form of lottery)
using the pattern of gas bubbles passing through a liquid.

16.1.2   Pseudo-random numbers

In this and the following few sections we look at computer algorithms for generating pseudo-random numbers. Most of the common techniques for doing this work with integers rather than real numbers. Typically they generate an integer i between zero and some maximum value $m - 1$. In order to turn these integers into random real numbers between zero and one, we simply divide by m.

The simplest and most widely used method of generating a sequence of pseudo-random integers i_n is to act with some function f on the previous value in the sequence:

$$i_n = f(i_{n-1}). \tag{16.3}$$

The function f should incorporate only integer arithmetic, such as addition or multiplication or integer division, in order that it always produce integers. In order to get the process started we also need to provide the first random number i_0 for the sequence. This first number is called the **seed** for the random number generator. It is in fact extremely useful that the generator requires a seed. It means that we can produce different sequences of pseudo-random numbers by feeding different seeds into Equation (16.3), or we can generate the same sequence more than once by reusing a single seed value. This second possibility may sound like an odd thing to do, but it can come in very handy for debugging Monte Carlo algorithms. If the algorithm does something strange on a particular run of the program, it may not repeat this strange behaviour on another run with a different set of random numbers. However, we can always duplicate the strange behaviour by using the same seed to reproduce the same set of random numbers, and we can do this as many times as we like until we track down the source of the problem.

Even before we start considering particular forms for the function f in Equation (16.3), one thing which is clear is that random number generators of this form will always produce a cycle of integers. If at any point during the sequence we produce a number i_n which is the same as one we have produced before i_{n-k}, then Equation (16.3) tells us that the sequence will obey $i_n = i_{n-k}$ for all subsequent values of n and hence the generator will fall into a cycle of length k. Since there are only m different values which i_n can take, the longest we will have to wait before this happens is m steps of the sequence, and this is therefore the longest possible length for the cycle. A large part of the task of designing a good random number generator is to find functions f which have a long cycle.

More sophisticated generators make use of two or more of the previous values in the sequence:

$$i_n = f(i_{n-1}, i_{n-2} \dots). \tag{16.4}$$

Generators of this form are also ultimately guaranteed to fall into a repeating cycle of values, for essentially the same reasons as before, although for a generator which uses k previous values of the sequence to generate the next, the cycle can be up to m^k steps long, which could be a very large number indeed for quite modest values of k.

It is not our intention here to go into the mathematics of random number generators in detail, since this is a book about Monte Carlo methods and not about random numbers. In this chapter we merely describe some of the most widely used and best-behaved functions for generating pseudo-random sequences, without attempting to prove their properties with any rigour. The information we give should be enough for you to choose a suitable random number generator for your Monte Carlo simulations and implement it on a computer in the language of your choice. For the reader who is interested in investigating the subject more thoroughly, we recommend the discussion by Knuth in Volume 2 of his classic work *The Art of Computer Programming* (1981).

16.1.3   Linear congruential generators

By far the most widely used type of random number generator is the **linear congruential generator**, first introduced by Lehmer (1951), which is a generator of exactly the type described by Equation (16.3). The function used to produce the sequence is

$$i_n = (ai_{n-1} + c) \bmod m, \tag{16.5}$$

where $p \bmod q$ represents the modulo operation, which returns the remainder after p is divided by q. (In C it is written as p%q, in FORTRAN as MOD(P,Q).) The highest random number that this equation can produce is $m - 1$, and so the longest sequence of numbers that can be produced has m elements, all of them different. Thus in order to make the sequence long we need to make m large. The maximum possible value of m is equal to the largest value that your computer can store in an integer variable, plus one. Let us call this number w. On modern computers integers are normally stored in 32 bits, which means that $w = 2^{32}$. Choosing $m = w$ is very convenient, because we can then perform arithmetic modulo w simply by adding and multiplying integers in variables of one word each and ignoring any overflow that arises during the operations. Unfortunately, however, it turns out that the random sequences produced by Equation (16.5) when $m = w$ are rather poor in one respect: the low order bits in the binary representation of the numbers will not be very random. An extreme example is the lowest order bit, which for $m = w$ will, depending on the values of a and c, either be a constant or will alternate between 1 and 0 for successive numbers i_n in the series. For applications which make use of the values of the bits in our random numbers (such as the multispin coding methods of Chapter 15) this could be disastrous. Thus it is advisable to make some choice other than $m = w$. However, this in turn leads us to another problem. If we make some other choice of m then we have to perform the modulo operation explicitly ourselves. But in order to do this correctly, we must be certain not to let the quantity $ai_{n-1} + c$ in Equation (16.5) overflow our integer variables. Since all the numbers i_n produced by our generator are less than m, we will be safe if we ensure that $a(m-1) + c < w$, but this places limits on the size of m, making it considerably smaller than w for all sensible choices of a and c.

There are other considerations in the choice of values for the constants. The length of the sequence generated is not always equal to m; the value of m merely provides an upper bound on the length. To see this consider for example the case of a, c and m all even. In this case the algorithm will produce only even integers, and clearly the sequence can then be no more than $\frac{1}{2}m$ in length. In addition to the length of the sequence there are more subtle problems associated with poor choices for the constants. For bad choices the numbers generated may simply be "not very random". A lengthy discussion is given by Knuth (1981), who notes that there are certain rules of thumb governing good choices of the constants, such as ensuring that a and m are coprime (i.e., have no factors in common other than 1), but to a large extent the choice of good values is simply a matter of trying out many of them. Knuth also gives a set of tests which may be applied to sequences of numbers to help judge how random they are. However, rather than go into the subject in more detail, we prefer to just give you some good values for use in Equation (16.5). Generators using these values have passed Knuth's tests and are acceptable for purposes not particularly sensitive to the quality of the random numbers generated. (For more sensitive applications, you should use one of the more sophisticated generators described in the rest of this chapter.)

The choice of constants does depend on whether one is working in a language like C or Pascal which provides facilities for performing unsigned integer arithmetic. For such languages an integer of 32 bits really can store any number up to 2^{32}. In that case we recommend the values $a = 2416$, $c = 374\,441$ and $m = 1\,771\,875$. For languages such as FORTRAN, which only provide signed integer arithmetic, a 32-bit integer can only store a positive number up to 2^{31}, since one bit is used to hold the sign. In this case we recommend the values $a = 9301$, $c = 49\,297$ and $m = 233\,280$. Both these sets of numbers achieve the maximum possible sequence length of m, and satisfy moderately stringent tests of their quality. For computers which use integers of lengths other than 32 bits, a more comprehensive list of good values for a, c and m is given in Press *et al.* (1988).

There is however an alternative to using these relatively small values of m. Schrage (1979) has demonstrated a trick which can be used to multiply two 32-bit integers together, modulo a third 32-bit integer, without overflowing our 32-bit variables. This allows us, without too much computational overhead, to implement a linear congruential generator with the maximum useful value of $m = w - 1$. The trick requires us to use signed integer arithmetic however, so we have to take $m = 2^{31} - 1$. As it turns out, one of the most thoroughly tested and widely used random number generators happens to use this value of m. (It's no coincidence in fact, given the prevalence of 32-bit integer arithmetic on computers.) This generator was first given by Lewis *et al.* in 1969, and makes use of the values $a = 16\,807$ and $c = 0$. For technical reasons Knuth recommends the choice $c = 0$ for all generators with $m = 2^n \pm 1$, and in this case it is also quite convenient, since it speeds up the arithmetic.

Schrage's trick works like this. Given m, we can define two numbers q and r thus:

$$q = \lfloor m/a \rfloor, \qquad r = m \bmod a, \tag{16.6}$$

so that

$$m = aq + r. \tag{16.7}$$

For the values of a and m given above, we have $q = 127\,773$ and $r = 2836$. It is important for the trick to work that $r < q$, but that is indeed the case here, so we are safe. Now it is clear that

$$ai_n \bmod m = (ai_n - \lfloor i_n/q \rfloor m) \bmod m. \tag{16.8}$$

This is true regardless of the value of q, since we can add or subtract any integer multiple of m from ai_n without changing the value of the remainder when we divide it by m. But now, using (16.7), we can write this as

$$
\begin{aligned}
ai_n \bmod m &= [ai_n - \lfloor i_n/q \rfloor(aq + r)] \bmod m \\
&= [a(i_n - \lfloor i_n/q \rfloor q) - \lfloor i_n/q \rfloor r] \bmod m \\
&= [a(i_n \bmod q) - \lfloor i_n/q \rfloor r] \bmod m.
\end{aligned}
\tag{16.9}
$$

The term $\lfloor i_n/q \rfloor r \le i_n(r/q)$, and since $r < q$, this number is less than i_n, which in turn is less than m. Furthermore, given that the number $i_n \bmod q$ lies between zero and $q - 1$, the term $a(i_n \bmod q) < aq$. The definition of q, Equation (16.6), then ensures that this term is also less than m. Thus both of these terms fit into our 32-bit signed integers, and, since neither of them can be negative, so does their difference. The difference can be negative however, which is why we need to use signed arithmetic to evaluate it. Taking the result modulo m now becomes easy: if the difference of the two terms is positive, we don't have to do anything; if it is negative we just need to add m. In other words

$$
ai_n \bmod m = \begin{cases} a(i_n \bmod q) - \lfloor i_n/q \rfloor r & \text{if this is not negative} \\ a(i_n \bmod q) - \lfloor i_n/q \rfloor r + m & \text{otherwise.} \end{cases}
\tag{16.10}
$$

With the given values of q and r we can use this equation to implement the random number generator of Lewis *et al.* An example, written in C, is given in Appendix B. The code is a little more complex than that for the straightforward linear congruential generator, but it has the significant advantage of a longer period (about 2 billion numbers, which is adequate for many smaller Monte Carlo simulations). It has another advantage too. All linear congruential generators can produce only a certain number of different possible integer values. If we convert these into real numbers between zero and one, we can only get a certain number of different possible reals. Of course, the available number of different reals between zero and one is always finite, limited as it is by the precision with which the computer represents real numbers. However, with the values for a, c and m we suggested for the simple 32-bit generator, for example, the number of different values generated by the algorithm is about 1.7 million, which is far less than the number allowed by the machine precision on most modern computers. (Double precision floating-point numbers typically have at least 48 bits of mantissa, which gives about 10^{14} different values in each decade, and significantly more than this between zero and one.) For many applications, this would not matter a great deal, but we can imagine cases in which it would. One of the authors for example, has performed extensive Monte Carlo simulations on one-dimensional systems with several million lattice sites. If one were not careful, it would be very easy to ignore the fact that not all the sites in such a system can be reached by choosing a site number using a single random number r and Equation (16.2), if r has only a million or so possible values. The result would be a Monte Carlo algorithm which violates the condition of ergodicity in a very bad way.

The generator described above still suffers from this drawback, but the situation is a lot less severe than in the case of the straightforward linear congruential generator. By using Schrage's trick, we have created a generator that can generate $m - 1 = 2\,147\,483\,646$ different numbers, which is enough for most Monte Carlo applications. The method does have one slight problem (as indeed do all generators which use $c = 0$) that if we seed the generator with the number $i_0 = 0$, we will just generate zeros forever afterwards, since ai_n will always be zero. Thus we should avoid using zero as the seed. The algorithm also never produces zero once seeded with any other number,[3] though for the particular choice of a and m given here, it does produce every other integer between 1 and $m - 1$. (This is why the number of different values produced is $m - 1$ and not m.) If we are worried about an algorithm which never produces zero, we can always subtract 1 from each number generated.

A different solution to the problem of writing a good linear congruential

---

[3]It can only produce zero if a and m have a common factor, which in this case they don't. Actually, m doesn't have any factors; $2^{31} - 1$ is a prime number.

random number generator is to write it in the machine language of the computer. This not only allows the generator to work faster, since the type of integer operations embodied in Equation (16.5) can be performed very efficiently in machine languages, but it also obviates the problem of overflowing integer variables, since most computers will allow you to perform the arithmetic using double words. This allows you to choose any value of m you like, and the usual choice is the same one we used above, $w-1$. Of course, most of us have no interest in learning the machine language of the computer we are using, which makes writing a generator in it an unappealing prospect. However, almost all machines these days come with a linear congruential random number generator already written in machine code which can be called from any high-level language such as FORTRAN or C. While these packaged routines certainly benefit from the advantages described above, it is prudent to exercise some caution in their use. Since we don't know what values of a, c and m the programmer chose in writing them, we have no way of telling whether they are actually good generators, unless we go to the trouble of running some tests ourselves (for example those described by Knuth). It may be that these generators are perfectly good, but there are also some horror stories about machines which were released with disastrous random number generators which could make a complete mess of your Monte Carlo simulation. So our advice is to be very careful. If you have any doubt about the quality of a random number generator supplied with your computer, you should consider either writing one of your own using one of the algorithms described in this chapter, or using the "shuffling" scheme we describe in the next section to improve the randomness of the manufacturer's algorithm.

16.1.4   Improving the linear congruential generator

Although adequate for many applications, linear congruential random number generators have problems. First there is the length of the sequence. Even the best of the generators described in the last section produces a sequence of only about 2 billion numbers. For a Metropolis algorithm simulation of the Ising model, for instance, which uses one random number to choose the spin to be flipped and another to decide whether to flip it, this means we could perform about 1 billion Monte Carlo steps before the cycle came round and repeated itself. Typical Metropolis simulations these days run into tens or hundreds of billions of Monte Carlo steps, so there is definitely the possibility of introducing errors into the results through correlations of this sort in the random numbers.

Another, more subtle problem concerns hidden correlations in the number sequences produced by linear congruential random number generators. It turns out (see Knuth (1981)) that if you take the numbers produced by one of these generators in groups of k (which can be any number you like) and regard these groups as the coordinates of points in a k-dimensional space, the points produced are constrained to lie on hyperplanes of co-dimension one in the space. Or to put it more simply, the points only fill a very small proportion of the k-space available to them, and are therefore obviously not really random. These correlations are certainly well hidden and might not matter for your particular application, but one must be cautious; as Press *et al.* (1988) point out, the papers whose results are in question because of poor random number generators form a significant portion of the Monte Carlo literature. If you are considering spending a substantial amount of time on a Monte Carlo calculation, you would do well to guard yourself against criticism of this kind by employing a random number generator you trust.

So how can we improve on the linear congruential generator? The most commonly used method is simply to take the random numbers it produces and shuffle them around a bit, to reduce their correlations and increase the repeat period of the generator. This technique requires very little extra computational effort after the numbers have been generated, and much improves the quality of the sequence. If you want to use a machine-coded generator which came with your computer, we would also suggest that you employ a shuffling scheme since this will reduce the risk of your getting poor results because someone messed up when they wrote the generator.

The standard shuffling scheme, due to Bays and Durham (1976), is this. We create an array of N integers $\{j_n\}$ where N is typically around 100 (the exact number is not important) and initially we fill it with random integers between zero and $m - 1$ generated by our linear congruential generator. We also generate one additional random integer, which we store in a variable which we will call y. Then the procedure goes as follows:

1. We calculate the index $k = \lfloor yN/m \rfloor$, which is an integer between zero and $N - 1$.

2. We return the kth element j_k from our array as the new random number generated by our shuffling algorithm, and in addition we set $y = j_k$.

3. We generate a new random integer using our linear congruential generator and we set j_k equal to this new number.

The numbers generated by this scheme depend on average on the last N numbers generated by the linear congruential generator, and so can be expected to have a repeat cycle on the order of m^N. For our purposes this is essentially infinite and we can stop worrying about the cycle. Furthermore, the subtle correlations concerned with hyperplanes that we mentioned are destroyed by this shuffling of our random numbers, greatly reducing the chances of spurious systematic errors being introduced into our Monte Carlo results.

The authors employed a linear congruential generator and a shuffling scheme similar to the one described here in many of the calculations performed during the writing of this book, and the results are, as far as we have been able to tell, satisfactory.

16.1.5   Shift register generators

The linear congruential generators of Section 16.1.3, in combination with the Bays–Durham shuffle of Section 16.1.4, give random numbers of sufficient quality for most Monte Carlo simulations. However, we must always be aware of the possibility that correlations in the sequence of numbers produced by a particular generator may give rise to systematic errors in our results. One way to guard against this problem is to have another, different random number generator at hand. If you suspect that your random number generator may be causing you trouble, you can repeat your simulations with the other generator and see if the problems go away. In fact, we recommend, as a precautionary measure, that you should always repeat some part of your simulation with another random number generator just to check that the results do not change. To this end we describe two other widely used types of random number generators in the next two sections: shift register or Tausworthe generators, and lagged Fibonacci generators. Either of these types of generators is good enough to provide a check on the results of our simulations. Also, as it turns out, they both usually generate random numbers faster than our linear congruential generator. You might therefore wish to consider using one of these types of generator as your primary source of random numbers and the slower, but reliable, linear congruential generator to double check the results, instead of the other way around.

First then, we look at **shift register generators** (Tausworthe 1965), which generate a random integer i_n by manipulation of the binary digits in the previous integer i_{n-1} in the series. The standard algorithm consists of two stages, a right-shift and a left-shift, as follows:

1. We shift the bits of i_{n-1} first to the right a certain number of places s, introducing zeros in the leftmost s bits which have just been vacated. Then we exclusive-OR the result with the original number i_{n-1}. This leaves us with an intermediate result, which we will call i'_{n-1}. If we denote the right-shift operation by R, then

$$i'_{n-1} = i_{n-1} \oplus R^s i_{n-1}, \qquad (16.11)$$

were $\oplus$ represents the XOR operation.

2. Now we perform the same procedure using a left-shift L, shifting some number t places to the left, introducing zeros in the rightmost t bits,

and then XORing the result with i'_{n-1} thus:

$$i_n = i'_{n-1} \oplus L^t i'_{n-1}. \tag{16.12}$$

In C, for example, the left- and right-shift operations are performed by the operators `<<` and `>>` and the exclusive-OR operation by `^`. All of these are fairly quick operations, so the generator is usually quite fast even when implemented in a high-level language. Furthermore, there is no problem using all 32 bits of an integer with this algorithm, or however many bits are available on your computer, since Equations (16.11) and (16.12) never produce an overflow.

Since shift-register random number generators generate each number from only the single preceding number in the sequence, the arguments given in Section 16.1.2 immediately tell us that the longest sequence they can generate has length w, i.e., one greater than the maximum integer which your computer can store. (In fact, as with the linear congruential generator, the longest sequence actually has length $w - 1$, since we cannot use any sequence which includes the number zero, because zero is mapped onto itself by the equations above.) As we pointed out in Section 16.1.4 this is inadequate for some longer Monte Carlo simulations. In this case, we can again employ the Bays–Durham shuffling scheme, as we did for the linear congruential generator, to increase the period and at the same time improve the quality of random numbers generated.

The only remaining question is what value the two parameters s and t should take. As with the linear congruential generators of Section 16.1.3, the choice of these numbers involves a certain amount of black magic. Here we merely state that for a 32-bit generator the values $s = 15$ and $t = 17$ are the most widely used. They also generate a sequence of numbers with the maximum possible period of $2^{32} - 1$. For a 31-bit generator, $s = 18$, $t = 13$ and $s = 28$, $t = 3$ are the commonest choices. For most purposes these choices will give adequate random numbers. A more thorough discussion of the possible choices is given by Marsaglia (1992).

16.1.6 Lagged Fibonacci generators

A third class of widely used random number generators are the **lagged Fibonacci generators**, which have the dual advantages of being fast and also of generating very long sequences of numbers, which obviates the need to employ a shuffling scheme, such as that of Section 16.1.4. In fact, as Knuth (1981) has stated, these generators "may well prove to be the very best source of random numbers for practical purposes". The only reason we cannot recommend them whole-heartedly is that their theoretical foundations are not as solid as those of the other generators we have discussed, so it is possible that there lurk correlations in the sequences they produce which have yet to be discovered. However, if used in combination with another generator to safeguard against such possibilities, they can be extremely useful. The authors have used one such generator extensively for some very large calculations in which efficiency of random number generation was crucial and obtained very satisfactory results.

Lagged Fibonacci generators are of the type described by Equation (16.4), and may be written in the general form

$$i_n = (i_{n-r} \circ i_{n-s}) \bmod m, \qquad (16.13)$$

where m, r and s are constants and $\circ$ can be any of the operations $+$ (plus), $-$ (minus), $\times$ (multiplication) or $\oplus$ (exclusive-OR). As before, the choice of the constants is crucial to producing random numbers of good quality. Knuth gives a thorough discussion of possible values. We will merely list some common ones.

Many people use the lagged Fibonacci generator with an exclusive-OR operation (Lewis and Payne 1973). (These generators are sometimes referred to as "generalized feedback shift register generators", for reasons which are quite obscure.) This has the advantage that the modulo operation in Equation (16.13) is not necessary, since the exclusive-OR of two n-bit integers can never itself exceed n bits. It turns out however that these generators produce poor random numbers as well as possessing rather short periods, unless the constants r and s are very large. A common choice for example is $r = 418$, $s = 1279$, which requires us always to store the last 1279 numbers generated. Considerably superior results can be obtained with much smaller values of r and s if we use other operators instead, and we recommend using (16.13) with either addition or multiplication. Addition usually gives a faster generator, but multiplication appears to give somewhat better random numbers.

Unlike the exclusive-OR case, both additive and multiplicative lagged Fibonacci generators require the use of the modulo operation in Equation (16.13). However, it turns out that the quality of random numbers generated is not highly sensitive to the value of the modulus m, the only restriction being that it should be an even number. Thus, it makes sense to choose m equal to the number w (i.e., one larger than the largest number which can be stored in an integer variable) so that we don't have to perform the modulo operation explicitly—we can just let the addition or multiplication operation overflow our integer variables and ignore the overflow.[4] As

---

[4]This happens automatically in machine languages, and also in the C programming language. Unfortunately, most implementations of FORTRAN and Pascal will not allow integers to overflow—the overflow will cause a run-time error which will halt the program. If you program in one of these languages it may be possible to disable this feature in your compiler, or you may be able to use a generator written in C while writing the rest of your program in the language of your choice.

for the other constants, the most common choice is $r = 24$, $s = 55$, first suggested by Mitchell and Moore in 1958 (unpublished). This choice appears to give excellent results for both additive and multiplicative generators. Knuth also suggests a number of other possibilities which you may like to play with.

These generators possess very long periods. In the additive case, for instance, it can be proved that the period is at least $2^{55} - 1$, which should be adequate for almost any purpose. They do have the disadvantage that we must store the last 55 numbers generated at all times. This however adds only a slight extra complexity to the algorithm and so is not a serious objection. The numbers can be stored in an integer array organized as a circular FIFO buffer of the kind described in Section 13.2.5. The values of the elements of this array need to be initialized before the first call to the generator—effectively the generator needs 55 seeds. These seeds could be generated, for example, using a linear congruential generator.

Both the additive and multiplicative generators produce random numbers of very high quality, although the multiplicative generator appears to be slightly better—it passes some extremely stringent tests of randomness for the numbers it generates (Marsaglia 1985) and we recommend it for calculations in which random numbers of high quality are important. It does have a couple of problems though. First, it is usually slower than the additive generator, since multiplication normally takes longer than addition. On modern RISC processors, however, this is not always true; many such processors can perform integer multiplication just as fast as addition. Second, it can only produce either even or odd integers, but not both. To see this, consider what happens if even a single one of our 55 seed integers is even. The product of this number with any other will produce an even result and, since 24 and 55 are coprime, this means that our buffer will, after some time, be entirely full of even numbers, and thereafter all numbers generated will be even. Only if all the seeds are chosen odd can the algorithm go on generating odd numbers for an arbitrary length of time, but in this case it will only generate odd numbers and no even ones. In fact, the normal practice is to take the latter course and seed the generator with 55 odd integers so that all numbers generated are odd. This reduces the number of different values which can be generated to $w/2$, rather than w. This is not normally a problem, but under certain special circumstances it could be, so it is as well to be aware of it.[5]

Note that with the additive generator we should be careful to ensure that not all the seeds are even, otherwise this generator will only ever produce even integers. In both cases, the seeds should be uniformly distributed over the entire range 0 to $w - 1$. If, for example, all the seeds were set equal to 1, then the first few numbers produced by either algorithm would not be good

---

[5]Note that it is only the number of different values which is reduced. The period of the generator is still long enough that you should not usually have to worry about it.

random numbers.

Although the theoretical foundations on which the lagged Fibonacci generators are built are less solid than those of the other methods we have discussed, they have produced excellent results in many tests. In combination with another generator they can prove very useful when speed and quality of random number generation are of the essence.

Armed with one of the random number generators described above, we can now generate random real numbers r_n between zero and one simply by dividing each integer generated by its maximum allowed value:

$$r_n = \frac{i_n}{m}. \tag{16.14}$$

And this is the fundamental random quantity which we use in Monte Carlo simulation. Sample programs in C implementing a number of the generators described here can be found in Appendix B.

16.2 Generating non-uniform random numbers

In the first part of this chapter we have seen how we can use linear congruential and other random number generators to generate pseudo-random floating-point numbers which are uniformly distributed between zero and one. Armed with these numbers and formulae such as those discussed at the beginning of Section 16.1, we can carry out such important tasks as choosing a site at random from a lattice and performing an operation with a certain probability. This is all we need to implement, for instance, the Metropolis algorithm for the Ising model. However, there are other random processes which require random numbers of different types. For example, the random-field Ising model of Section 6.1.1 has (at least in one version) random local fields on every site of the lattice which have values Gaussianly distributed about zero. How should we go about generating these Gaussian random numbers? In this section we discuss techniques which allow us to take our uniformly distributed random numbers and from them produce random numbers with any other distribution we desire. Thus, we do not need to have a different sort of random number generator for every distribution of random numbers. Just the one generator will do when coupled with the techniques outlined here for producing new distributions from the old ones.

16.2.1 The transformation method

The best method for producing non-uniformly distributed random numbers, given uniformly distributed ones, is the **transformation method**. It goes like this. Suppose we want to produce real random numbers x which lie between the limits x_{min} and x_{max} and are distributed according to some function $f(x)$. That is, the probability of producing a number in the interval between x and $x + dx$ should be $f(x)\,dx$ provided $x_{min} \le x < x_{max}$, or zero otherwise. The function $f(x)$ is a properly normalized probability distribution such that its integral is unity:

$$\int_{x_{min}}^{x_{max}} f(x)\,dx = 1. \tag{16.15}$$

The limits x_{min} and x_{max} are allowed to be $\pm\infty$ if we want.

The fraction $F(x)$ of these random numbers which lie below some value x is given by

$$F(x) = \int_{x_{min}}^{x} f(x')\,dx'. \tag{16.16}$$

The same fraction of our uniformly distributed random numbers lies in the interval $0 \le r < F(x)$. What we want to do is map numbers from our uniform random number generator which fall into this region onto the numbers between x_{min} and x in our new distribution. As it turns out, it's very easy to arrange this. All we have to do is observe that the largest numbers in each case should correspond to one another. In other words the number x in our non-uniform distribution should be generated when the number

$$r = F(x) \tag{16.17}$$

is produced by our uniform generator. Given the random numbers r, then, all we have to do to get the non-uniform ones is invert this equation, to give a formula for x. The only catch is that $F(x)$ is defined as a definite integral over the function f, so inverting the equation requires us first to perform the integral. For some functions f this is easy, but for others the integral cannot be done and in that case the transformation method will not work.

Let us see a couple of examples of cases in which the transformation method does work. First, suppose we want to generate random numbers between $-\infty$ and $+\infty$, drawn from the Lorentzian distribution

$$f(x) = \frac{1}{\pi}\frac{\Gamma}{\Gamma^2 + x^2}. \tag{16.18}$$

Here Γ is a free parameter, the "width" of the Lorentzian, and the factor of π^{-1} and the Γ on top of the fraction are required so that the distribution is normalized to 1 over the interval $-\infty$ to $+\infty$. Now Equations (16.16) and (16.17) tell us that the Lorentzianly distributed random number x should be generated every time our uniform random number generator produces the number r, where

$$r = \frac{1}{\pi}\int_{-\infty}^{x} \frac{\Gamma}{\Gamma^2 + x'^2}\,dx' = \frac{1}{\pi}\left[\tan^{-1}\frac{x'}{\Gamma}\right]_{-\infty}^{x}$$

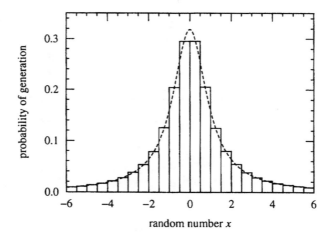

FIGURE 16.1 Demonstration of the the transformation method for generating non-uniform random numbers. The histogram shows the distribution of one million Lorentzianly distributed random numbers generated using Equation (16.20). The dotted line is the Lorentzian itself.

$$= \frac{1}{2} + \frac{1}{\pi} \tan^{-1} \frac{x}{\Gamma}. \tag{16.19}$$

Now we can invert this equation to get

$$x = \Gamma \tan\left[\pi\left(r - \tfrac{1}{2}\right)\right]. \tag{16.20}$$

All we have to do now is generate random numbers r uniformly in the range between zero and one using any of the techniques described in the first part of this chapter and then feed them into this equation, and what we get out is a set of numbers x which are also random, but now distributed according to the Lorentzian, Equation (16.18). Just to prove that it actually works, we have generated one million random numbers using this formula with $\Gamma = 1$, and made a histogram of them in Figure 16.1. The dashed line is the Lorentzian itself, and, as we can see, the agreement between line and histogram is excellent.

Another example: suppose we want to generate a unit vector which points in a random direction in three-dimensional space. We might want to do this, for example, in order to implement the Metropolis algorithm for the Heisenberg model as described in Section 4.5.3, which requires us to choose a new random direction for one of our Heisenberg spins (which are three-dimensional unit vectors) at each step of the simulation. We can represent a unit vector by the two angles θ and ϕ of spherical coordinates, such that the components of the unit vector are

$$x = \sin\theta\cos\phi,$$
$$y = \sin\theta\sin\phi, \qquad (16.21)$$
$$z = \cos\theta.$$

We know that the element of solid angle in these coordinates is $\sin\theta\,d\theta\,d\phi$. We want to generate random values of θ and ϕ, such that an equal number of the vectors they describe fall in equal divisions of solid angle. In other words we want to generate uniformly distributed values of ϕ between 0 and 2π (which is easy) and we want to generate values of θ between 0 and π distributed according to the frequency function

$$f(\theta) = \tfrac{1}{2}\sin\theta, \qquad (16.22)$$

where the factor of $\tfrac{1}{2}$ is required to ensure that $f(\theta)$ integrates to unity. Again employing Equations (16.16) and (16.17), this implies that the value θ should be generated every time our uniform random number generator produces the number r, where

$$r = \tfrac{1}{2}\int_0^\theta \sin\theta'\,d\theta' = \tfrac{1}{2}\Big[-\cos\theta'\Big]_0^\theta$$
$$= \tfrac{1}{2}(1 - \cos\theta). \qquad (16.23)$$

Now we can invert this equation to give us θ:

$$\theta = \cos^{-1}(1 - 2r). \qquad (16.24)$$

So, in order to generate our random, spherically symmetric unit vectors, all we need do is produce two uniformly distributed random numbers between zero and one, feed one of them into this equation to get θ, and the other into Equation (16.1) to produce a value for ϕ between 0 and 2π, and then feed these angles into Equation (16.21) in order to get x, y and z.

16.2.2 Generating Gaussian random numbers

An issue which arises often is the generation of random numbers which are distributed according to a Gaussian or normal distribution with standard deviation σ:

$$f(x) = \frac{1}{\sqrt{2\pi\sigma^2}}\exp\left(-\frac{x^2}{2\sigma^2}\right). \qquad (16.25)$$

(Again, the leading numerical factor ensures that $f(x)$ is normalized to unity over the range $-\infty$ to $+\infty$.) Let us see what happens if we apply our transformation method to this distribution function. Equations (16.16) and (16.17) tell us that the Gaussianly distributed random number x should be generated every time our uniform random number generator produces the number r, where

$$r = \frac{1}{\sqrt{2\pi\sigma^2}} \int_{-\infty}^{x} \exp\left(-\frac{x'^2}{2\sigma^2}\right) dx' = \frac{1}{2}\left[1 + \text{erf}\left(\frac{x}{\sqrt{2}\sigma}\right)\right], \qquad (16.26)$$

where $\text{erf}(x)$ is the error function, which is essentially just the definite integral of a Gaussian. Unfortunately, there is no known closed-form expression for the error function, which makes it impossible to invert this equation.[6]

Generating Gaussian random numbers is extremely important for many applications however, so other methods have been developed to tackle the problem. The standard way of doing it is a two-dimensional variation of the transformation method. Imagine we have two independent random numbers x and y, both drawn from a Gaussian distribution with the same standard deviation σ. The probability that the point (x, y) falls in some small element $dx\,dy$ of the xy plane is then

$$f(x, y)\,dx\,dy = \frac{1}{2\pi\sigma^2} \exp\left(-\frac{x^2 + y^2}{2\sigma^2}\right) dx\,dy. \qquad (16.27)$$

We could alternatively express this in polar coordinates as the probability that the point falls in the elemental area $r\,dr\,d\theta$ with radial coordinate between r and $r + dr$ and angular coordinate between θ and $\theta + d\theta$:

$$f(r, \theta)\,dr\,d\theta = \frac{1}{2\pi\sigma^2} \exp\left(-\frac{r^2}{2\sigma^2}\right) r\,dr\,d\theta. \qquad (16.28)$$

Thus, if we can generate random values of r and θ according to this distribution and then transform them back into Cartesian coordinates x and y, we will have two random numbers which are distributed according to a Gaussian with standard deviation σ. Since most applications require many random numbers, generating two in this way is not a problem; we simply save up one of them for use later.

Generating the θ variable according to Equation (16.28) is trivial—we just need to produce a uniformly distributed real number between zero and 2π, which we can do using Equation (16.1). The radial coordinate r can be generated using the transformation method. The normalized distribution function for r is

$$f(r) = \frac{1}{\sigma^2} \exp\left(-\frac{r^2}{2\sigma^2}\right) r. \qquad (16.29)$$

---

[6]Some computer languages, including C, provide a library function which can evaluate $\text{erf}(x)$ using an asymptotic series approximation. However, no such library functions exist for evaluating the inverse of the function.

The transformation method then says that we should produce a value r for this coordinate every time our uniform random number generator produces a number ρ (we use ρ this time to avoid confusion between variables called r) such that

$$\rho = \frac{1}{\sigma^2} \int_0^r \exp\left(-\frac{r'^2}{2\sigma^2}\right) r' \, dr' = 1 - \exp\left(-\frac{r^2}{2\sigma^2}\right). \tag{16.30}$$

Rearranging for r, this gives us

$$r = \sqrt{-2\sigma^2 \log(1 - \rho)}. \tag{16.31}$$

With this value for r and our random value for θ, the two numbers

$$x = r \sin\theta, \quad y = r \cos\theta \tag{16.32}$$

are independent Gaussianly distributed random numbers.

A couple of brief points are worth making. First, notice that the term inside the logarithm in Equation (16.31) is $1 - \rho$, where ρ is a uniformly distributed random number between zero and one, which comes from our standard random number generator. It is tempting to say that therefore $1 - \rho$ is also a random number between zero and one, so we can save ourselves the subtraction and just write $\log \rho$. This however would be a mistake, since, as you will recall, the standard random number generator produces numbers in the range $0 \leq \rho < 1$, so that it is possible, if unlikely, that ρ will be exactly zero. (Though slim, the chance is finite since the computer has only a limited precision with which it can represent real numbers.) Such a zero value of ρ will cause a program error when we try to evaluate the logarithm of zero. No such errors will occur if we use the correct formula, Equation (16.31), which never tries to evaluate the log of zero.

Second, we should point out that it is not really necessary that we be able to generate Gaussian random numbers for any value of σ, as we are here considering. If we can generate random numbers with standard deviation one, then multiplying each number by some constant σ will give us Gaussian random numbers with that value of σ as their standard deviation. For this reason most Gaussian random number generators use the algorithm above with $\sigma = 1$ and expect you to multiply the numbers yourself by whatever standard deviation you want. Notice that this property is not true of all distributions of random numbers. It only works for ones like the Gaussian for which the distribution function can be written as a function of x/σ.

16.2.3 The rejection method

If we want random numbers distributed according to some function $f(x)$, but we don't know the definite integral (16.16) of the function, then we can't use the transformation method described in the Section 16.2.1. In this section we describe another method for generating random numbers, the **rejection method**, which is simple to implement and can generate random numbers according to any distribution, whether it is integrable or not. The method does have some drawbacks, however, which make it inferior to the transformation method for integrable functions:

1. It is considerably less efficient than the transformation method. It requires us to generate at least two and often more than two random numbers uniformly distributed between zero and one for every non-uniform number returned.

2. It only works for distributions defined over a finite range. In other words we can't have $x_{min} = -\infty$ or $x_{max} = +\infty$ as we did in some of the examples in the last two sections.

These drawbacks are offset by the generality of the method, and certainly there are plenty of situations where the rejection method is the method of choice.

In its simplest form, the rejection method works like this. We want to generate random numbers x in the interval from x_{min} to x_{max} distributed according to some function $f(x)$. Let f_{max} be the maximum value which the function attains in the interval. We generate a random number x uniformly in the interval between x_{min} and x_{max} using Equation (16.1). (This only works if x_{min} and x_{max} are finite, which is the reason for condition 2 above.) Now we generate another random number r between zero and one and we keep the random number x if

$$r < \frac{f(x)}{f_{max}}.$$ (16.33)

Otherwise, we reject x and generate another number between x_{min} and x_{max}. The process continues until we accept one of our numbers, and that is the number which our random number generator returns. The factor of f_{max} on the bottom of Equation (16.33) ensures that the acceptance probability $f(x)/f_{max}$ has a maximum value of one, which makes the algorithm as efficient as possible.

Why does it work? Well, the probability that Equation (16.1) generates a number in some small range x to $x + dx$ is

$$p_{gen}(x) \, dx = \frac{dx}{x_{max} - x_{min}}.$$ (16.34)

The probability that it is accepted comes from Equation (16.33). It is just:

$$p_{accept}(x) = \frac{f(x)}{f_{max}}.$$ (16.35)

And the probability that we return a value of x in this interval is then the product of these two probabilities:

$$p(x)\,\mathrm{d}x = p_{\mathrm{gen}}(x)\,p_{\mathrm{accept}}(x)\,\mathrm{d}x = \frac{f(x)\,\mathrm{d}x}{f_{\mathrm{max}}(x_{\mathrm{max}} - x_{\mathrm{min}})}. \tag{16.36}$$

Clearly this is proportional to $f(x)\,\mathrm{d}x$, as we want it to be. Notice that there is no need in this case for the function $f(x)$ to be normalized to unity over the range of allowed values for x. The factor of f_{max} ensures that the results of the algorithm will be unchanged if we multiply $f(x)$ by any constant factor.

The inefficiency of the algorithm arises for two reasons. First, every time we try a new value of x we have to generate two new random numbers between zero and one. One of them is used to produce the value of x using Equation (16.1), and the other is used in Equation (16.33) to decide whether to accept or reject that value. Second, the very fact that we reject some candidate numbers reduces the algorithm's efficiency. It means that we waste time generating numbers and then throwing them away only to generate more. We can calculate the average effect of these inefficiencies as follows. The total probability of returning a value—any value—on one particular iteration of the algorithm is the integral of (16.36) over the allowed range of x, and the average number of iterations we have to perform to get one acceptance is the reciprocal of this integral. Two random numbers between zero and one are required for each iteration, so the total average number of calls we need to make to our basic random number generator for every random number we return is

$$\text{total calls} = \frac{2f_{\mathrm{max}}(x_{\mathrm{max}} - x_{\mathrm{min}})}{\int_{x_{\mathrm{min}}}^{x_{\mathrm{max}}} f(x)\,\mathrm{d}x}. \tag{16.37}$$

As an example, let us look at the generation of Gaussian random numbers again. In Section 16.2.2 we saw that it is not possible to use the normal transformation method to generate Gaussianly distributed random numbers, but we did describe another method, based on a variation of the transformation method, which can generate two independent Gaussian random numbers with the same standard deviation, given as input two independent random numbers between zero and one. Let us compare this method to the rejection method described here. First, we note that the rejection method cannot generate numbers with an infinite (or semi-infinite) range, so we have to choose finite values for x_{min} and x_{max}. What values we choose depends on the precise details of our application, but for now we note that if we choose the limits to lie at about 6 standard deviations from the centre of the distribution that will leave out less than one part in 10^7 of the area under the distribution, which for most applications will be good enough. For a distribution of random numbers centred at zero with $\sigma = 1$, the function $f(x)$ can then be written

$$f(x) = \exp\left(-\tfrac{1}{2}x^2\right), \tag{16.38}$$

and the total number of calls per Gaussian random number generated is

$$\text{total calls} = \frac{24}{\int_{-6}^{+6} \exp\left(-\tfrac{1}{2}x^2\right) \mathrm{d}x} = 9.57\ldots \tag{16.39}$$

In other words we will have to make about ten calls on average for each number we generate—ten times as many as for the method of Section 16.2.2. The rejection method is therefore a poor way of generating Gaussian random numbers. However it may still be useful for distributions for which no other method exists.

16.2.4   The hybrid method

The rejection method described in the preceding section is a simple alternative to the transformation method when we want to generate non-uniform random numbers with non-integrable distribution functions. As we saw, however, it is an inefficient method, requiring several calls to our basic random number generator for each number it returns. At the expense of a certain increase in programming complexity, we can reduce this inefficiency by using a hybrid method which combines features of both the transformation and rejection methods, while still being applicable to any distribution function $f(x)$. The method works like this.

First we must find a function $g(x)$ which approximates to $f(x)$, but which is integrable, and which is greater than or equal to $f(x)$ for all x in the range between $x_{\min}$ and $x_{\max}$. Then we apply the transformation method to generate a random number distributed according to $g(x)$. Assuming $g(x)$ is normalized, the probability of generating a number between x and $x + \mathrm{d}x$ is

$$p_{\text{gen}}(x)\,\mathrm{d}x = g(x)\,\mathrm{d}x. \tag{16.40}$$

Now we accept or reject this number in a manner similar to the rejection method, but with acceptance probability

$$p_{\text{accept}}(x) = \frac{f(x)}{g(x)}. \tag{16.41}$$

Thus the total probability of generating a number in the given interval is

$$p(x)\,\mathrm{d}x = p_{\text{gen}}(x)\,p_{\text{accept}}(x)\,\mathrm{d}x = f(x)\,\mathrm{d}x, \tag{16.42}$$

as it should be. The normal rejection method could be regarded as a special case of this algorithm in which $g(x)$ is just equal to $f_{\max}$ for all x. Clearly

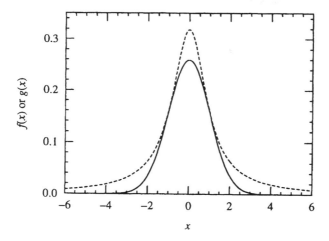

FIGURE 16.2 An example of suitable functions $f(x)$ and $g(x)$ for use in the hybrid method described in the text. In this case $f(x)$ is a Gaussian (solid line) and $g(x)$ is a Lorentzian (dotted line). The Gaussian is scaled so that $f(x) \leq g(x)$ for all x.

however, the better $g(x)$ approximates to $f(x)$ the more efficient the algorithm will become. Note that $f(x)$ does not have to be normalized to unity in order for the method to work. In fact, it cannot be normalized to unity, since we know that $g(x)$ is normalized, and $f(x) \leq g(x)$ for all x. Thus the integral of (16.42) between $x_{\min}$ and $x_{\max}$ must be less than one, implying that the method has less than perfect efficiency. However, it can still be a lot more efficient than the simple rejection method. The hybrid method also does not require that the limits $x_{\min}$ and $x_{\max}$ be finite.

As an example, let's take Gaussian random numbers again. Suppose we choose for $g(x)$ the Lorentzian function, Equation (16.18), with $\Gamma = 1$, which has a bell shape similar to that of the Gaussian, but decays to zero slower in the tails. In Figure 16.2 we show the Lorentzian, normalized to unity, with the Gaussian $f(x)$ scaled so that it lies below it for all values of x. In this example, we found by simple trial and error that multiplying the normalized Gaussian by 0.65 brought us very close to touching the Lorentzian.

Now we can apply Equation (16.20) to generate random numbers between plus and minus infinity with a Lorentzian distribution, and then Equation (16.41) to decide whether to accept them or not. The integral over Equation (16.42) then gives us simply 0.65 for the fraction of proposed numbers which are accepted (since this is the factor by which we scaled the normalized Gaussian curve to get $f(x)$). Given that we have to generate two random numbers between zero and one for each proposed number, that means that on average about three random numbers are generated for each Gaussian number returned, which is much better than the ten numbers we had to generate with the simple rejection method.

16.3   Generating random bits

Occasionally the need arises for the generation of random binary digits—bits—which are 0 or 1 with a certain probability. The principal context in which we run across this problem is in multispin-coded algorithms of the type discussed in Chapter 15. For example, in Section 15.1.1 we described a multispin-coded Monte Carlo algorithm for carrying out the single-spin-flip Metropolis algorithm on the Ising model. This algorithm requires a source of random words (usually 32 or 64 bits) in which each bit is 1 with probability $\exp(-4\beta J)$. How do we go about generating such random words?

Suppose we want to generate words of n bits in which each bit is 1 with probability p. The simplest case is when $p = \frac{1}{2}$, in which case the words are uniformly distributed random integers between zero and $2^n - 1$. We have already seen a number of ways of generating such integers in Section 16.1. Any of the generators described in that section can be used for multispin coding applications and will produce words in which the bits are 0 or 1 with probability $\frac{1}{2}$, although the linear congruential generator of Section 16.1.3 will probably have too short a period for many applications. The improved linear congruential generator of Section 16.1.4 which makes use of the Bays–Durham shuffle is a better choice and is adequate in many cases, although you should note that the implementation which we described only generates 31-bit random numbers; the top bit generated is always zero and so cannot be used in a Monte Carlo calculation. The shift register generator described in Section 16.1.5 will also work fine in most cases and generates true 32-bit numbers. However, the best choices are probably the lagged Fibonacci generators of Section 16.1.6, which are both quick and produce independent random bits of quality suitable for most Monte Carlo applications. If you use the multiplicative Fibonacci generator, you should bear in mind that it always generates odd numbers, which means that the lowest order bit is always one and hence cannot be used. If you use the additive Fibonacci generator this problem does not arise.

Once we have a way of generating words made up of random bits which are one with probability $p = \frac{1}{2}$, we can derive other values of p from them. If, for instance, we take the bitwise AND of two such words, the probability that any bit will be a one is $p = \frac{1}{4}$. If we take the bitwise OR, we get $p = \frac{3}{4}$. We can then combine these words to generate further values of p. In general, if we have a word W_p composed of random bits which are one with probability p, and we use our random number generator to produce another r with bits which are one with probability $\frac{1}{2}$, then by taking the bitwise

AND of the two we get a new word $W_{p'} = r \wedge W_p$ in which each bit is one with probability

$$p' = \tfrac{1}{2}p. \tag{16.43}$$

If we take the bitwise OR of the two $W_{p'} = r \vee W_p$ we get

$$p' = \tfrac{1}{2}p + \tfrac{1}{2}. \tag{16.44}$$

These relations tell us that by the judicious combination using AND and OR operations of the appropriate number of independent random words in which each bit is one with probability $\tfrac{1}{2}$, we can produce independent bits whose probability p of being one is any rational fraction with denominator a power of two.

In practice, given a desired value of p, the best way to find the appropriate sequence of operations to generate it is to work backwards from the result we want until we get $p = \tfrac{1}{2}$. As an example, here is how we would generate a word $W_{19/32}$ using, in this case, five different random words $r_1 \ldots r_5$ with $p = \tfrac{1}{2}$:

$$\begin{aligned}
W_{19/32} &= r_1 \vee W_{3/16}, \\
W_{3/16} &= r_2 \wedge W_{3/8}, \\
W_{3/8} &= r_3 \wedge W_{3/4}, \\
W_{3/4} &= r_4 \vee W_{1/2}, \\
W_{1/2} &= r_5.
\end{aligned} \tag{16.45}$$

Putting these all together, we can write $W_{19/32}$ as

$$W_{19/32} = r_1 \vee (r_2 \wedge (r_3 \wedge (r_4 \vee r_5))). \tag{16.46}$$

It is not hard to see that in order to produce a value of p with a denominator of 2^k in this fashion we will have to combine k different random words r_i using AND and OR operations.

This is clever, but it is not enough. We need to be able to produce bits which are one with probability p being any floating-point number we specify. It is true that we can get arbitrarily close to any real number with a rational fraction, or even with a rational fraction whose denominator is a power of two. However, if the denominator required to get sufficiently close to a particular real number is very large, we are going to have to combine a lot of different random words r in order to generate just one with the desired value of p, and this could take a lot of time. So we use a different trick.

Suppose we want to generate a random word in which the bits are one with probability p. And suppose we are able to generate two different random words, one W_{p_1} in which the bits are one with probability $p_1 < p$, and one W_{p_2} in which they are one with probability $p_2 > p$. We can do this, for example, using the techniques described above. Now let us choose between these two random words at random, with probability q that we choose W_{p_1} and $1 - q$ that we choose W_{p_2}. Then the probability that any bit in the chosen word will be one is $qp_1 + (1 - q)p_2$. Setting this equal to the desired probability p and solving for q we get

$$q = \frac{p_2 - p}{p_2 - p_1}. \tag{16.47}$$

This then allows us to generate words with any real value of p we want.

In practice, the most efficient way to use this method is to decide first with probability q whether we are going to choose W_{p_1} or W_{p_2} and then to generate the appropriate word. This is obviously more efficient than generating both words first and throwing one away. Nonetheless, we still have to generate one such random word as well as generating a random floating-point number in order to make the choice. There are other, more subtle problems as well. In particular, this method tends to generate random words in which there are correlations between the bits. Any particular bit generated does indeed have exactly the desired probability of being a one. But if one such bit is a one, then there is a statistical bias in favour of the other bits in the same word being ones also. This is most obvious in the case where we choose the two words W_{p_1} and W_{p_2} to be composed of all zeros and all ones respectively. This choice satisfies the conditions we specified for the two words regardless of the value of p, since $p_1 = 0$ and $p_2 = 1$. However, it would be rather a silly choice, since the correlations between the bits are extremely bad. If any bit in a word generated in this way is a one, then all the others will be ones as well. And if any bit is a zero, all the others will be zeros.

In practice, we get the best results if we make the values of p_1 and p_2 as close as possible to the desired value of p, but there is clearly a compromise to be struck here between the time taken to do this and the amount of correlation between the bits which we are willing to tolerate. In fact, in the case of the "asynchronous update" multispin algorithms we studied in Chapter 15 in which variables from several different systems are packed into the bits of one word, some correlation between random bits is acceptable and quite widely spaced values of p_1 and p_2 such as, say, 0 and $\frac{1}{2}$, usually work fine. (The example program for the two-dimensional Ising model given in Appendix B uses this choice, and gives perfectly satisfactory results.) The correlation between the bits may give rise to some correlation between the values of observable quantities measured in the different systems simulated, and we should be careful that this does not lead us to underestimate the statistical errors on such quantities. Other than this, however, a little correlation between bits will do no harm.

Problems

16.1 For a linear congruential generator, Equation (16.5), with $c = 0$ derive a "skip formula" which gives the value of i_{n+k} in terms of i_n for any k, without our having to actually carry out k iterations of the generator.

16.2 An alternative to the Bays–Durham shuffling scheme of Section 16.1.4 is as follows. We store an array of N integers $\{j_n\}$ just as before and each time we wish to generate a new random number we produce a random integer i using our linear congruential generator. We use this integer to calculate an index $k = \lfloor iN/m \rfloor$. We return j_k as our new random number and put i in its place in the array. Unfortunately, this scheme would be a very poor one. Why?

16.3 How would you generate floating-point numbers $x \geq 0$ according to the distribution $p(x) \propto e^{-x}$?

References

Ashcroft, N. W. and Mermin, N. D. 1976 *Solid State Physics*, Saunders College Publishing, Orlando.

Badger, L. 1994 *Mathematics Magazine* **64**, 83.

Barkema, G. T. and MacFarland, T. 1994 *Phys. Rev. E* **50**, 1623.

Barkema, G. T. and Marko, J. F. 1993 *Phys. Rev. Lett.* **71**, 2070.

Barkema, G. T., Marko, J. F. and Widom, B. 1994 *Phys. Rev. E* **49**, 5303.

Barkema, G. T. and Newman, M. E. J. 1997 *Physica A* **244**, 25.

Barkema, G. T. and Newman, M. E. J. 1998 *Phys. Rev. E* **57**, 1155.

Barkema, G. T. and Newman, M. E. J. 1999 in *Monte Carlo Methods in Chemical Physics*, D. Ferguson, J. I. Siepmann and D. G. Truhlar (eds.), Wiley, New York.

Barnes, W. J. 1929 *Proc. R. Soc. London A* **125**, 670.

Baxter, R. J. 1971 *Phys. Rev. Lett.* **26**, 832.

Baxter, R. J. 1982 *Exactly Solved Models in Statistical Mechanics*, Academic Press, London.

Bays, C. and Durham, S. D. 1976 *ACM Trans. Math. Software* **2**, 59.

Berg, B. A. 1993 *Int. J. Mod. Phys. C* **4**, 249.

Bernal, J. D. and Fowler, R. H. 1933 *J. Chem. Phys.* **1**, 515.

Bevington, P. R. and Robinson, D. K. 1992 *Data Reduction and Error Analysis for the Physical Sciences*, McGraw-Hill, New York.

Binder, K. and Heermann, D. W. 1992 *Monte Carlo Simulation in Statistical Physics*, Springer-Verlag, Berlin.

Binney, J. J., Dowrick, N. J., Fisher, A. J. and Newman, M. E. J. 1992 *The Theory of Critical Phenomena*, Oxford University Press.

Blöte, H. W. J and Swendsen, R. H. 1979 *Phys. Rev. B* **20**, 2077.

Booth, T. L. 1971 *Digital Networks and Computer Systems*, Wiley, New York.

Bortz, A. B., Kalos, M. H. and Lebowitz, J. L. 1975 *J. Comp. Phys.* **17**, 10.

Breeman, M., Barkema, G. T. and Boerma, D. O. 1995 *Surf. Sci.* **323**, 71.

Breeman, M., Barkema, G. T. and Boerma, D. O. 1996 *Thin Solid Films* **272**, 195.

Coddington, P. D. and Baillie, C. F. 1991 *Phys. Rev. B* **43**, 10617.

Coddington, P. D. and Baillie, C. F. 1992 *Phys. Rev. Lett.* **68**, 962.

Coniglio, A. and Klein, W. 1980 *J. Phys. A* **13**, 2775.

Cooley, J. W. and Tukey, J. W. 1965 *Math. Computation* **19**, 297.

Dörrie, H. 1965 *One Hundred Great Problems of Elementary Mathematics*, Dover, New York.

Eckert, J. P., Jr. 1980 in *A History of Computing in the Twentieth Century*, N. Metropolis, J. Howlett and G.-C. Rota (eds.), Academic Press, New York.

Edwards, S. F. and Anderson, P. W. 1975 *J. Phys. F* **5**, 965.

Efron, B. 1979 *SIAM Review* **21**, 460.

Ferrenberg, A. M. and Landau, D. P. 1991 *Phys. Rev. B* **44**, 5081.

Ferrenberg, A. M., Landau, D. P. and Wong, Y. J. 1992 *Phys. Rev. Lett.* **69**, 3382.

Ferrenberg, A. M. and Swendsen, R. H. 1988 *Phys. Rev. Lett.* **61**, 2635.

Ferrenberg, A. M. and Swendsen, R. H. 1989 *Phys. Rev. Lett.* **63**, 1195.

Feynman, R. P. 1985 *Surely You're Joking, Mr Feynman*, Norton, New York.

Fischer, K. H. and Hertz, J. A. 1991 *Spin Glasses*, Cambridge University Press.

Fisher, M. E. 1974 *Rev. Mod. Phys.* **46**, 597.

Fortuin, C. M. and Kasteleyn, P. W. 1972 *Physica* **57**, 536.

Futrelle, R. P. and McGinty, D. J. 1971 *Chem. Phys. Lett.* **12**, 285.

de Gennes, P. G. 1971 *J. Chem. Phys.* **55**, 572.

Giauque, W. F. and Stout, J. W. 1936 *J. Am. Chem. Soc.* **58**, 1144.

Gibbs, J. W. 1902 *Elementary Principles in Statistical Mechanics*. Reprinted 1981, Ox Bow Press, Woodridge.

Gottlob, A. P. and Hasenbusch, M. 1993 *Physica A* **201**, 593.

Grandy, W. T., Jr. 1987 *Foundations of Statistical Mechanics*, Reidel, Dordrecht.

Heller, C., Duke, T. A. J. and Viovy, J.-L. 1994 *Biopolymers* **25**, 431.

Hukushima, K. and Nemoto, K. 1996 *J. Phys. Soc. Japan* **65**, 1604.

Jayaprakash, C. and Saam, W. F. 1984 *Phys. Rev. B* **30**, 3916.

Kalos, M. H. and Whitlock, P. A. 1986 *Monte Carlo Methods, Volume 1: Basics*, Wiley, New York.

Kandel, D. 1991 Ph.D. thesis, Weizmann Institute of Science, Rehovot.

Kandel, D., Domany, E. and Brandt, A. 1989 *Phys. Rev. B* **40**, 330.

Kawasaki, K. 1965 *Phys. Rev.* **145**, 224.

Knuth, D. E. 1981 *The Art of Computer Programming, Volume 2: Semi-numerical Algorithms,* Addison-Wesley, Reading.

Landau, D. P., Tang, S. and Wansleben, S. 1988 *J. Phys. Colloq.* **49**, 1525.

Lee, J. 1993 *Phys. Rev. Lett.* **71**, 211.

Lehmer, D. H. 1951 in *Proceedings of the Second Symposium on Large-Scale Digital Calculating Machinery,* Harvard University Press, Cambridge.

Lenard, A. 1961 *J. Math. Phys.* **2**, 682.

Lewis, P. A., Goodman, A. S. and Miller, J. M. 1969 *IBM Syst. J.* **8**, 136.

Lewis, T. G. and Payne, W. H. 1973 *Journal ACM* **20**, 456.

Lieb, E. H. 1967 *Phys. Rev.* **162**, 162; *Phys. Rev. Lett.* **18**, 1046; *Phys. Rev. Lett.* **19**, 108.

Lieb, E. H. and Wu, F. Y. 1970 in *Phase Transitions and Critical Phenomena,* Vol. 1, C. Domb and J. L. Lebowitz (eds.), Academic Press, London.

Lifshitz, I. M. and Slyozov, V. V. 1961 *J. Phys. Chem. Solids* **19**, 35.

Machta, J., Choi, Y. S., Lucke, A., Schweizer, T. and Chayes, L. V. 1995 *Phys. Rev. Lett.* **75**, 2792.

Machta, J., Choi, Y. S., Lucke, A., Schweizer, T. and Chayes, L. V. 1996 *Phys. Rev. E* **54**, 1332.

Marinari, E. and Parisi, G. 1992 *Europhys. Lett.* **19** 451.

Marsaglia, G. 1985 in *Computer Science and Statistics: Sixteenth Symposium on the Interface,* Elsevier, Amsterdam.

Marsaglia, G. 1992 *Proc. Symposia Appl. Math.* **46**, 73.

Mattis, D. C. 1976 *Phys. Lett. A* **56**, 421.

Matz, R., Hunter, D. L. and Jan, N. 1994 *J. Stat. Phys.* **74**, 903.

Metropolis, N. 1980 in *A History of Computing in the Twentieth Century,* N. Metropolis, J. Howlett and G.-C. Rota (eds.), Academic Press, New York.

Metropolis, N., Rosenbluth, A. W., Rosenbluth, M. N., Teller, A. H. and Teller, E. 1953 *J. Chem. Phys.* **21**, 1087.

Metropolis, N. and Ulam, S. 1949 *J. Am. Stat. Assoc.* **44**, 335.

Mouritsen, O. G. 1990 *Int. J. Mod. Phys. B* **4**, 1925.

Müller-Krumbhaar, H. and Binder, K. 1973 *J. Stat. Phys.* **8**, 1.

Nagle, J. F. 1966 *J. Math. Phys.* **7**, 1484.

Newman, M. E. J. and Barkema, G. T. 1996 *Phys. Rev. E* **53**, 393.

Newman, M. E. J. and Barkema, G. T. 1997 *Phys. Rev. E* **56**, 3468.

Niedermayer, F. 1988 *Phys. Rev. Lett.* **61**, 2026.

Nightingale, M. P. and Blöte, H. W. J. 1996 *Phys. Rev. Lett.* **76**, 4548.

Onsager, L. 1944 *Phys. Rev.* **65**, 117.

Onsager, L. 1949 *Nuovo Cimento (suppl.)* **6**, 261.

Pathria, R. K. 1972 *Statistical Mechanics*, Butterworth-Heinemann, Oxford.

Pauling, L. 1935 *J. Am. Chem. Soc.* **57**, 2680.

Prähofer, M. and Spohn, H. 1996 *Physica A* **233**, 191.

Press, W. H., Flannery, B. P., Teukolsky, S. A. and Vetterling, W. T. 1988 *Numerical Recipes: The Art of Scientific Computing*, Cambridge University Press.

Rahman, A. and Stillinger, F. H. 1972 *J. Chem. Phys.* **57**, 4009.

Rieger, H. 1995 *Phys. Rev. B* **52**, 6659.

Rubinstein, M. 1987 *Phys. Rev. Lett.* **59**, 1946.

Rys, F. 1963 *Helv. Phys. Acta* **36**, 537.

Saleur, H. 1991 *Nucl. Phys. B* **360**, 219.

Schrage, L. 1979 *ACM Trans. Math. Software* **5**, 132.

Schülke, L. and Zheng, B. 1995 *Phys. Lett. A* **204**, 295.

Sedgewick, R. 1988 *Algorithms*, Addison-Wesley, Reading.

Segrè, E. 1980 *From X-Rays to Quarks*, Freeman, San Francisco.

Shinozaki, A. and Oono, Y. 1991 *Phys. Rev. Lett.* **66**, 173.

Slater, J. C. 1941 *J. Chem. Phys.* **9**, 16.

Sweeny, M. 1983 *Phys. Rev. B* **27**, 4445.

Swendsen, R. H. 1979 *Phys. Rev. Lett.* **42**, 859.

Swendsen, R. H. 1982 in *Topics in Current Physics 30*, T. W. Burkhardt and J. M. J. van Leeuwen (eds.), Springer-Verlag, Berlin.

Swendsen, R. H. and Wang, J.-S. 1987 *Phys. Rev. Lett.* **58**, 86.

Tausworthe, R. C. 1965 *Math. Comp.* **19**, 201.

Valleau, J. P. and Card, D. N. 1972 *J. Chem. Phys.* **57**, 5457.

Wagner, C. 1961 *Z. Elektrochem.* **65**, 581.

Wang, J.-S., Swendsen, R. H. and Kotecky, R. 1990 *Phys. Rev. B* **42**, 2465.

Wansleben, S. and Landau, D. P. 1991 *Phys. Rev. B* **43**, 6006.

Wilson, K. G. and Fisher, M. E. 1972 *Phys. Rev. Lett.* **28**, 240.

Wilson, K. G. and Kogut, J. 1974 *Phys. Rep.* **12C**, 75.

Wolff, U. 1989 *Phys. Rev. Lett.* **62**, 361.

Wulff, G. 1901 *Z. Kristallogr.* **34**, 449.

Yanagawa, A. and Nagle, J. F. 1979 *Chem. Phys.* **43**, 329.

Yang, C. N. 1952 *Phys. Rev.* **85**, 808.

Appendices

A

Answers to problems

Chapter 1

1.1 False. The probability of being in any *particular* state with energy E is indeed proportional to $e^{-\beta E}$, as we stated in Section 1.2. However, if there are several states with energy E then the system is more likely to have that energy than if there are few such states. Often this is expressed in terms of a "density of states" $\rho(E)$ which is the number of states with energy E. Then the probability of having a certain energy varies in proportion to $\rho(E)\, e^{-\beta E}$. The density of states is discussed in more detail in Sections 4.5.3 and 6.3.

1.2 For this simple two state system, the sum rule (1.2) tells us that $w_1 = 1 - w_0$. Using this relation we can write the master equation in the form

$$\frac{\mathrm{d}w_0}{\mathrm{d}t} = R_0\left\{1 - w_0\left[1 + e^{-\beta(E_1 - E_0)}\right]\right\}.$$

Since E_0 and E_1 are constant, we can directly integrate this equation to give w_0, with the integration constant being set by the initial conditions:

$$w_0 = \frac{1 - e^{-R_0(1 + e^{-\beta(E_1 - E_0)})t}}{1 + e^{-\beta(E_1 - E_0)}}.$$

Taking the limit $t \to \infty$, the exponential in the numerator vanishes, and by rearranging we can show that the probability $p_0 = w_0(\infty)$ is

$$p_0 = \frac{e^{-\beta E_0}}{e^{-\beta E_0} + e^{-\beta E_1}},$$

as in Equation (1.5). The solutions for w_1 and p_1 follow from the sum rule.

1.3 If we have n out of our N particles in state 1 then the total energy of the system is $H = E_1 n$. The number of states with this energy is the number of ways of choosing n from N, and hence the partition function is

$$Z = \sum_n \binom{N}{n} e^{-\beta E_1 n} = \left[1 + e^{-\beta E_1}\right]^N.$$

Using Equation (1.9), the internal energy is then

$$U = -\frac{1}{Z}\frac{\partial Z}{\partial \beta} = \frac{E_1 N}{1 + e^{\beta E_1}}.$$

1.4 In one dimension, the Hamiltonian for the Ising model, Equation (1.30), can be written as $H = -J\sum_i s_i s_{i+1}$. Making the proposed change of variables this becomes

$$H = -JN + 2J\sum_i \sigma_i = -JN + 2Jn,$$

where n is the number of variables σ_i which are equal to $+1$. With periodic boundary conditions n must be even, but we can ignore this constraint in the limit of large N, and the problem then becomes the same as the previous one with $E_1 = 2J$, except for the additive constant $-JN$. Thus the internal energy of the system is

$$U = -JN + \frac{2JN}{1 + e^{2\beta J}} = -JN\tanh\beta J.$$

Chapter 2

2.1 The appropriate generalization of (1.1) to the case of a discrete time variable is

$$w_\nu(t+1) - w_\nu(t) = \sum_\mu [w_\mu(t)P(\mu \to \nu) - w_\nu(t)P(\nu \to \mu)].$$

Using Equation (2.5) we can show that the second term is just

$$\sum_\mu w_\nu(t)P(\nu \to \mu) = w_\nu(t),$$

and the required result follows immediately.

2.2 We should choose the transition rates to be inversely proportional to the amount of time we want the system to spend in each state. In other words, $P(\mu \to \nu) \propto \exp(\beta E_\mu)$. The most efficient choice is then to set

$$P(0 \to 1) = e^{-\beta(E_2 - E_0)}, \quad P(1 \to 2) = e^{-\beta(E_2 - E_1)}, \quad P(2 \to 0) = 1.$$

Chapter 3

3.1 Using the full Ising Hamiltonian, Equation (3.1), the generalization of Equation (3.8) to the case of finite B is

$$E_\nu - E_\mu = -J \sum_{i \text{ n.n. to } k} s_i^\mu (s_k^\nu - s_k^\mu) - B \sum_i s_i^\nu + B \sum_i s_i^\mu.$$

The first sum gives exactly the same thing as in the $B = 0$ case. The second two give $-B(s_k^\nu - s_k^\mu)$, since all the other spins are unchanged and cancel out. Using Equation (3.9) we then get

$$E_\nu - E_\mu = 2Js_k^\mu \sum_{i \text{ n.n. to } k} s_i^\mu + 2Bs_k^\mu.$$

3.2 The expression for our estimate of the error on the mean square of a set of numbers is a generalization of Equation (3.37). Our best estimate of the mean square is

$$\overline{x^2} = \frac{1}{n} \sum_{i=1}^n x_i^2,$$

and our estimate of the variance on this quantity is

$$\text{var } \overline{x^2} = \frac{1}{n} \sum_{i=1}^n (x_i^2 - \overline{x^2})^2 = \overline{x^4} - \overline{x^2}^2.$$

The error is then

$$\sigma = \sqrt{\frac{1}{n-1}(\overline{x^4} - \overline{x^2}^2)}.$$

Applying this to the twelve numbers given in the problem, we get a value of $\sigma = 5.05$. Directly applying the jackknife method to the same set of numbers, we get a value of $\sigma = 5.28$.

3.3 There are a number of ways we might estimate the partition function. Perhaps the most direct is to measure the internal energy and make use of Equation (1.9) to derive the following formula:

$$Z(\beta) = \mathcal{N} \exp\left[-\int_0^\beta U(\beta) \, d\beta\right].$$

If we measure U over a range of temperatures from $\beta = 0$ (which is the same as $T = \infty$) to the temperature we are interested in, we can calculate Z by numerically integrating in much the same way as we did for the entropy in Section 3.5. The integration constant $\mathcal{N}$ is not very important since it drops out of most calculations of real quantities. If we want to fix its value however, we can do so by noting that at $\beta = 0$ all the terms in the partition function, Equation (1.6), are equal to 1, so that $Z(\beta = 0)$ is just equal to the number of states of the system. This then is also the value of the constant $\mathcal{N}$.

3.4 There are many possible solutions to this problem of course. Here is one simple program to perform the desired calculation for a system of N spins. It is written in C and relies on a function drandom() which returns a random real number between zero and one. This is just the guts of the program. The code given here performs one Monte Carlo step of the simulation. If you want to actually use it, you'll have to add some extra lines to make it into a complete program. In particular, you'll need to set up the initial values of the array s[] which stores the spins, and the variables beta and J which store the corresponding quantities from the model.

```
int s[N];
double beta,J;

void move()
{
   int i;
   int n1,n2,delta;

   i = N*drandom();

   if ((n1=i+1)>=N) n1 -= N;
   if ((n2=i-1)<0) n2 += N;

   delta = s[i]*(s[n1]+s[n2]);

   if (delta<=0) {
     s[i] = -s[i];
   } else if (drandom()<exp(-4*beta*J)) {
     s[i] = -s[i];
   }
}
```

This is not the most efficient implementation of this algorithm. It could be made slightly faster by storing the value of $e^{-4\beta J}$ in a separate variable so that we don't have to recalculate it at every step.

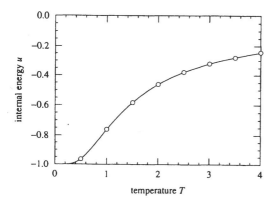

The points in this figure are our results for the internal energy per spin of an $N = 1000$ system simulated using the program above with $J = 1$. The solid line is the exact solution from Problem 1.4.

Chapter 4

4.1 In order to produce a configuration of the system which is statistically independent of the current one, the domain walls have to diffuse an average distance on the order of the correlation length ξ. If the domain walls perform a random walk across the system, then the mean square distance moved increases linearly with time $\langle x^2 \rangle \sim t$. Thus the time τ taken to diffuse a distance ξ—the correlation time—goes as $\tau \sim \xi^2$. Comparing this result with Equation (4.6) we get $z = 2$. In fact, the measured values of z for the two- and three-dimensional Ising models, Table 4.1, are close to 2 for the Metropolis algorithm, as indeed they are for all single-spin-flip algorithms, indicating that the domain walls do approximately perform a random walk for such algorithms. For the cluster algorithms however, the domain walls are more mobile, and this is the fundamental reason why these algorithms achieve lower values of z.

4.2 Consider as we did in Section 4.2.1 two states separated by the flipping of a single cluster, and suppose that that cluster contains k spins. The appropriate generalization of Equation (4.11) to the case of non-zero magnetic field is

$$E_\nu - E_\mu = 2J(m - n) + 2SBk,$$

where $S = \pm 1$ is the value of the spins in the cluster before they are flipped over. Substituting this into Equation (4.10) and rearranging, we find that Equation (4.12) is now modified thus:

$$\frac{A(\mu \to \nu)}{A(\nu \to \mu)} = [e^{2\beta J}(1 - P_{\text{add}})]^{n-m} e^{-2\beta SBk}.$$

The simplest way to satisfy this equation is to choose $P_{\text{add}} = 1 - e^{-2\beta J}$, as in the normal Wolff algorithm, so that the factor inside the square brackets is 1. Then the ratio of the acceptance probabilities for forward and backward moves is $e^{-2\beta SBk}$. The most efficient choice of acceptance ratio to satisfy this constraint is

$$A(\mu \rightarrow \nu) = \begin{cases} e^{-2\beta SBk} & \text{if } SB > 0 \\ 1 & \text{otherwise.} \end{cases}$$

Notice that this reduces to the usual Wolff result $A(\mu \rightarrow \nu) = 1$ when $B = 0$.

4.3 The Swendsen–Wang algorithm is just the same as the Wolff algorithm except that it covers the entire lattice with clusters, rather than just creating one. In fact, we already worked out the equivalent of Equation (4.24) for this case in Section 4.3.2 when we showed (Equation (4.21)) that

$$\langle m^2 \rangle = \frac{1}{N^2} \Big\langle \sum_i n_i^2 \Big\rangle.$$

Thus, using Equation (1.36), the susceptibility above T_c is given by

$$\chi = \frac{\beta}{N} \Big\langle \sum_i n_i^2 \Big\rangle.$$

Chapter 5

5.1 The algorithm satisfies ergodicity for the same reason that the Kawasaki algorithm does; all exchanges of up- and down-spins in the Kawasaki algorithm are also possible moves in this algorithm, so that if one of them satisfies ergodicity then the other does as well. The proof of detailed balance is also the same as for the Kawasaki algorithm except that we need to show that the selection probabilities $g(\mu \rightarrow \nu)$ and $g(\nu \rightarrow \mu)$ are the same in either direction. This however is simple since the probability of picking a particular minority spin is always the same because the number of such spins is conserved, and the probability of picking a particular one of its neighbours is constant at $1/z$, where z is the lattice coordination number. To see why the algorithm is more efficient than the Kawasaki one, consider the case $\rho < \frac{1}{2}$. In this case the particles (rather than the vacancies) are in the minority. Then there are two ways to pick any particular particle–particle pair, one way of picking a particle–vacancy pair, and no ways to pick a vacancy–vacancy pair. When $\rho = \frac{1}{2}$, particle–particle and vacancy–vacancy pairs are equally common on average, which means that the Kawasaki algorithm and this new one are equally efficient. But when $\rho < \frac{1}{2}$ the number of particle–particle pairs goes down relative to the number of vacancy–vacancy ones, and hence the new algorithm does better than the Kawasaki one. However, the algorithm is also more complicated to program, since it requires us to maintain an up-to-date list of where all the particles in the system are. This makes the program slower than in the Kawasaki case. Thus in practice we have to have a density ρ which is substantially less than $\frac{1}{2}$ in order for the algorithm to pay dividends. A similar argument can be made to show that we get an efficiency gain when ρ is significantly greater than $\frac{1}{2}$ also.

5.2 The energy difference between the two states can be written

$$E_\nu - E_\mu = -J \sum_{\langle ij \rangle} \left(s_i^\nu s_j^\nu - s_i^\mu s_j^\mu \right).$$

Only the spins k and k' which are exchanged change their values. All others stay the same, so the only contributions which don't cancel out are the ones involving these spins. Noting that the interaction between the two spins themselves doesn't change when they are exchanged, we then get

$$E_\nu - E_\mu = -J \left[\sum_{\substack{i \neq k' \text{ and} \\ \text{n.n. to } k}} s_i^\mu \left(s_k^\nu - s_k^\mu \right) + \sum_{\substack{j \neq k \text{ and} \\ \text{n.n. to } k'}} s_j^\mu \left(s_{k'}^\nu - s_{k'}^\mu \right) \right].$$

In the case where $s_k \neq s_{k'}$, which is the only case we are really interested in, we can simplify this using Equation (3.9) to give

$$E_\nu - E_\mu = 2J \left[s_k^\mu \sum_{\substack{i \neq k' \text{ and} \\ \text{n.n. to } k}} s_i^\mu + s_{k'}^\mu \sum_{\substack{j \neq k \text{ and} \\ \text{n.n. to } k'}} s_j^\mu \right],$$

which is the answer we were looking for.

5.3 Suppose that $\rho \leq \frac{1}{2}$, so that it is the particles rather than the vacancies which form the droplet in configuration (a). In terms of its width and height, W and H, the number of unsatisfied bonds around the perimeter of the droplet is $2W + 2H$. For a droplet of a given area, this number is minimized when the droplet is a square. The area of the droplet is ρL^2, which means that the square has dimensions $W = H = \sqrt{\rho}L$, and hence the number of unsatisfied bonds is $4\sqrt{\rho}L$. In configuration (b), the total interface length is simply the length of two straight interfaces which stretch across the whole system, i.e., $2L$. The lengths of the interfaces in cases (a) and (b) are thus equal when $\rho = \frac{1}{4}$. Below this density the square droplet has the lowest possible interface length and so is the preferred configuration. Above it the banded configuration is preferred. When $\rho > \frac{1}{2}$ a similar argument shows that the droplet configuration is only energetically favourable above $\rho = \frac{3}{4}$.

Chapter 6

6.1 The condition of detailed balance, Equation (2.14), tells us that the transition probabilities for transitions between state 1 and the state at the top of the barrier must satisfy

$$\frac{P(1 \rightarrow B)}{P(B \rightarrow 1)} = e^{-\beta(E_B - E_1)}.$$

Since $P(B \rightarrow 1)$ is a probability, the largest value it can have is 1, which means that the largest value which $P(1 \rightarrow B)$ can have is $\exp[-\beta(E_B - E_1)] = \exp[-\beta B_1]$, as claimed in the problem. (This of course is just the Metropolis algorithm for this transition.) The exponential increase in the time taken to cross a barrier with barrier height is called the Arrhenius law, and is discussed in more detail in Chapter 11.

6.2 The Hamiltonian of the Mattis model can be written as

$$H = -\sum_{\langle ij \rangle} \xi_i \xi_j s_i s_j.$$

Defining a new set of variables $t_i = \xi_i s_i$, which take the values ± 1, this becomes

$$H = -\sum_{\langle ij \rangle} t_i t_j,$$

which is the Hamiltonian of the normal Ising model with $J = 1$ and is therefore not glassy. In other words, there are possible choices of the values of J_{ij} which are half $+1$ and half -1 and appear random to the eye, but nonetheless are not glassy. However, the number of choices of the J_{ij} in the Mattis model is 2^N where N is the number of sites on the lattice. If we choose J_{ij} purely at random, there are $2^{Nz/2}$ choices, where z is the lattice coordination number. Since $z > 2$ for all lattices in more than one dimension, this means that the number of possible choices in the general case becomes exponentially larger than the number for the Mattis model as N becomes large.

6.3 Using the expression $H = -JN + 2Jn$ for the Hamiltonian from Problem 1.4 and noting that the number of states with a certain value of n is just the number of ways of choosing n spins out of N, we can write the number of states with energy E as

$$\rho(E) = \binom{N}{n} = \binom{N}{\frac{1}{2}N + \frac{E}{2J}}.$$

The most efficient choice of acceptance ratio is then given by Equation (6.20) with this expression substituted in.

6.4 As explained in Section 1.2.1, the range of energies which the system passes through, measured in terms of energy per spin, decreases as $1/\sqrt{N}$ with the size of the system (or as $L^{-d/2}$ with the linear dimension). The

simulation with the highest temperature in a simulated tempering calculation should be above the glass temperature T_g in order that we sample a
representative selection of states. The number of simulations necessary to
cover the energy range from the energy of this simulation $u(T_g)$ down to the
energy we are interested in therefore increases as $\sqrt{N}$ with the system size.
For a three-dimensional system, for instance, it will go as $L^{3/2}$.

Chapter 7

7.1 The number of possible configurations of arrows at a vertex is the number of ways of choosing the three outgoing arrows out of the six possible
directions, which is 20. If we ignore the first ice rule, this gives us a total
of 20^N possible states of the lattice, where N is the number of vertices. By
Pauling's argument however, there is a 50% probability that any bond in
one of these configurations will be doubly occupied, or not occupied at all.
There are a total of $3N$ bonds, and hence when we reinstate the first ice rule
we reduce the number of states by a factor of 2^{3N}. Thus the entropy per
vertex is approximately $N^{-1} \log(20^N / 2^{3N}) = \log \frac{5}{2} \simeq 0.916$.

7.2 If we again take a $q = 3$ Potts model, we can ensure that nearest-
neighbour spins have different values if we set $J = -\infty$. In addition, from
Equation (7.20) we see that the energy of the F model is proportional to
the number of next-nearest-neighbour spin pairs which have the same value,
and we can incorporate this by giving our Potts model finite ferromagnetic
next-nearest-neighbour interactions. This makes the Potts model equivalent
to the F model, except for an (infinite) additive shift in the energy.

7.3 The simplest way to do this is to create a loop algorithm similar to that
for the six-vertex model, except that when the loop passes through a vertex
it can leave along *any* bond with equal probability, regardless of the state of
the arrows at the vertex. It is not hard to show that this will produce only
configurations containing the eight allowed vertex types. Both short and
long loop versions of this algorithm are possible, and both obey ergodicity
and detailed balance for the same reasons that they did in the six-vertex
model.

Chapter 8

8.1 For a finite sized system the specific heat per site at the critical temperature scales with the size L of the system as $c \sim L^{\alpha/\nu}$ (see Equation (8.43)). This means that the specific heat for the whole system scales
as $C \sim L^{d+\alpha/\nu}$. Thus the temperature range ΔT varies with system size as
$\Delta T \sim L^{-d/2-\alpha/2\nu}$. This expression can be simplified using Equation (8.93)
to give $\Delta T \sim L^{-1/\nu}$. Interestingly, as Equation (8.54) shows, the range of

values of the reduced temperature t to which a particular range of the scaling variable x corresponds in a finite size scaling calculation also scales as $L^{-1/\nu}$. This means that the range of values of x over which we can estimate a scaling function like $\tilde{\chi}(x)$ using a single simulation at T_c does not change as the system simulated gets larger. This makes the single histogram method very convenient for many finite size scaling calculations.

8.2 The criterion which we applied to derive Equation (8.10) was that in order to extrapolate to a temperature T, the number of samples in the histogram at energy $U(T)$ should be significantly greater than 1. Let us say that we require there to be at least k samples at this energy. Then, assuming that the distribution of energies sampled is roughly Gaussian, our criterion can be written

$$\frac{n}{\sqrt{2\pi\sigma_E^2}} \exp\left\{ -\frac{[U(T) - U(T_0)]^2}{2\sigma_E^2} \right\} = k.$$

Using Equation (8.9) to eliminate the difference $U(T) - U(T_0)$ in favour of ΔT and then rearranging, we find the desired result that $\Delta T \sim \sqrt{\log n}$. Note that this is a very slow increase in ΔT with n. In terms of extrapolation range, increasing the length of our Monte Carlo simulation does us very little good, although it does still increase the accuracy of our estimate of the quantity of interest in the usual $\sqrt{n}$ fashion.

8.3 The answer to this problem is exactly the same as the answer to Problem 6.4. The multiple histogram method requires there to be an overlap in the ranges of energy sampled by the separate simulations, and these ranges decrease with system size N as $1/\sqrt{N}$. Thus the number of simulations required to cover a given range goes up as $\sqrt{N}$. The fact that the simulated tempering and multiple histogram methods place exactly the same requirements on the overlap of simulations is actually quite useful. It tells us that we can always use the multiple histogram method to interpolate between the different temperatures used in a simulated tempering calculation.

8.4 In systems of finite size, the effective temperature of the critical point is higher than in infinite systems. This is because the length scale of the fluctuations ξ diverges as we approach T_c from above until it reaches the system size, which it does before we get to T_c. From there on the system behaves as it would below T_c since the fluctuations fill the entire system. Thus we expect the peak in the susceptibility to occur at a temperature $T > T_c$ in a finite system. With a scaling form like Equation (8.51) this means that the scaling function must have its peak at $t > 0$.

8.5 The equivalent of Equation (8.46) for τ is

$$\tau = \xi^z \tau_0(L/\xi).$$

Defining the scaling function

$$\tilde{\tau}(x) = x^{-z\nu}\tau_0(x^\nu),$$

this becomes

$$\tau = L^z\tilde{\tau}(L^{1/\nu}|t|),$$

which is the appropriate scaling equation for the finite size behaviour of the correlation time.

Chapter 10

10.1 The answer to this problem comes from Problem 5.3, where we showed that the lowest energy state of such a system is one in which the up- and down-pointing spins form two bands stretching across the system, with two straight domain walls dividing them. For a phase-separating system, the final state after running the simulation for a long time will be just such a state. At finite temperature the domain walls will usually not be straight and will fluctuate in position, but the overall geometry of the state will be as described in Problem 5.3.

10.2 The probability of choosing spin i as our first spin is proportional to $(z - n_i)\exp(-4\beta n_i)$. Spin i has $z - n_i$ anti-aligned neighbours, so the probability of choosing a particular neighbour j out of these is $1/(z-n_i)$. The total probability of picking the pair i,j is thus proportional to $\exp(-4\beta n_i)$. The same pair of spins could also have been picked in the opposite order, starting with spin j. The probability of this happening is proportional to $\exp(-4\beta n_j)$. Thus the total probability of picking the pair i,j is proportional to $\exp(-4\beta n_i) + \exp(-4\beta n_j)$, as desired.

Chapter 11

11.1 The ratio of the rates for the exchange and hopping processes is $10e^{-\beta\Delta B}$. Since the exchange process moves atoms $\sqrt{2}$ times as far as the hopping process, the diffusion constants for the two processes will be equal when this ratio is $\frac{1}{2}$ (because the diffusion constant is proportional to the mean square distance moved by an atom in a given time). Plugging in the value of ΔB, this means that the exchange process dominates for temperatures above about 770 K. This is below the melting point of copper, but still sufficiently far above typical laboratory temperatures that we don't usually have to worry about it.

11.2 As pointed out in Section 11.2.2, the time taken per Monte Carlo step increases logarithmically with the number of possible moves. Since the number of moves scales on average as the number of sites on the lattice, each move will take about a factor $\log 250^2 / \log 100^2 = 1.20$ longer on the larger system. The total number of Monte Carlo steps we have to perform in order to simulate a given interval of real time increases linearly with the number of lattice sites, so that we have to perform about a factor of $250^2/100^2 = 6.25$ more steps in the larger simulation. Overall, therefore, the simulation will take about 7.5 times as much CPU time on the larger lattice, or 7.5 hours in this case.

Chapter 12

12.1 The simplest expression which gives v in terms of only one combination of the independent variables N and E is

$$vN^{3/2} = c_1 N^{1/2} E + c_3 [N^{1/2} E]^3.$$

12.2 (a) Since the chain is not allowed to move transversely to its own length its only mode of translation is the longitudinal reptation mode. (b) For a chain of N reptons there are $N - 2$ adjacent pairs of links and two end links giving a total of N overall. In two dimensions each one can move to three other positions, so the total number of possible moves is $3N$. As in the repton model, each move should be attempted once per unit time, so the appropriate time increment is $\Delta t = 1/(3N)$ per move. (c) The six possible moves of the first and last links on the chain are always allowed. The probability of two adjacent links in the interior of the chain being anti-parallel is $\frac{1}{4}$, and each such anti-parallel pair has three possible moves. Thus the average number of allowed moves in the chain is $6 + \frac{3}{4}(N - 2) = \frac{9}{2} + \frac{3}{4}N$. The average probability of a move being allowed is the ratio of allowed moves to total moves or $(\frac{9}{2} + \frac{3}{4}N)/(3N) = (N + 6)/(4N)$.

Chapter 13

13.1 (a) On the square lattice the disks cover $\frac{1}{4}\pi \simeq 78.5\%$ of the space. (b) On the triangular lattice they cover $\pi/(2\sqrt{3}) \simeq 90.7\%$. The triangular packing is the most dense packing of circles in two-space, which is the main reason why the triangular lattice crops up so often in physical systems.

13.2 Before we stretch the lattice, the neighbours of the point (h, k, l) are $(h \pm 1, k \pm 1, l \pm 1)$, all located at a distance of $\sqrt{3}$. After stretching, these neighbours are located at a distance of 2. In addition the sites $(h \pm 2, k, l)$ and $(h, k \pm 2, l)$ are now also located at the same distance from our point, so that the lattice has become 12-fold coordinated. Since it is still a Bravais lattice, it must be the fcc lattice, which is the only 12-fold coordinated such lattice.

13.3 This can be accomplished quite easily by keeping a record at each node of the number of values stored in the leaves below that node. Then we can choose a random leaf as follows. Starting at the root node we examine its first child node. Suppose this child has n_1 values stored in leaves below it. If $r < n_1$ we proceed to the child. Otherwise we subtract n_1 from r and proceed to the other child. We repeat this process all the way down the tree until we come to a leaf. Since at each node we have chosen a child in proportion to the number of leaves below it, every leaf has exactly the same chance as every other of being chosen.

13.4 Let us write the average number of levels of the tree which we have to search through as $\alpha \log_2 N$. The first level takes us one unit of time to search through. At the second level we have a probability of $\frac{2}{3}$ that the remaining subtree below us contains $\frac{2}{3}N$ values, and $\frac{1}{3}$ that it contains $\frac{1}{3}N$ values. Thus the average total number of levels we have to search through can also be written as $1 + \frac{2}{3}\alpha \log_2\left(\frac{2}{3}N\right) + \frac{1}{3}\alpha \log_2\left(\frac{1}{3}N\right)$. Equating this with our first expression and solving for the multiplicative prefactor α, we find $\alpha = \left[\log_2(3) - \frac{2}{3}\right]^{-1} \simeq 1.09$. In other words, the tree can be quite badly unbalanced and the search is still almost as fast is it is in the balanced tree.

Chapter 14

14.1 (a) Yes, it is possible. If we tilt the lines separating the domains 30° away from the vertical, fixing the spins $\left(\frac{1}{2}i, i \bmod 22\right)$ for all $0 \le i < 66$, the lattice forms a single strip which wraps three times around the periodic boundary conditions. Cutting this strip into six equal pieces, each cut requiring us to fix an additional six spins, we achieve our goal. The total number of fixed spins is $66 + 36 = 102$. (b) The most economical way of dividing up the lattice is to fix spins along lines set at 45° to the vertical. If we fix the set of spins (i, i) and $(i, L-i-1)$ for all $0 \le i < L$, and another set exactly the same but translated by $\frac{1}{2}L$ horizontally (with periodic boundary conditions), then we can divide an $L \times L$ lattice into eight equal domains. Each line fixes L spins, but 8 spins are shared between lines, so the total number of fixed spins is $4L - 8$. For $L = 252$ this gives us our 1000 fixed spins. Thus we can divide a 252×252 lattice into eight domains by fixing 1000 spins. (If you can find a larger lattice which can be divided with this many fixed spins, we want to hear about it.)

14.2 No. The spins located at the points where the two sets of boundaries cross belong to both sets and hence will never get changed. We must use at least three different sets of boundaries to achieve ergodicity.

14.3 The volume V of the region covered by each processor scales as n^{-1}, where n is the number of processors used. On average the values of about half of the spins on the lattice have to be transmitted every time the boundaries are shifted, so each processor will have to send and receive a number of values which also scales as n^{-1}. The amount of CPU time spent simulating each region for one correlation time scales as L^{d+z} (see Equation (4.8)) where $L \sim V^{1/d}$ is the linear dimension of the region. Thus the CPU time scales with the number of processors as $n^{-1-z/d}$. The ratio of communication time to calculation time therefore scales as $n^{-1}/n^{-1-z/d} = n^{z/d}$. For the alternative algorithm we have to transmit a message every time we change a spin on the boundary of a region. The boundaries cover a fraction of the system which scales as L^{-1}. Using the relations given above this means that the fraction of time spent on communication goes as $n^{1/d}$. This implies that the second algorithm will, at least in theory, have superior performance if we use a sufficiently large number of processors provided the dynamic exponent z is greater than one. Since $z \simeq 2$ for the Metropolis algorithm, does this mean that we should be using the second algorithm? No, we should not, for a number of reasons. First there are the practical reasons to do with the synchronization of the different processors given at the beginning of Section 14.2.1. In addition however, the calculations above do not really give the whole picture, since they assume that the time spent on communication is proportional to the amount of data to be communicated. In practice, the time spent usually does increase linearly with the amount of data, but it also has an offset—an extra time penalty associated with transmitting any message, no matter how small—whose size is independent of the amount of data. Depending on the type of parallel computer used for the computation, this offset can range from the negligible to being much greater than the linear term. For some computers therefore, the scaling arguments above may be completely irrelevant (although for some they are not).

Chapter 15

15.1 (a) Writing the occupation variables of the two sites after the move as s_i' and s_j', the truth table is

s_i	s_j	s_i'	s_j'
0	0	0	0
0	1	1	0
1	0	0	1
1	1	1	1

Thus the master expressions are $s_i' = s_j$ and $s_j' = s_i$.

(b) The values of s_i' and s_j' now depend on r as well as s_i and s_j. The full truth table is

s_i	s_j	r	s_i'	s_j'
0	0	0	0	0
0	0	1	0	0
0	1	0	1	0
0	1	1	1	0
1	0	0	1	0
1	0	1	0	1
1	1	0	1	1
1	1	1	1	1

Suitable master expressions are

$$s_i' = s_j \vee (s_i \wedge \overline{r}), \qquad s_j' = s_i \wedge (s_j \vee r).$$

15.2 (a) The truth table for h is

a_i	b_i	a_j	b_j	h
0	0	0	0	0
0	0	0	1	0
0	0	1	0	0
0	0	1	1	1
0	1	0	0	0
0	1	0	1	0
0	1	1	0	1
0	1	1	1	0
1	0	0	0	0
1	0	0	1	1
1	0	1	0	0
1	0	1	1	0
1	1	0	0	1
1	1	0	1	0
1	1	1	0	0
1	1	1	1	0

Using the additive normal form introduced in Section 15.3 we can then write h as

$$\begin{aligned} h = \; & [\overline{a_i} \wedge \overline{b_i} \wedge a_j \wedge b_j] \vee [\overline{a_i} \wedge b_i \wedge a_j \wedge \overline{b_j}] \\ & \vee [a_i \wedge \overline{b_i} \wedge \overline{a_j} \wedge b_j] \vee [a_i \wedge b_i \wedge \overline{a_j} \wedge \overline{b_j}]. \end{aligned}$$

A much simpler expression for the same quantity which makes use of the XOR operation is

$$h = (a_i \oplus a_j) \wedge (b_i \oplus b_j).$$

Note however, that there is no easy way of finding this simplification using a Karnaugh map. Karnaugh maps never generate Boolean expressions containing the XOR operation.

(b) Appropriate master expressions for the interior moves are

$$a'_i = (h \wedge r_0) \vee (\overline{h} \wedge a_i), \qquad a'_j = (h \wedge \overline{r_0}) \vee (\overline{h} \wedge a_j),$$
$$b'_i = (h \wedge r_1) \vee (\overline{h} \wedge b_i), \qquad b'_j = (h \wedge \overline{r_1}) \vee (\overline{h} \wedge b_j),$$

where r_0 and r_1 are random bits which are one with probability $\frac{1}{2}$.

Chapter 16

16.1 It is simple to show that

$$i_{n+1} = ai_n \bmod m = ai_n - \lfloor ai_n/m \rfloor m,$$

and hence that

$$i_{n+2} = a(ai_n - \lfloor ai_n/m \rfloor m) \bmod m = a^2 i_n \bmod m.$$

Iterating the same argument we can then derive the skip formula

$$i_{n+k} = a^k i_n \bmod m.$$

16.2 The position in the array at which we store the number i depends only on the value of i itself, so that each time a particular value is generated it is stored in the same position. Thus, after the generator has been running for a while, the number pulled out of the array for a particular value of i will also be the same every time we generate that value of i. Hence the period of this generator will be the same as that of the original linear congruential generator, which makes this a poor shuffling scheme.

16.3 Using the transformation method, the number x should be generated every time our uniform random number generator produces a number

$$r = \int_0^x e^{-x} \, dx = 1 - e^{-x}.$$

Rearranging for x this gives us

$$x = -\log(1 - r).$$

Note that, as with Equation (16.31), we should keep the $1 - r$ here. We cannot get rid of the subtraction because if we did we would end up trying to calculate the logarithm of zero every so often, which would probably cause our program to crash.

B

Sample programs

In this appendix we give a number of examples of computer programs which implement algorithms described in this book. We have not attempted to give programs for all the algorithms discussed—there are too many of them. Neither have we given whole programs. Most of the repetitive details such as creating variables and initializing them, seeding random number generators and printing out results have been omitted. What remains is the kernel of code which makes the Monte Carlo algorithm run. By studying these programs you may get an idea of some of the tricks which can be used to make a Monte Carlo simulation fast.

The programs here are written in C. Many of them call a function drandom() which returns a random double-precision floating point number in the range $0 \leq r < 1$. Three different possible forms for this function are given in Section B.5.

B.1 Algorithms for the Ising model

In this section we give three different algorithms for simulating the ordinary Ising model introduced in Section 1.2.2 on a square lattice in two dimensions.

B.1.1 Metropolis algorithm

Here we give two functions which implement the single-spin-flip Metropolis algorithm of Section 3.1 for the Ising model with $J = 1$. The first, initialize(), sets up a lookup table of the possible values of the acceptance ratio $\exp(-\beta\Delta E)$. The second, sweep(), performs one "sweep" of the lattice, i.e., one Monte Carlo step per spin, using the Metropolis algorithm. The program makes use of helical boundary conditions for speed (see Section 13.1.1).

```
/* beta    = inverse temperature
 * prob[] = array of acceptance probabilities
 * s[]     = lattice of spins with helical boundary conditions
 * L       = constant edge length of lattice
 */

#include <math.h>

#define N (L*L)
#define XNN 1
#define YNN L

int s[N];
double prob[5];
double beta;

void initialize()
{
  int i;
  for (i=2; i<5; i+=2) prob[i] = exp(-2*beta*i);
}

void sweep()
{
  int i,k;
  int nn,sum,delta;

  for (k=0; k<N; k++) {

    /* Choose a site */

    i = N*drandom();

    /* Calculate the sum of the neighbouring spins */

    if ((nn=i+XNN)>=N) nn -= N;
    sum = s[nn];
    if ((nn=i-XNN)<0) nn += N;
    sum += s[nn];
    if ((nn=i+YNN)>=N) nn -= N;
    sum += s[nn];
    if ((nn=i-YNN)<0) nn += N;
    sum += s[nn];
```

```
    /* Calculate the change in energy */

    delta = sum*s[i];

    /* Decide whether to flip spin */

    if (delta<=0) {
      s[i] = -s[i];
    } else if (drandom()<prob[delta]) {
      s[i] = -s[i];
    }

  }
}
```

B.1.2   Multispin-coded Metropolis algorithm

The following function performs one sweep of the lattice using the multispin-coded version of the Metropolis algorithm described in Section 15.1.2. The random bit patterns needed by the algorithm are generated using Equation (16.47) with $p_1 = 0$ and $p_2 = \frac{1}{2}$. In addition to the random floating-point function drandom(), this program also calls a function lrandom() which returns a random unsigned long integer. One possible implementation of this function is given in Section B.5.3.

```
/* p0   = exp(-8*beta*J)
 * p1   = (exp(-4*beta*J) - p0)/(1 - p0)
 * s[]  = lattice of spins with helical boundary conditions
 * L    = constant edge length of lattice
 */

#define N (L*L)
#define XNN 1
#define YNN L
#define ZERO 0x00000000

unsigned long s[N];
double p0,p1;

sweep()
{
  int i,k;
  int n1,n2,n3,n4;
```

```
unsigned long spin;
unsigned long a1,a2,a3,a4;
unsigned long R1,R2;
unsigned long r0,r1;

for (k=0; k<N; k++) {

  /* Choose a spin and find its four neighbours */

  i = N*drandom();
  spin = s[i];

  if ((n1=i+XNN)>=N) n1 -= N;
  if ((n2=i-XNN)<0) n2 += N;
  if ((n3=i+YNN)>=N) n3 -= N;
  if ((n4=i-YNN)<0) n4 += N;

  /* Calculate the quantities a1 to a4 */

  a1 = spin^s[n1];
  a2 = spin^s[n2];
  a3 = spin^s[n3];
  a4 = spin^s[n4];

  /* Now calculate R1 and R2 */

  R1 = a1|a2|a3|a4;
  R2 = ((a1|a2)&(a3|a4))|((a1&a2)|(a3&a4));

  /* Generate the random bit patterns */

  if (drandom()<2*p0) r0 = lrandom();
  else r0 = ZERO;
  if (drandom()<2*p1) r1 = lrandom();
  else r1 = ZERO;

  /* Update the spin */

  s[i] ^= R2|(R1&r1)|r0;
}
}
```

B.1.3 Wolff algorithm

Here we give a function to perform one Monte Carlo step of the Wolff cluster algorithm (Section 4.2) for the two-dimensional Ising model. See Section 13.2.5 for a description of how this program works. In this implementation we use a LIFO buffer or stack to hold the spins in the cluster.

```
/* padd = 1 - exp(-2*beta*J)
 * s[]  = lattice of spins with helical boundary conditions
 * L    = constant edge length of lattice
 */

#define N (L*L)
#define XNN 1
#define YNN L

int s[N];
double padd;

void step()
{
  int i;
  int sp;
  int oldspin,newspin;
  int current,nn;
  int stack[N];

  /* Choose the seed spin for the cluster,
   * put it on the stack, and flip it
   */

  i = N*drandom();

  stack[0] = i;
  sp = 1;
  oldspin = s[i];
  newspin = -s[i];
  s[i] = newspin;

  while (sp) {

    /* Pull a site off the stack */

    current = stack[--sp];
```

```
/* Check the neighbours */

if ((nn=current+XNN)>=N) nn -= N;
if (s[nn]==oldspin)
  if (drandom()<padd) {
    stack[sp++] = nn;
    s[nn] = newspin;
  }

if ((nn=current-XNN)<0) nn += N;
if (s[nn]==oldspin)
  if (drandom()<padd) {
    stack[sp++] = nn;
    s[nn] = newspin;
  }

if ((nn=current+YNN)>=N) nn -= N;
if (s[nn]==oldspin)
  if (drandom()<padd) {
    stack[sp++] = nn;
    s[nn] = newspin;
  }

if ((nn=current-YNN)<0) nn += N;
if (s[nn]==oldspin)
  if (drandom()<padd) {
    stack[sp++] = nn;
    s[nn] = newspin;
  }

  }
}
```

B.2   Algorithms for the COP Ising model

In this section we give two algorithms for the conserved-order-parameter
(COP) Ising model of Chapter 5 on a square lattice in two dimensions.

B.2.1   Non-local algorithm

Here we give two functions to implement the non-local algorithm described
in Section 5.2 with $J = 1$. The first, initialize(), creates a lookup table of

acceptance probabilities, just as in the Metropolis program of Section B.1.1. The second, sweep(), does the actual simulation. Before we use these functions, we also need to set up the arrays up[] and down[] with the initial coordinates of all the up- and down-pointing spins on the lattice respectively, and set the global variables nup and ndown equal to the number of elements in each of these arrays.

```c
/* beta    = inverse temperature
 * prob[]  = array of acceptance probabilities
 * s[]     = lattice of spins with helical boundary conditions
 * up[]    = list of up-pointing spins
 * down[]  = list of down-pointing spins
 * nup     = number of up-pointing spins
 * ndown   = number of down-pointing spins
 * L       = constant edge length of lattice
 */

#include <math.h>

#define N (L*L)
#define XNN 1
#define YNN L

int s[N];
int up[N],down[N];
int nup,ndown;
double prob[17];
double beta;

void initialize()
{
  int i;
  for (i=0; i<17; i++) prob[i] = exp(-beta*i);
}

void sweep()
{
  int i;
  int delta;
  int iup,idown;
  int xup,xdown;
  int upnn1,upnn2,upnn3,upnn4;
  int downnn1,downnn2,downnn3,downnn4;
  int term1,term2;
```

```
for (i=0; i<N; i++) {

  /* Choose a pair of spins */

  iup = nup*drandom();
  idown = ndown*drandom();

  xup = up[iup];
  xdown = down[idown];

  /* Calculate their nearest neighbours */

  if ((upnn1=xup+XNN)>=N) upnn1 -= N;
  if ((upnn2=xup-XNN)<0) upnn2 += N;
  if ((upnn3=xup+YNN)>=N) upnn3 -= N;
  if ((upnn4=xup-YNN)<0) upnn4 += N;

  if ((downnn1=xdown+XNN)>=N) downnn1 -= N;
  if ((downnn2=xdown-XNN)<0) downnn2 += N;
  if ((downnn3=xdown+YNN)>=N) downnn3 -= N;
  if ((downnn4=xdown-YNN)<0) downnn4 += N;

  /* Calculate the local energies before swapping */

  term1 = s[upnn1]+s[upnn2]+s[upnn3]+s[upnn4];
  term2 = -s[downnn1]-s[downnn2]-s[downnn3]-s[downnn4];

  /* Swap the spins over */

  s[xup] = -1;
  s[xdown] = +1;

  /* Calculate the difference in the local energies after */

  term1 -= -s[upnn1]-s[upnn2]-s[upnn3]-s[upnn4];
  term2 -= s[downnn1]+s[downnn2]+s[downnn3]+s[downnn4];

  /* Calculate total change in energy */

  delta = term1 + term2;

  /* Accept or reject the move */
```

```
    if (delta<=0) {
      up[iup] = xdown;
      down[idown] = xup;
    } else if (drandom()<prob[delta]) {
      up[iup] = xdown;
      down[idown] = xup;
    } else {
      s[xup] = +1;
      s[xdown] = -1;
    }
  }
}
```

B.2.2   Continuous time algorithm

The following functions perform one sweep of the lattice using the continuous time algorithm described in Section 5.2.1. In this algorithm the spins on the lattice are stored in several different lists, divided according to whether they are pointing up or down and the value of their spin coordination number, Equation (5.16). We also maintain two arrays which tell us in which list each spin appears and where, so that we can find it quickly. For convenience we have split the code into three functions. The function sweep() actually carries out the simulation. It is aided by spincoord(), which returns the spin coordination number of a given spin, and by update(), which updates the entries in the lists for the neighbours of spins which are flipped. Notice that in order to avoid having any gaps in our lists, we move the last spin in a list to fill the space left when we delete a spin.

```
/* s[]      = lattice of spins with helical boundary conditions
 * coord[]  = spin coordination of each spin on the lattice
 * up[][]   = lists of all up spins by coordination number
 * nup[]    = number of spins in each of these lists
 * down[]   = lists of all down spins by coordination number
 * ndown[]  = number of spins in each of these lists
 * loc[]    = location in relevant list of each spin on lattice
 * time     = real time variable
 * prob[]   = lookup table of values of exp(-2*beta*J*n)
 */

#define N (L*L)
#define XNN 1
#define YNN L
```

```
int s[N];
int coord[N];
int loc[N];
int nup[5],ndown[5];
int up[5][N],down[5][N];
double time;
double prob[5];

void sweep()
{
  int i,j;
  int c1,c2;
  int l1,l2;
  int x1,x2;
  int nn;
  double rd,sum;
  double sumup,sumdown;

  for (i=0; i<N; i++) {

    /* Calculate the sums */

    sumup = sumdown = 0.0;
    for (j=0; j<5; j++) {
      sumup += nup[j]*prob[j];
      sumdown += ndown[j]*prob[j];
    }

    /* Update the time variable */

    time += 1.0/(sumup*sumdown);

    /* Choose the spins */

    rd = sumup*drandom();
    for (c1=0,sum=0.0; c1<5; c1++) {
      sum += nup[c1]*prob[c1];
      if (sum>rd) break;
    }
    l1 = nup[c1]*drandom();
    x1 = up[c1][l1];

    rd = sumdown*drandom();
```

```
for (c2=0,sum=0.0; c2<5; c2++) {
  sum += ndown[c2]*prob[c2];
  if (sum>rd) break;
}
l2 = ndown[c2]*drandom();
x2 = down[c2][l2];

/* Exchange their values */

s[x1] = -1;
s[x2] = +1;

/* Update the entries in the lists for the two spins */

up[c1][l1] = up[c1][--nup[c1]];
loc[up[c1][l1]] = l1;
coord[x1] = c1 = spincoord(x1);
loc[x1] = ndown[c1];
down[c1][ndown[c1]++] = x1;

down[c2][l2] = down[c2][--ndown[c2]];
loc[down[c2][l2]] = l2;
coord[x2] = c2 = spincoord(x2);
loc[x2] = nup[c2];
up[c2][nup[c2]++] = x2;

/* Now do the same for each of the eight nearest
 * neighbours of the spins which have been exchanged
 */

if ((nn=x1+XNN)>=N) nn -= N;
update(nn);
if ((nn=x1-XNN)<0) nn += N;
update(nn);
if ((nn=x1+YNN)>=N) nn -= N;
update(nn);
if ((nn=x1-YNN)<0) nn += N;
update(nn);

if ((nn=x2+XNN)>=N) nn -= N;
update(nn);
if ((nn=x2-XNN)<0) nn += N;
update(nn);
```

```
    if ((nn=x2+YNN)>=N) nn -= N;
    update(nn);
    if ((nn=x2-YNN)<0) nn += N;
    update(nn);

  }
}

int spincoord(int spin)
{
  int nn1,nn2,nn3,nn4;

  /* Find the four neighbours of the spin */

  if ((nn1=spin+XNN)>=N) nn1 -= N;
  if ((nn2=spin-XNN)<0) nn2 += N;
  if ((nn3=spin+YNN)>=N) nn3 -= N;
  if ((nn4=spin-YNN)<0) nn4 += N;

  return (s[spin]*(s[nn1]+s[nn2]+s[nn3]+s[nn4])+4)/2;
}

void update(int spin)
{
  int c,l;

  c = coord[spin];
  l = loc[spin];

  if (s[spin]==+1) {
    up[c][l] = up[c][--nup[c]];
    loc[up[c][l]] = l;
    coord[spin] = c = spincoord(spin);
    loc[spin] = nup[c];
    up[c][nup[c]++] = spin;
  } else {
    down[c][l] = down[c][--ndown[c]];
    loc[down[c][l]] = l;
    coord[spin] = c = spincoord(spin);
    loc[spin] = ndown[c];
    down[c][ndown[c]++] = spin;
  }
}
```

B.3 Algorithms for Potts models

Here we give a function which implements the heat-bath algorithm for high-q Potts models on a square lattice in two dimensions. As in previous algorithms, the function makes use of a lookup table prob[] of the values of the exponentials needed by the algorithm. Note also the section in which the variable states is calculated. This variable holds the number of different states of the nearest neighbours of the selected spin. (For example, if all the nearest neighbours of the selected spin are the same, this variable should be 1, because there is only one nearest-neighbour spin state.) The code which calculates this number involves some trickery; you may like to take a moment to understand how it works.

```
/* prob[] = array of acceptance probabilities
 * s[]    = lattice of spins with helical boundary conditions
 * L      = constant edge length of lattice
 */

#define N (L*L)
#define XNN 1
#define YNN L

int s[N];
double prob[5];

void sweep()
{
  int i,n;
  int states;
  int nright,nleft,nup,ndown;
  int sright,sleft,sup,sdown;
  int zright,zleft,zup,zdown;
  double bright,bleft,bup,bdown;
  double sum,rn;

  for (n=0; n<N; n++) {

    /* Choose a spin at random */

    i = N*drandom();

    /* Find the neighbouring spins */

    if ((nright=i+XNN)>=N) nright -= N;
```

```
    if ((nleft=i-XNN)<0) nleft += N;
    if ((nup=i+YNN)>=N) nup -= N;
    if ((ndown=i-YNN)<0) ndown += N;

    sright = s[nright];
    sleft = s[nleft];
    sup = s[nup];
    sdown = s[ndown];

    /* Calculate the number of nearest-neighbour states */

    zright = zleft = zup = zdown = 1;
    if (sright==sleft) zright++, zleft++;
    if (sup==sdown) zup++, zdown++;
    if (sright==sup) zright++, zup++;
    if (sup==sleft) zup++, zleft++;
    if (sleft==sdown) zleft++, zdown++;
    if (sdown==sright) zdown++, zright++;

    states = ((zup+zdown)*zright*zleft +
              (zright+zleft)*zup*zdown)/
              (zup*zdown*zleft*zright);

    /* Calculate the weights for moves which set the spin
     * equal to one of its nearest neighbours
     */

    bright = prob[zright]/zright;
    bleft = prob[zleft]/zleft;
    bup = prob[zup]/zup;
    bdown = prob[zdown]/zdown;

    /* Choose the new state for the spin using the heat-bath
     * algorithm
     */

    sum = bright + bleft + bup + bdown;
    rn = (sum+q-states)*drandom();

    /* Find which state this corresponds to.  Check first to
     * see if it's one of the states which is aligned with a
     * nearest-neighbour spin
     */
```

```
if (rn<bright) {
  s[i] = sright;
} else {
  rn -= bright;
  if (rn<bleft) {
    s[i] = sleft;
  } else {
    rn -= bleft;
    if (rn<bup) {
      s[i] = sup;
    } else {
      rn -= bup;
      if (rn<bdown) {
        s[i] = sdown;
      } else {
        rn -= bdown;

        /* If not, it must be a non-aligned state */

        do {
          if (sright<=rn) {
            rn += 1.0/zright;
            sright = q;
            continue;
          }
          if (sleft<=rn) {
            rn += 1.0/zleft;
            sleft = q;
            continue;
          }
          if (sup<=rn) {
            rn += 1.0/zup;
            sup = q;
            continue;
          }
          if (sdown<=rn) {
            rn += 1.0/zdown;
            sdown = q;
            continue;
          }
          break;
        } while (TRUE);
```

```
            s[i] = rn;
          }
        }
      }
    }
  }
}
```

B.4   Algorithms for ice models

Here we give a function which performs one step of the short loop Monte Carlo algorithm for the square ice model. The square ice model is described in Section 7.1.2. The short loop algorithm is described in Section 7.3. The program works by tilting the square lattice by 45°, so that the bonds fall on a square lattice themselves, and storing this lattice in the elements of an ordinary array using helical boundary conditions. The system is then a checkerboard of bonds which go upward to the left and bonds which go upward to the right. For helical boundary conditions this means that the dimensions of the lattice should be of the form $L \times (L+1)$ so that everything fits together properly.

```
/* arrow[]        = direction of the arrows
 * step[]         = sequence of arrows in this move
 * lasttime[]     = last move in which this arrow was changed
 * iter           = number of this Monte Carlo step
 * L              = edge length of lattice (must be odd)
 * LU, LD, RU, RD = possible directions of arrows (left-up,
 *                  left-down, right-up, right-down)
 */

#define N (L*(L+1))
#define XNN 1
#define YNN L
#define LU 2
#define RD (-LU)
#define RU 3
#define LD (-RU)

int arrow[N];
int step[N];
int lasttime[N];
int iter;
```

```
move()
{
  int i;
  int first,curr;
  int currarrow;
  int nc,cand;
  int clist[2];
  int len=0,noloop=1;

  /* Choose the first arrow */

  first = N*random();
  lasttime[first] = iter;
  step[len++] = curr = first;
  currarrow = arrow[curr];

  /* Generate the loop */

  do {

    /* Find candidate arrows for next step of loop */

    nc = 0;
    switch (currarrow) {

    case LU:
      if ((cand=curr-XNN)<0) cand += N;
      if (arrow[cand]==LD) clist[nc++] = cand;
      if ((cand=curr-XNN-YNN)<0) cand += N;
      if (arrow[cand]==LU) clist[nc++] = cand;
      if ((cand=curr-YNN)<0) cand += N;
      if (arrow[cand]==RU) clist[nc++] = cand;
      break;

    case LD:
      if ((cand=curr-XNN)<0) cand += N;
      if (arrow[cand]==LU) clist[nc++] = cand;
      if ((cand=curr-XNN+YNN)>=N) cand -= N;
      if (arrow[cand]==LD) clist[nc++] = cand;
      if ((cand=curr+YNN)>=N) cand -= N;
      if (arrow[cand]==RD) clist[nc++] = cand;
      break;
```

```
case RU:
  if ((cand=curr+XNN)>=N) cand -= N;
  if (arrow[cand]==RD) clist[nc++] = cand;
  if ((cand=curr+XNN-YNN)<0) cand += N;
  if (arrow[cand]==RU) clist[nc++] = cand;
  if ((cand=curr-YNN)<0) cand += N;
  if (arrow[cand]==LU) clist[nc++] = cand;
  break;

case RD:
  if ((cand=curr+XNN)>=N) cand -= N;
  if (arrow[cand]==RU) clist[nc++] = cand;
  if ((cand=curr+XNN+YNN)>=N) cand -= N;
  if (arrow[cand]==RD) clist[nc++] = cand;
  if ((cand=curr+YNN)>=N) cand -= N;
  if (arrow[cand]==LD) clist[nc++] = cand;
  break;

}

/* Check to see if we have returned to a previous site */

if (lasttime[clist[0]]==iter) {
  step[len++] = clist[0];
  noloop = 0;
} else if (lasttime[clist[1]]==iter) {
  step[len++] = clist[1];
  noloop = 0;
} else {

  /* If not, choose randomly between the two candidates */

  if (random()<0.5) curr = clist[0];
  else curr = clist[1];
  lasttime[curr] = iter;
  step[len++] = curr;
  currarrow = arrow[curr];

}

} while (noloop);
```

```
/* Follow the path up to the loop and then reverse all the
 * arrows in it
 */

for (i=0; step[i]!=step[len-1]; i++);
for ( ; i<len-1; i++) arrow[step[i]] *= -1;
}
```

B.5 Random number generators

In this section we give code to implement three of the random number generators discussed in Section 16.1. Each of these generates random long integers and then multiplies by a conversion factor to give a random double-precision floating-point number in the range $0 \leq r < 1$. It is a simple matter to remove the conversion if we want instead to return the random integers themselves for use in, for example, multispin coding (see Chapter 15).

B.5.1 Linear congruential generator

This function generates a random number using the linear congruential generator of Lewis *et al.* (1969), implemented using the modulus trick given by Schrage (1979) to prevent integer overflows in 32 bits (see Section 16.1.3). The global variable i should be seeded with a non-zero long integer before the generator is used for the first time.

```
#define a 16807
#define m 2147483647
#define q 127773
#define r 2836
#define conv (1.0/(m-1))

long i;

double drandom()
{
  long l;

  l = i/q;
  i = a*(i-q*l) - r*l;
  if (i<0) i += m;

  return conv*(i-1);
}
```

B.5.2   Shuffled linear congruential generator

This next function uses the same linear congruential generator as the previous example, but employs the shuffling scheme of Bays and Durham (1976) which we described in Section 16.1.4 to improve the quality of the numbers generated and increase the repeat period of the generator. Again, the global variable i should be seeded with a non-zero long integer before the generator is used for the first time, and in addition we also need to seed the shuffling array j [] and the variable y with random long integers, which we could do, for example, using the basic linear congruential generator of Section B.5.1.

```
#include <math.h>

#define a 16807
#define m 2147483647
#define q 127773
#define r 2836
#define conv (1.0/(m-1))

#define N 64

long i;
long y;
long j[N];

double drandom()
{
  long l;
  long k;

  l = i/q;
  i = a*(i-q*l) - r*l;
  if (i<0) i += m;

  k = floor((double) y*N/m);
  y = j[k];
  j[k] = i;
  return conv*(y-1);
}
```

B.5.3   Lagged Fibonacci generator

The following three functions implement the additive lagged Fibonacci generator of Mitchell and Moore (unpublished), as described by Knuth (1981) and discussed in Section 16.1.6. The first function, seed(), takes a single unsigned long integer as a seed and seeds the array used by the generator with a simple linear congruential generator. The second, drandom(), is the generator itself. As you can see, the code for the generator is very brief, which makes it a particularly fast way of generating random numbers. The third function, lrandom(), is a version of the same generator to generate random 32-bit integers which could be used, for example, in multispin coding applications, such as the Ising model algorithm given in Section B.1.2. In fact, the exact same generator could equally well be used to generate 64-bit integers on a computer which had 64-bit words. Only the conversion factor conv2 would have to be changed and the seeding routine would have to be modified to initialize the array ia[] with 64-bit integers.

```
#define a 2416
#define c 374441
#define m 1771875
#define conv1 2423.9674
#define conv2 (1/4294967296.0)

unsigned long ia[55];
long p,pp;

void seed(unsigned long i)
{
   int n;

   for (n=0; n<55; n++) ia[n] = conv1*(i=(a*i+c)%m);
   p = 0;
   pp = 24;
}

double drandom()
{
   if (--p<0) p = 54;
   if (--pp<0) pp = 54;
   return conv2*(ia[p]+=ia[pp]);
}

long lrandom()
{
   if (--p<0) p = 54;
   if (--pp<0) pp = 54;
   return ia[p]+=ia[pp];
}
```

Index

Page numbers in bold denote definitions or principal references.

Made in the USA
Middletown, DE
25 February 2016